Design, Modeling and Evaluation of Protective Relays for Power Systems

AF714527

Mladen Kezunovic • Jinfeng Ren • Saeed Lotfifard

Design, Modeling and Evaluation of Protective Relays for Power Systems

Springer

Mladen Kezunovic
Department of ECEN
Texas A&M University
College Station
TX, USA

Jinfeng Ren
Alstom Grid
Redmond, WA, USA

Saeed Lotfifard
School of Electrical Engineering
and Computer Science
Washington State University
Pullman, WA, USA

ISBN 978-3-319-36741-5 ISBN 978-3-319-20919-7 (eBook)
DOI 10.1007/978-3-319-20919-7

Springer Cham Heidelberg New York Dordrecht London
© Springer International Publishing Switzerland 2016
Softcover re-print of the Hardcover 1st edition 2016
This work is subject to copyright. All rights are reserved by the Publisher, whether the whole or part of the material is concerned, specifically the rights of translation, reprinting, reuse of illustrations, recitation, broadcasting, reproduction on microfilms or in any other physical way, and transmission or information storage and retrieval, electronic adaptation, computer software, or by similar or dissimilar methodology now known or hereafter developed.
The use of general descriptive names, registered names, trademarks, service marks, etc. in this publication does not imply, even in the absence of a specific statement, that such names are exempt from the relevant protective laws and regulations and therefore free for general use.
The publisher, the authors and the editors are safe to assume that the advice and information in this book are believed to be true and accurate at the date of publication. Neither the publisher nor the authors or the editors give a warranty, express or implied, with respect to the material contained herein or for any errors or omissions that may have been made.

Printed on acid-free paper

Springer International Publishing AG Switzerland is part of Springer Science+Business Media (www.springer.com)

Preface

This book has been written for university students, professionals in the area of protective relaying, and other interested individuals with minimum engineering skills to study the material on their own. To achieve this goal, the book has been written in an unconventional way: It uses a simulation tool called MERIT 2000, based on widely known MATLAB software, to offer hands-on experience in understanding and implementing protective relaying designs.

Many books on protective relaying have been published over the years, which provide an excellent background on power system faults and protective relaying principles. However, most of these books assume that the reader is interested in learning about relaying principles and on how protective relay products can be used to implement protective approaches for various power system apparatus.

This book takes a different approach: It assumes that the reader is interested in learning how the relays work, what the basic design principles are, and how an implemented design of a relay may be evaluated. With this goal in mind, the book tries to make the learning process a design experience wherein the reader starts using the software engineering tools (MATLAB) from the very beginning as the basic relay design principles are introduced. The book is not a substitute for a reference on the fundamentals of relaying but is rather a complementary source on the topic.

To provide a hands-on experience, the authors have provided MERIT 2000 software developed in MATLAB as a supplement to this book. The software has been in use since 2000 in laboratory assignments in courses at Texas A&M University.

College Station, TX, USA — Mladen Kezunovic
Redmond, WA, USA — Jinfeng Ren
Pullman, WA, USA — Saeed Lotfifard

Acknowledgments

The effort to create exercises that provide a hands-on experience in this book is the result of collaboration in specifying, implementing, testing, and evaluating the MERIT 2000 software. This effort was financially supported at its beginning by the National Science Foundation Grant ECS-96-19294 awarded to Texas A&M University in 1996 and in part by EPRI under contract WO 8618-02 awarded to Texas A&M University and Washington State University in 2002.

Texas A&M faculty, Dr. Mladen Kezunovic, and Dr. Garng Huang, as well as Dr. Ali Abur who was with Texas A&M University at the time the NSF grant was awarded, worked on different aspects of the power system analysis and developed MATLAB-based software to facilitate the experimental part of the respective analysis issues. Dr. Kezunovic was responsible for the protective relaying area and development of the MERIT 2000 software used in this book.

Several former graduate students, including the co-authors of the book, as well as scholars in Dr. Kezunovic's group, have contributed to the development of the MERIT 2000 software. Dr. Bogdan Kasztenny is acknowledged for his key role in this regard while he was a postdoc in Dr. Kezunovic's lab at Texas A&M University in 1997–1999.

Final thanks go to the many undergraduate and graduate students who undertook courses in protective relaying at Texas A&M University since 2000. They all provided feedback, and some contributed with new developments over the years making the MERIT 2000 software a very valuable teaching and learning tool. For those who want to engage in the learning experience, the software will be made available through a third party as a supplement to our book. The authors will keep updating the software as new developments are reported.

Contents

List of Figures

List of Tables

Chapter 1
Introduction

1.1 Scope

In the protection of electric power systems, a protective relay is defined as "a relay whose function is to detect defective lines or apparatus or other power system conditions of an abnormal or dangerous nature and to initiate appropriate control circuit action" [1]. The theory and application of the protective relaying is an important part of the education for an electrical engineer who specializes in power systems [2]. Multidisciplinary and integrative nature of the protection solutions requires new modeling, simulation, and testing tools to understand the diverse device behavior.

In this book, the hands-on experiences with modeling, simulation, and testing tools provide needed insights into the integration concepts associated with design and application of protective relaying to modernize the power engineering education. This book can be used by professionals and students (undergraduate and graduate levels) for better understanding the design concept and related applications for protective relaying. The modeling and simulation techniques that provide an ability to study interaction between various devices facilitate understanding of new standards for Intelligent Electronic Devices (IEDs). Using simulation tools that demonstrate interaction between power systems and IEDs allows in-depth studying of the relay dependability and security performance when exposed to the stressed power system conditions.

The contents cover the following topics:

- Basics of symmetrical components and their application in power system fault analysis
- Fundamentals of protective relaying and design principles for overcurrent, impedance (distance), and differential relaying
- Library of modeling modules for digital relays and their interfacing to power systems

© Springer International Publishing Switzerland 2016

M. Kezunovic et al., *Design, Modeling and Evaluation of Protective Relays for Power Systems*, DOI 10.1007/978-3-319-20919-7_1

- Use of software models for designing and implementing specific protection solutions including line, transformer, and busbar protection. The line protection covers overcurrent scheme, differential relaying, zone protection, and pilot protection.
- Testing methodology and tools for validating protection solutions modeled in software as well as protection device products.

1.2 Basics of Protection Relaying

Faults in electric power systems usually, but not always, provide significant changes in the system measurements, which can be used to distinguish between tolerable and intolerable system conditions. The conditions include overcurrent, over- or under-voltage, power factor or phase angle, power or current direction, impedance, frequency, temperature, physical movements, pressure, and contamination of the insulation [2, 3].

Protection is the science, skill, and art of applying and setting relays to provide maximum sensitivity to faults and undesirable conditions, but to avoid their operation under unwanted conditions. Both failure to operate (dependability) and incorrect operation (security) can result in major system upsets involving increased equipment damage, increased personnel and public hazards, and possible long interruption of service [4]. To minimize the potential catastrophic problems that can result in the power system failing due to a protection failure, the practice is to use several relays or relay systems operating in parallel. These can be at the same location (primary backup), at the same station (local backup), or at various remote stations (remote backup). The various protective devices must be properly coordinated. For a protection system, an adequate redundancy capability is important. However, additional redundancy may have a negative impact on security. Thus, an optimal balance should be achieved through specific protection design.

The component representation of a typical protective relay is shown in Fig. 1.1. The components can be electromechanical, electronic, or both. Originally, all protective relays were of the electromechanical type. Analog type electronic relays using discreet electronic components were introduced in the 1970s. Nowadays, microprocessor-based electronic relays, also referred as numerical relays or digital relays, have been widely used in protecting electric power systems.

With digital relays, the protection principles and fundamentals are essentially unchanged from the basic ideas introduced before, but many benefits, such as higher accuracy, reduced space, lower equipment and installation costs, wider application and setting capabilities, and various other desirable supplemental

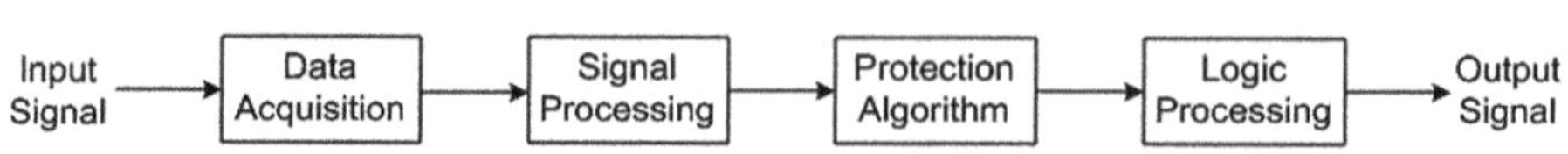

Fig. 1.1 Component representation of a typical protective relay

features, are provided. These include control logic, remote and peer-to-peer communications, event recording, fault location, and self-monitoring and checking.

Design and application of a protective relay and the overall protection solution rely on many factors: features of target power system being protected, fault characteristics, protection algorithms, configuration and setting, communication and control methods. The approach throughout this book is to utilize relaying and power system models to facilitate the protection design, implementation, and validation.

1.3 Modeling and Simulation Methodology and Tools

In this book, the simulation models that represent digital relays and power system components are run in MATLAB/SIMULINK environment [4, 5]. Libraries of protective relay modules, power system elements, and protection schemes are provided for an easy use by readers when learning the principles of protective relay design and application. The relay elements library, signal source library and analysis tools, protective relay models, and power network elements are briefly described here as examples of the comprehensive tools. The details and other models, such as the various protection algorithms, communication schemes, transformer and busbar protection system and their related power network models, will be provided in following chapters.

1.3.1 Relay Elements Library

The library of relay elements includes the major modules for designing a protective relay function such as data acquisition, measurement, decision making, and communication. As shown in Fig. 1.2 and Table 1.1, this library allows one to design a variety of relays by selecting, connecting, and configuring models of appropriate elements from the library.

1.3.2 Signal Source Library and Analysis Tools

The signal source library generates various input signals for protective relays. As given in Fig. 1.3, it includes transient modeling of power system for closed-loop simulation (link to ATP/EMTP [6] and the SimPowerSystem in Simulink) as well as the analytical way of generating the signals. In addition, a file format conversion for data from major transient file formats (COMTRADE [7] and MATLAB's native format) is also provided.

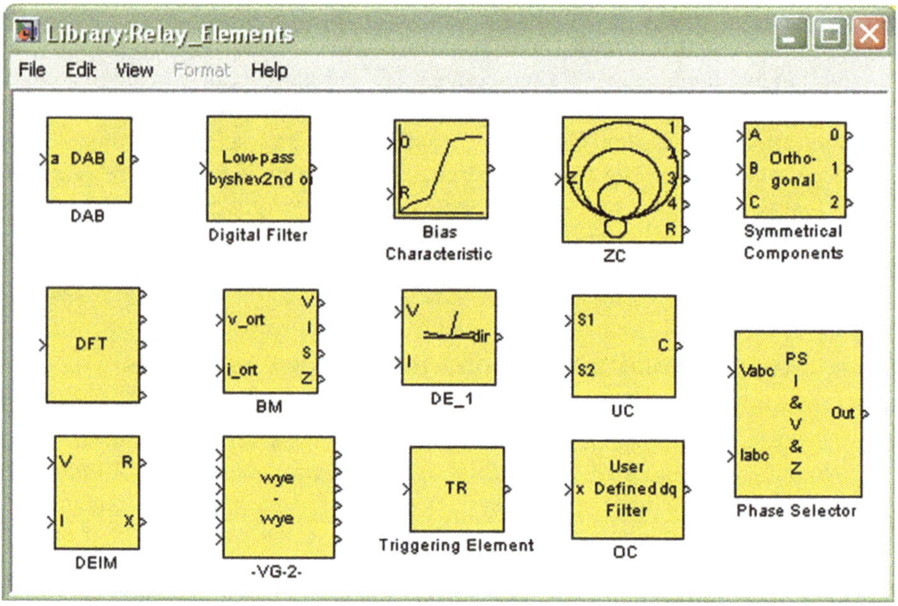

Fig. 1.2 Library of protective relay elements

Analysis tools contain functional elements that facilitate the analysis and testing procedures. Take the output phasor display as an example (see Fig. 1.4). This module can be fed by up to six phasors which are multiplexed into a single input vector. It displays them in both the graphical and numerical forms. This block can be disabled. That is presented in the model but not active. The phasor display can be of the steady state type (the values are displayed when the simulation terminates) or of the transient type (the plot refreshed every given time interval).

1.3.3 Relay Models and Power Network Elements Library

The relay models represent complete digital relays and protection terminals such as overcurrent, impedance (distance), and differential relays. Those models are composed of the blocks of the relay elements library and the general SIMULINK libraries. Some of the relay models emulate actual relays to the extent possible using the publicly available design details. Figure 1.5 presents an example of a three-phase directional overcurrent relay. It features three-phase directional element, three-phase inverse-time, and residual overcurrent elements. In this model:

- The DAB and OC module takes analog inputs and simulates the analog filter, signal conditioner, A/D converter, and phasor measurement estimator.

Table 1.1 Summary of relay elements library

Element	Description
Data Acquisition Board (DAB)	An analog input signal is filtered for anti-aliasing, conditioned, sampled, and forwarded as a data window of signal samples. Options include analog filtering, type, order and cut-off frequency; number of bits of A/D converter; conditioning gain and length of data window
Digital Filter (DF)	An input signal is filtered digitally. The predefined filters include various types of IIR filters, and Walsh, Fourier FIR. "Free-expression" digital filter can be set
Digital Fourier Transform (DFT)	An input data window is captured and the phasors of up to five harmonics are calculated using the DFT technique. The options include selection of the frequency of the reference (first) harmonics and other requested harmonics
Basic Measurement (BM)	The voltage and current phasors are captured and the amplitudes, impedance components, and power are calculated. The post-filtering may be applied using either mean or median filters
Differential Equation-based Impedance Measurement (DEIM)	The block measures the impedance based on the differential equation approach. The pre-filtering using either Walsh or Fourier filters of selectable window length may be applied. The post-filtering using either mean or median filters may be applied. Either Euler or trapezoidal method of numerical differentiation may be used
Universal Comparator (UC)	The block is fed by two signals and doses a comparison between the signal or a threshold, the signal and time or the two signals. The direction of comparison may be alternated and the standard time characteristics are included
Zone Comparator (ZC)	The block is fed by the resistance and reactance values and emulates four forward impedance zones and a reverse one. Either the mho or "free-expression" shapes may be set
Triggering Element (TR)	The block is fed by the data window of a signal and acts as a transient detector. Implemented methods include sample-to-sample, cycle-to-cycle, and value-to-threshold checking
Symmetrical Components (SC)	The block is fed by three phase signals and produces three symmetrical components signals. Either phasors or instantaneous values are utilized
Bias Characteristic	The block is fed by the operating and restraining signal and applies a "free-expression" bias characteristic

- The time overcurrent, residual and directional protection elements take three-phase current, zero-sequence current, and phase-to-phase voltages as inputs and perform corresponding functions.
- The logic block analyzes digital inputs and initiates trip signal according to the logic configuration.

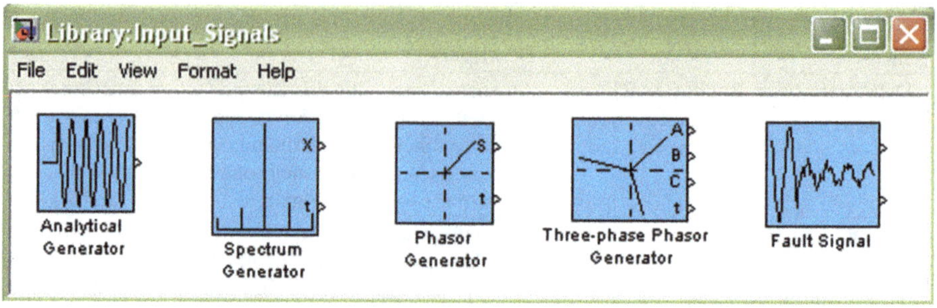

Fig. 1.3 Signal source library

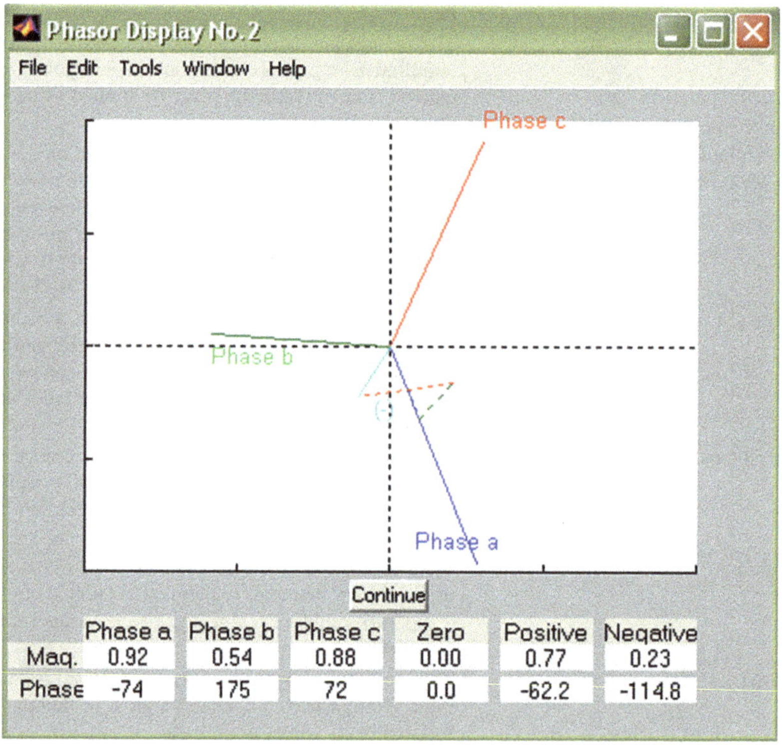

Fig. 1.4 Output phasor display block

From this example the relay model can be assembled from the available libraries in a very convenient way. Modifications of its structures and parameters are intuitive. Once developed, the model can be packed into a subsystem, seen in Fig. 1.6, and categorized into a library of relay elements for the future use. This

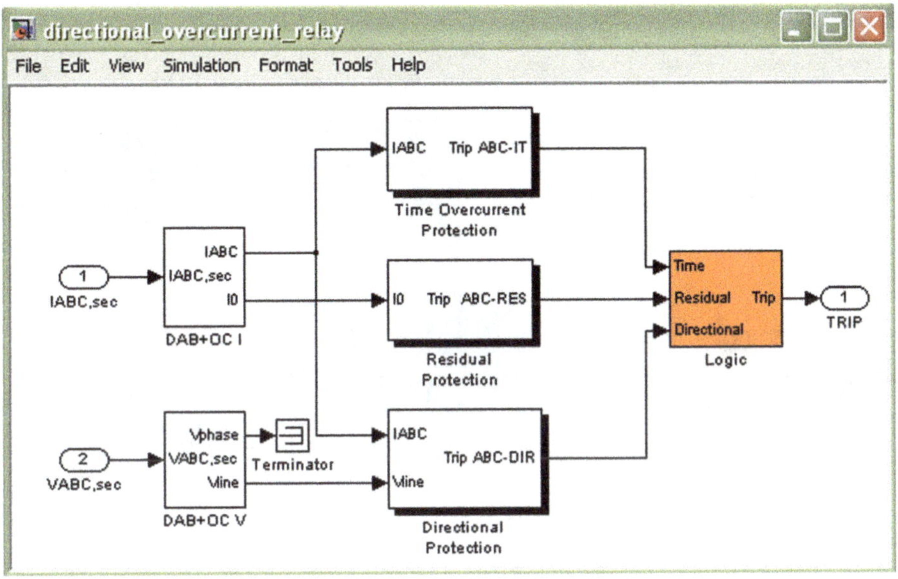

Fig. 1.5 Structure of directional overcurrent relay model

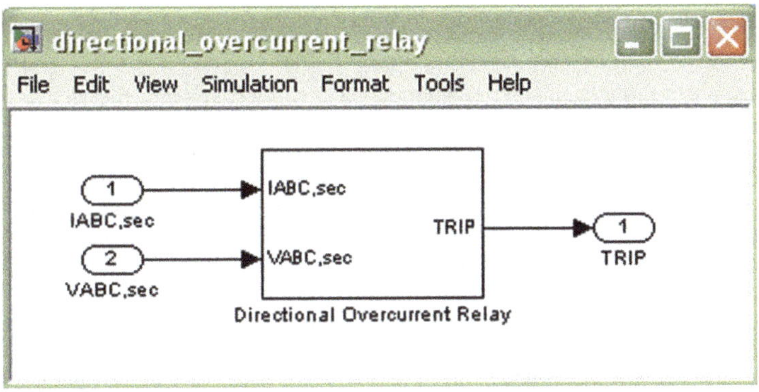

Fig. 1.6 Directional overcurrent relay block

approach can accelerate the model development phase significantly. In the same way, a larger structure of protection system can be built out of relays and relay element libraries.

Once the developed relay model has been verified through testing, it can be applied to the power network protection. Power network library (Fig. 1.7) contains the major elements for constructing an application scenario, such as generator, load, transmission line circuit breaker as well as the instrument transformers. Disturbances can be simulated for generating voltages and/or currents fed to relays.

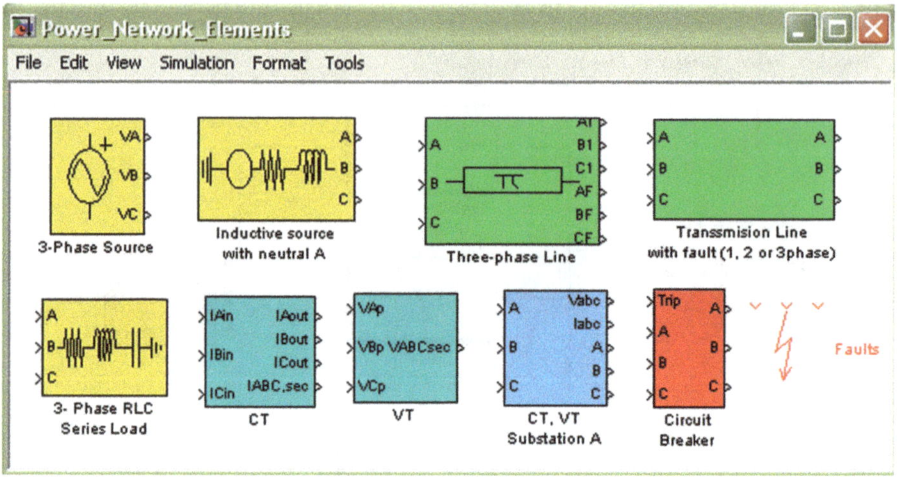

Fig. 1.7 Power network elements library

References

1. IEEE standard for relays and relay systems associated with electric power apparatus, IEEE Standard C37.90-2005, January 31, 2006
2. Kezunovic M (2005) Fundamentals of power system protection. In: Chen W-K (ed) The electrical engineering handbook. Chapter on electric power systems. Elsevier Academic, San Diego, pp 787–804. ISBN 978-0-12-170960-0
3. Lewis Blackburn J, Domin TJ (2006) Protective relaying: principles and applications, 3rd edn. CRC Press, Boca Raton
4. IEEE PSRC Working Group D5 (chaired by E.A. Udren) (1997) Proposed statistical performance measures for microprocessor-based transmission-line protective relays: explanations of the statistics. IEEE Trans Power Delivery 12(1):134–143
5. Using SIMULINK, The Mathworks Inc., 2000
6. CanAm EMTP User Group (2001) Alternative Transient Program (ATP) Rule Book, Portland, OR [Online]. http://www.eeug.org/files/secret/ATP_RuleBook
7. Communication networks and systems in substation—Part 9-2: specific communication service mapping (SCSM)—sampled analogue values over ISO 8802-3, IEC Std. 61850

Chapter 2
Power System Fault Analysis and Short-Circuit Computations

2.1 Introduction

This chapter contains the material for learning basics of power system fault analysis and short-circuit calculation at the elementary level.

First, the basic theory of symmetrical components and sequence networks is presented with the software (the exercise: "Unbalanced System Operation" and "Short-Circuit Analysis") illustrating the following issues:

- Converting phase signals into symmetrical components including both numerical and graphical approaches
- Representing a three-phase unbalanced system by 3 three-phase balanced symmetrical component systems
- Representing power system three-phase elements by their symmetrical networks
- Connecting symmetrical networks for single line to ground faults

By dealing with an original (physical) three-phase power system and its various representations using symmetrical components and networks, and having the tools to compare the input/output signals, one can learn how the concept of symmetrical components and networks helps in the analysis of unbalanced power systems.

Next, the basics of creating sequence networks for the common power system elements are presented. The following three-phase devices are covered by examples: transmission lines, loads, transformers, synchronous machines, and induction motors.

The software exercises are organized to include the model of a given three-phase device and related three single-phase sequence networks that represent the device in the symmetrical component domain. Both the original three-phase element and its sequence networks are excited with the corresponding external signals and the behavior of the two network representations is compared to prove and illustrate by simulation that the power system elements can be represented by appropriate sequence networks.

© Springer International Publishing Switzerland 2016

M. Kezunovic et al., *Design, Modeling and Evaluation of Protective Relays for Power Systems*, DOI 10.1007/978-3-319-20919-7_2

Last, the matrix methods for performing short-circuit studies in large power systems are provided at different learning levels listed below.

Elementary level: An educational software and accompanying classroom material (exercise: "Short-Circuit Studies using Matrix Methods") are available for illustrating the basics of forming both the impedance matrices and using the matrices for short-circuit calculations.

Advanced level: Projects are designed to develop a simple short-circuit program from scratch using both the admittance and impedance methods. Samples of project assignments and selected solutions are attached at the end of this chapter.

2.2 Symmetrical Components

Under normal conditions, a power system operates under balanced conditions and per-phase (single-phase representation) analysis can be used to analyze the system. Under some conditions, the system may become unbalanced and per-phase analysis can no longer be used. This exercise shows how to analyze unbalanced systems in steady state, using symmetrical components.

The use of symmetrical components will be explained briefly in the following paragraphs. A more detailed explanation may be found in [1].

The phasors of the currents and voltages in a power system are given usually in the three-phase domain, i.e., I_a, I_b, I_c, V_a, V_b, and V_c. Under normal operation, the currents and voltages, respectively, are balanced; i.e., they have the same magnitude and are separated by 120° from each other. Under these circumstances, each phase current and voltage can be decomposed in symmetrical components as follows:

$$
\begin{aligned}
I_a &= I_a^0 + I_a^+ + I_a^- \qquad & V_a &= V_a^0 + V_a^+ + V_a^- \\
I_b &= I_b^0 + I_b^+ + I_b^- \qquad & V_b &= V_b^0 + V_b^+ + V_b^- \\
I_c &= I_c^0 + I_c^+ + I_c^- \qquad & V_c &= V_c^0 + V_c^+ + V_c^-
\end{aligned}
\tag{2.1}
$$

where I_a^0, I_b^0, I_c^0 and V_a^0, V_b^0, V_c^0 are the zero-sequence sets, I_a^+, I_b^+, I_c^+ and V_a^+, V_b^+, V_c^+ are the positive-sequence sets, and I_a^-, I_b^-, I_c^- and V_a^-, V_b^-, V_c^- are the negative-sequence sets.

A zero sequence set has the following property:

$$I_a^0 = I_b^0 = I_c^0$$

The components of the positive- and negative-sequence sets also have the same magnitude, but they are rotated by 120° as follows. For the positive sequence set, I_a^+ is leading I_b^+ by 120°, and I_c^+ by 240°. In the negative sequence set, I_a^- is lagging I_b^- by 120°, and I_c^- by 240°.

Of the nine symmetrical components mentioned above, only three can be chosen independently. The common convention is that the I_a^+, I_a^-, and I_a^0 are chosen as the independent variables and the remainder are then expressed in terms of the lead variables. Then we can express the currents (or voltages) in the phase domain as function of the symmetrical components of phase a as follows:

$$\begin{bmatrix} I_a \\ I_b \\ I_c \end{bmatrix} = \begin{bmatrix} 1 & 1 & 1 \\ 1 & a^2 & a \\ 1 & a & a^2 \end{bmatrix} \begin{bmatrix} I_a^0 \\ I_a^+ \\ I_a^- \end{bmatrix} \tag{2.2}$$

where:

$$a = e^{j2\pi/3} = 1\angle 120^{\circ}$$
$$a^2 = 1\angle - 120^{\circ}$$

Or in compact form:

$$I_{abc} = A \cdot I_{012} \tag{2.3}$$

where, $A = \begin{bmatrix} 1 & 1 & 1 \\ 1 & a^2 & a \\ 1 & a & a^2 \end{bmatrix}$.

On the other hand:

$$I_{012} = A^{-1} \cdot I_{abc} \tag{2.4}$$

where, $A^{-1} = \frac{1}{3}\begin{bmatrix} 1 & 1 & 1 \\ 1 & a & a^2 \\ 1 & a^2 & a \end{bmatrix}$.

Matrix A is the symmetrical components transformation matrix.

In Fig. 2.1, a representation of the single line to ground (SLG) fault is shown. The symmetrical components for the case of a SLG fault are shown in Fig. 2.2. In this case if the phasors that represent the symmetrical components for each phase current are added, we have:

$$\begin{aligned} I_a &= I_a^0 + I_a^+ + I_a^- = I_f \\ I_b &= I_b^0 + I_b^+ + I_b^- = 0 \\ I_c &= I_c^0 + I_c^+ + I_c^- = 0 \end{aligned} \tag{2.5}$$

The exercise provided in the following sections illustrates the use of symmetrical components for the analysis of unbalanced networks. Unbalanced system sources and a single-line-to-ground fault in a simple power system are analyzed. After the completion of this exercise, one can understand how the symmetrical components are used to analyze an unbalanced condition in a power system.

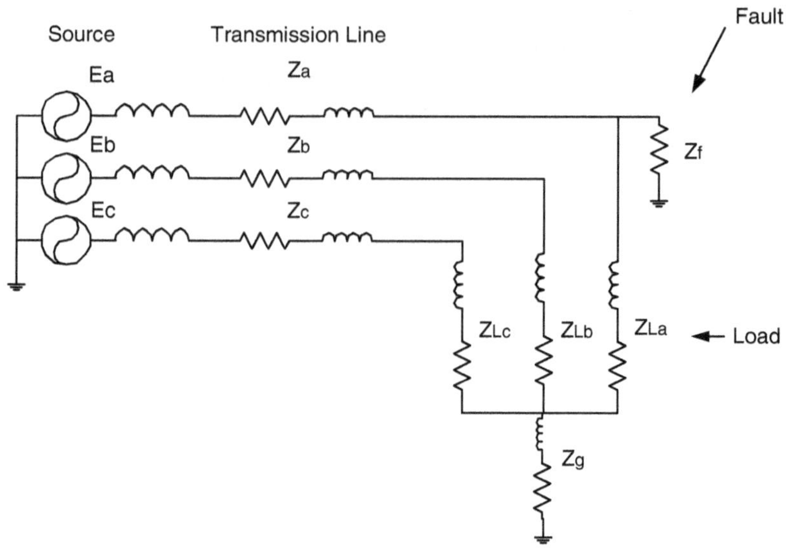

Fig. 2.1 A SLG fault representation

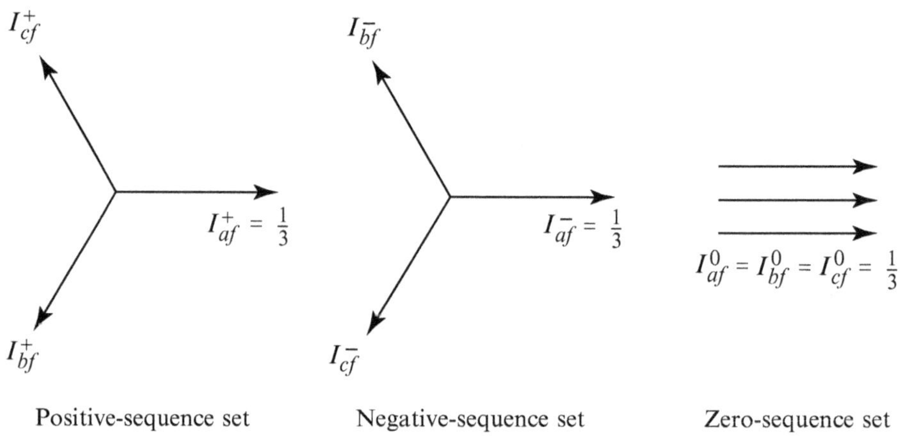

Fig. 2.2 The sequence current phasors for a single-line-to-ground fault

The simulation modules provided in this section will help one understand the principle of the symmetrical component method for solving unbalanced power system conditions.

There are two Simulink modules developed to demonstrate the unbalanced networks: SCo1.mdl and SCo2.mdl. Both models demonstrate the use of the symmetrical components and networks to analyze unbalanced networks. In the first module, a system with an unbalanced source is analyzed, while in the second one a system with a single-line-to-ground fault is analyzed. Do the following procedure to execute the models in MERIT 2000 software provided as a supplement to this book:

1. Open MATLAB
2. Type SCo1 or SCo2 to run the first or second module, respectively

In each case, a new window will appear in the screen. To run simulation, just click on the run button of the top bar.

There are also two MATLAB scripts named abc2pnz.m and pnz2abc.m, for converting phasors between the three-phase domain and the three-sequence (symmetrical component) domain.

2.2.1 Module 1: Analysis of a System with an Unbalanced Source Using Symmetrical Components

Module 1 in its main window consists of two options as shown in Fig. 2.3. To see the contents of each of the options, position the mouse pointer on the desired box and double click.

The option on the left hand side contains a three-phase network with unbalanced source as shown in Fig. 2.4. The block labeled "Measuring block" measures the phase currents and voltages in the time domain (output ports labeled i and v, respectively) and also their corresponding phasors (output ports I_{abc} and V_{abc}). A block labeled Display of *abc* or 012 phasors is used to display the phasors in *abc* domain and also shows graphically how the positive, negative, and zero sequence phasors for phase a (i.e., I_a^+, I_a^-, and I_a^0) are obtained from phasors I_a, I_b and I_c. A figure with the mentioned phasors is shown at the end of the simulation, if the parameter "Block active" in its dialog box is checked (see Fig. 2.5). When running simulations, be sure that only one of the block types is active.

At the end of the simulation, a figure like the one shown in Fig. 2.6 will be displayed. Press the "continue" button in the figure to see how the positive, negative, and zero sequence phasors are formed from the *abc* phasors.

The option indicated on the right hand side of Fig. 2.3 contains five sub-options shown in Fig. 2.7. The option "Network with source decomposed in symmetrical

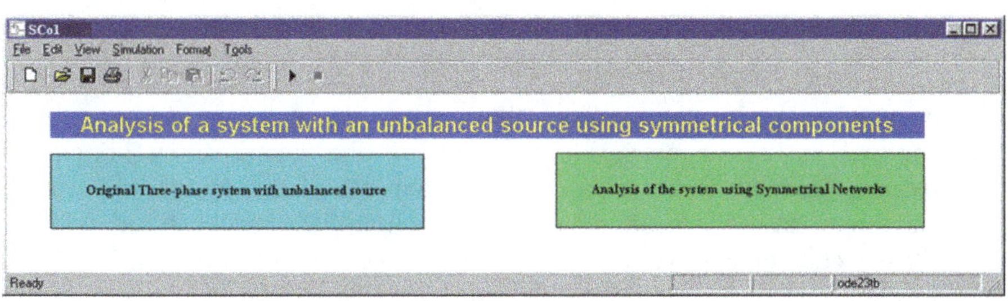

Fig. 2.3 Main window for Module 1

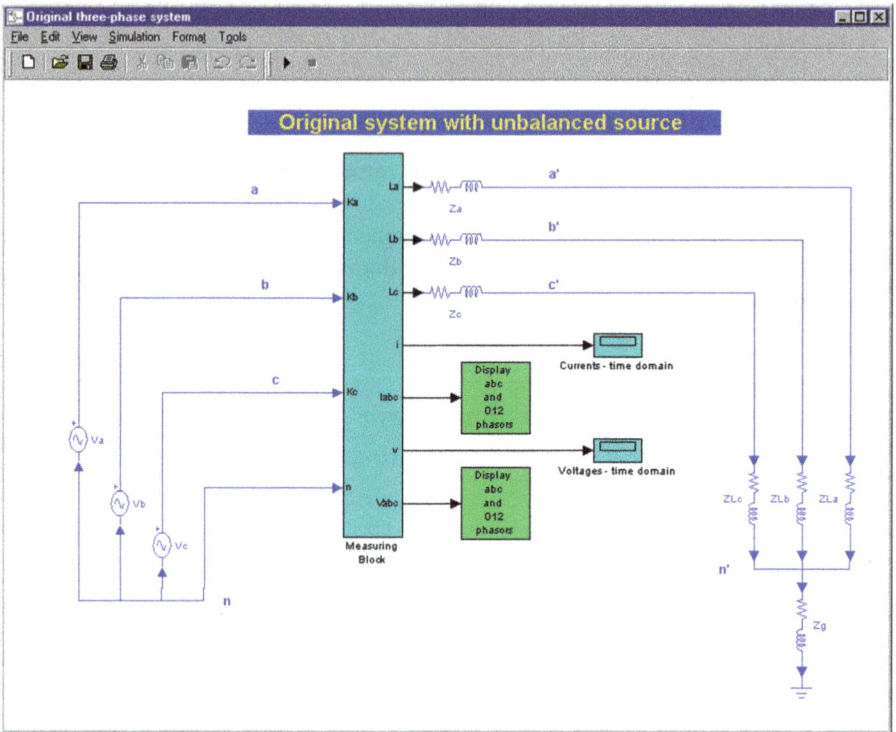

Fig. 2.4 Three-phase network with unbalanced source

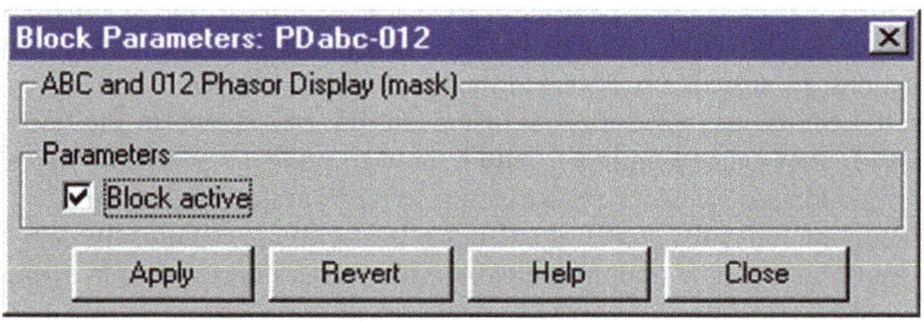

Fig. 2.5 Dialog box for the *abc* and 012-phasor display block

components" contains the original three-phase system, but with the source decomposed in its symmetrical components. If the superposition principle is applied and only the positive sequence components of the source are kept, we have a three-phase balanced network. This circuit can be analyzed with the per-phase technique and we have positive sequence network shown in Fig. 2.8. If we do the same, but for the negative sequence sources, we obtain the negative sequence network of Fig. 2.9.

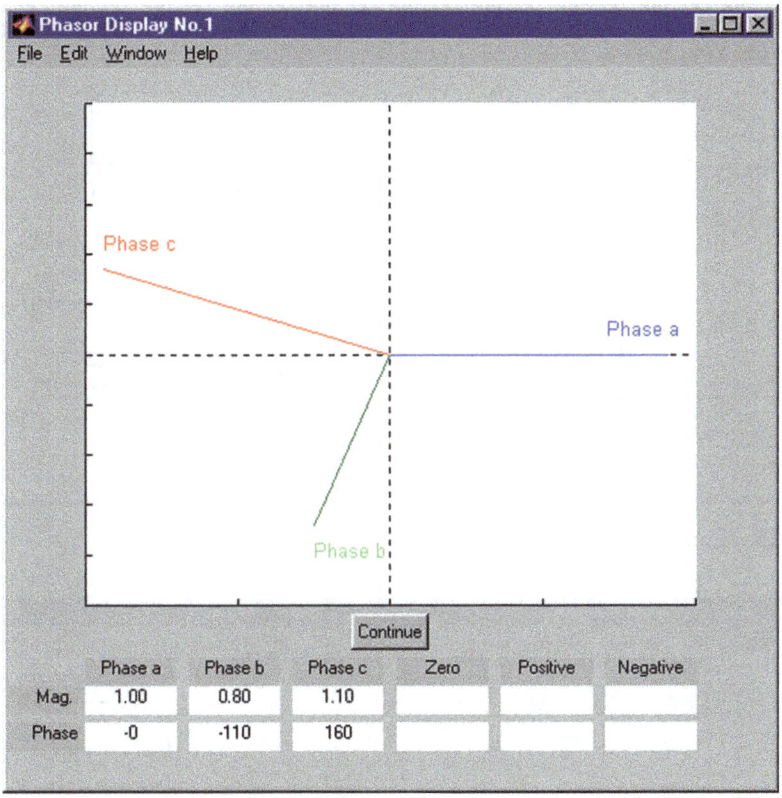

Fig. 2.6 *abc* or 012 phasor display

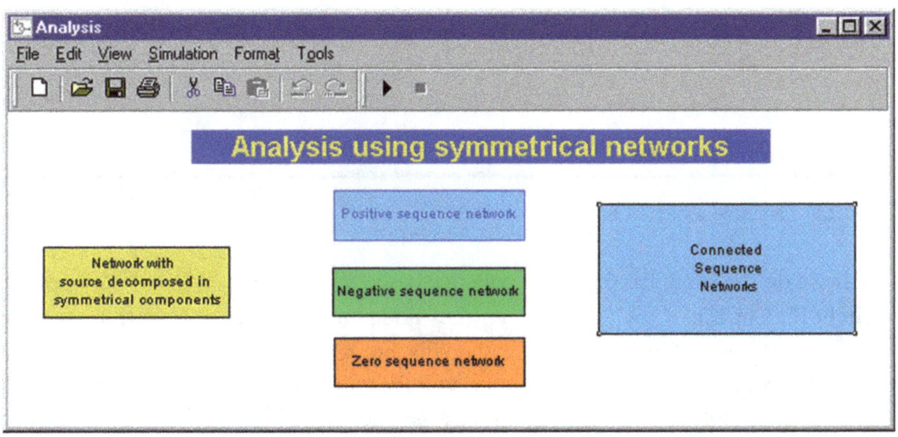

Fig. 2.7 Analysis using symmetrical components

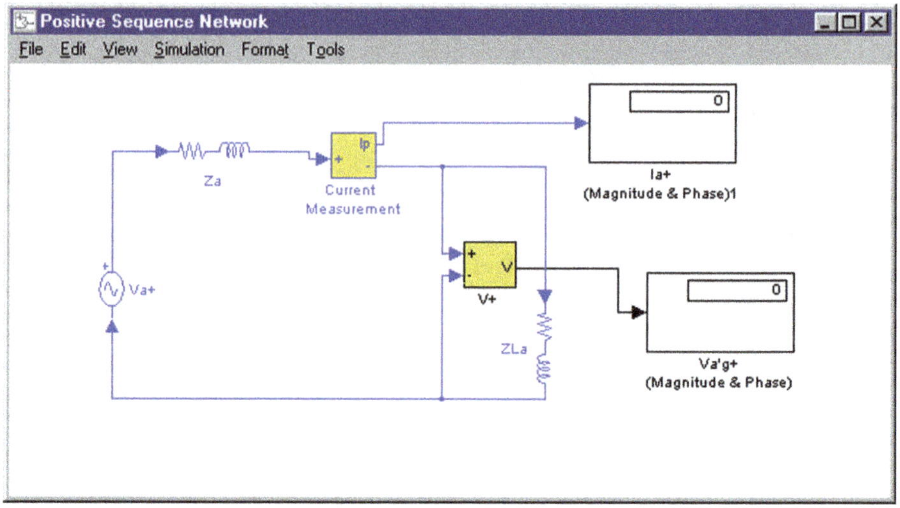

Fig. 2.8 Positive sequence network

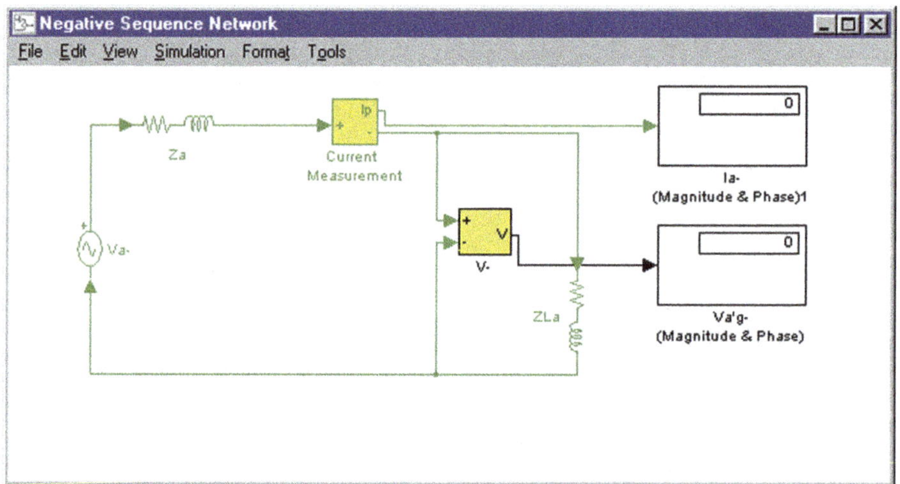

Fig. 2.9 Negative sequence network

If we do the same for the zero sequence case, we have $V_a^0 = V_b^0 = V_c^0$. Applying Kirchoff Current Law (KCL) at the node n ($I_a^0 + I_b^0 + I_c^0 = 0$), we get $I_a^0 = I_b^0 = I_c^0 = 0$. This means that there is no zero-sequence current flowing in the network. As all the zero-sequence voltages and currents are equal, the circuit can be represented with only one phase as shown in Fig. 2.10. This is the zero sequence network. Note that Z_g is multiplied by three in that network. This is to take into account the fact that in the original network, the magnitude of the current that circulates through Z_g is three times I_a^0. In the network as shown in Fig. 2.10, only I_a^0

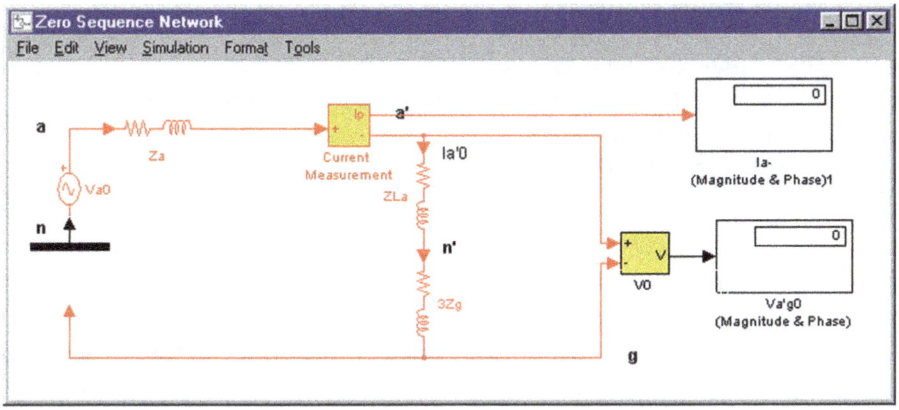

Fig. 2.10 Zero sequence network

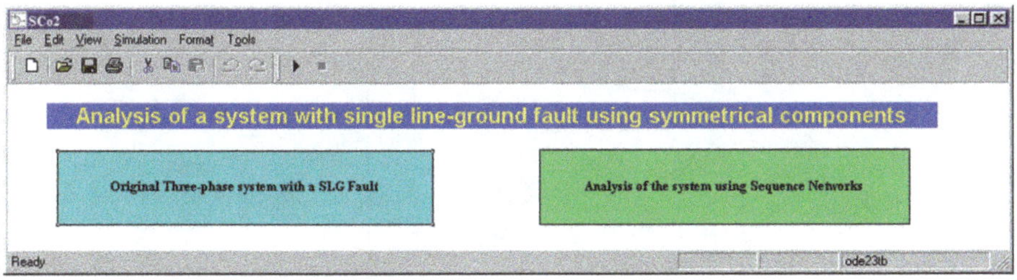

Fig. 2.11 Connected sequence networks

circulates through Z_g, then, Z_g must be multiplied by three to represent the same voltage drop as in the original network.

These displays are selected by clicking on the appropriate boxes on the display shown in Fig. 2.7.

In the first model, as there is no fault, the symmetrical networks are independent of each other. In Fig. 2.11, the three sequence and the three phase quantities obtained using the outputs of this circuit are shown to confirm that the analysis of the unbalanced network is valid by using symmetrical components. This display is brought up by clicking on the far right box on the display shown in Fig. 2.7.

2.2.2 Module 2: Analysis of a System with Single Line-Ground Fault Using Symmetrical Components

Module 2 has a main window that contains two options shown as boxes in Fig. 2.12. The box on the left hand side shows a simple system with a fault to ground in phase *a* in three-phase *abc* domain (Fig. 2.13). The fault impedance is Z_f. The box on the

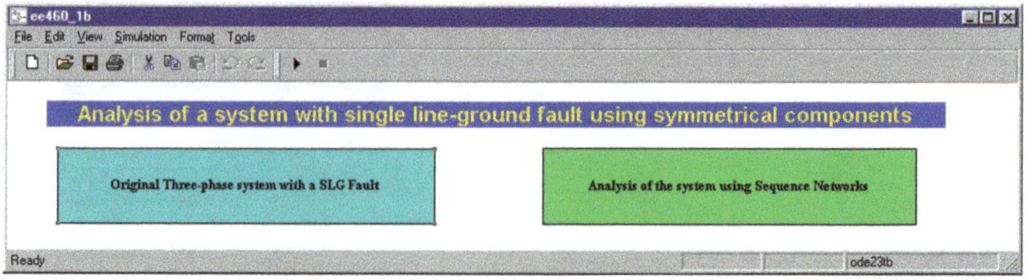

Fig. 2.12 Main window for Module 2

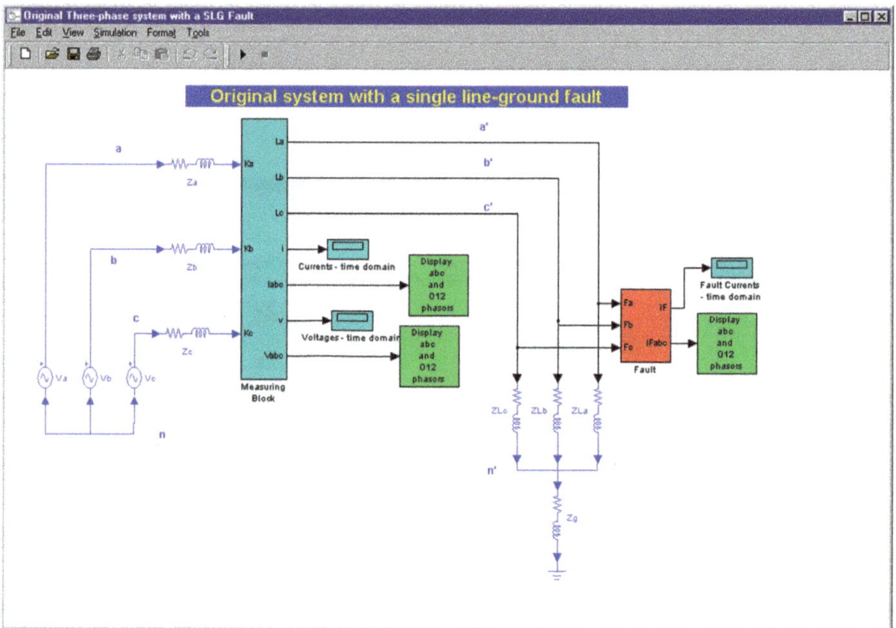

Fig. 2.13 System with a single line-ground fault

right hand side of the main window contains five sub-options that are similar to Fig. 2.7. To see the contents of each of the options, position the mouse pointer in the desired box and double click.

2.2.2.1 Fault Connections Represented Using Symmetrical Components

In this model, another representation equivalent to the three-phase model in the main window is shown in Fig. 2.14. For a phase *a* to ground fault, the current that

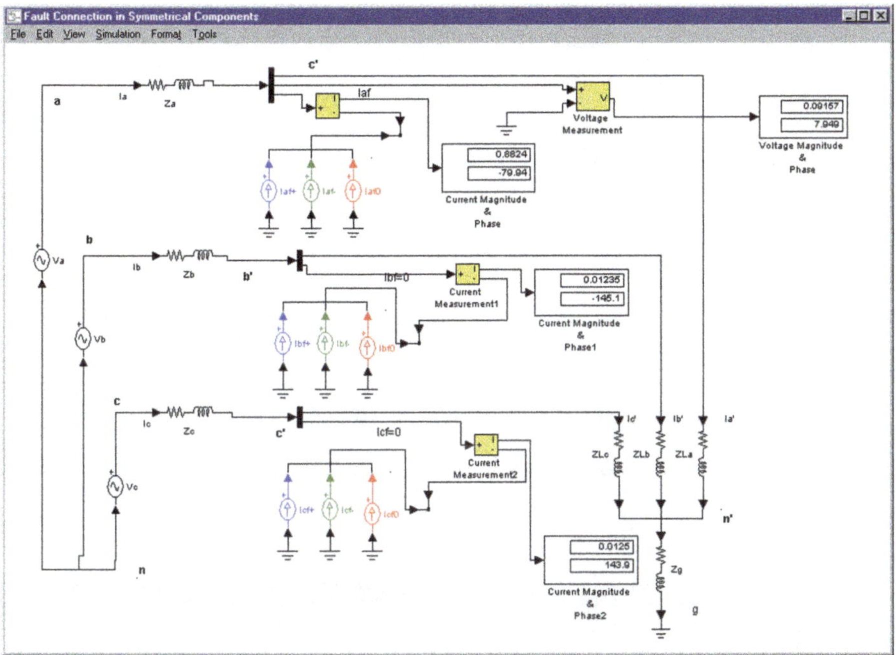

Fig. 2.14 Replacement of fault currents by symmetrical components

flows through the fault impedance in phase *a* is substituted by current sources that represent its symmetrical components. In phase *b* and *c* there is no fault, therefore the fault current is zero. Current sources that represent the symmetrical components of the fault currents in these phases are added to the model. As the fault currents in these non-faulted phases are zero, the sum of the symmetrical components should equal to zero. In Fig. 2.14, it can be checked that in fact, the sum of the symmetrical components of the fault current in phase *b* is zero. The same happens for phase *c*.

As the system is linear, superposition can be applied to the system shown in Fig. 2.13 and it can be decomposed in the positive-, negative-, and zero-sequence networks. The three-phase voltage source is balanced, so it only has components in the positive-sequence network.

Note that in Figs. 2.14, 2.15, 2.16, and 2.17, the currents and voltages are displayed using measuring blocks (yellow boxes), in which, if connected as shown, the current and voltage phasors are obatined, respectively.

2.2.2.2 Positive Sequence Network

If only the positive-sequence sources are left in the system of Fig. 2.13, the remaining system is balanced. Therefore, per-phase analysis can be used to obtain

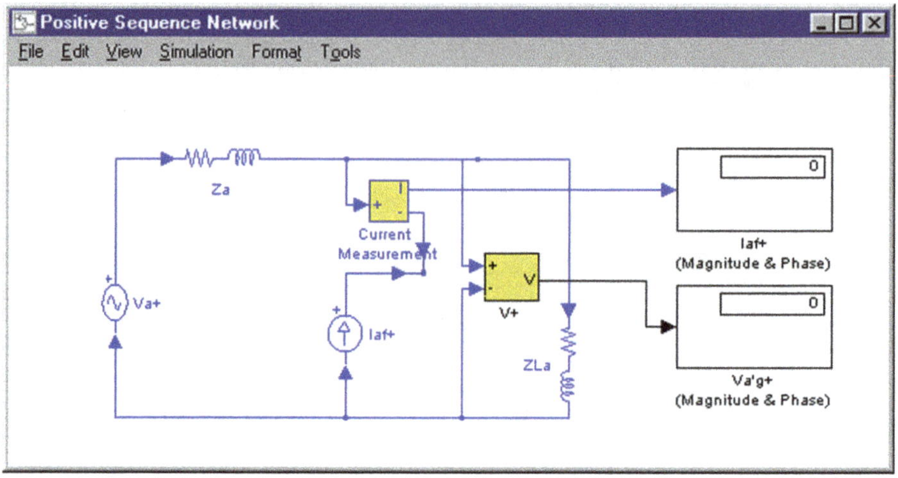

Fig. 2.15 Positive-sequence network

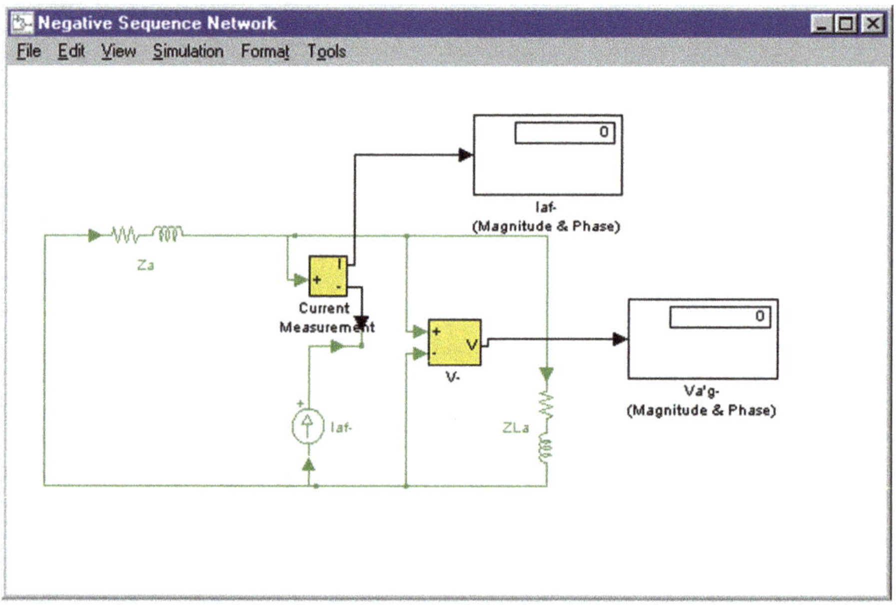

Fig. 2.16 Negative-sequence network

the variables of the phase a. The subsystem in Fig. 2.15 shows the corresponding network. As a result of analyzing the circuit, we get:

$$V_{a'g}^{+} = \frac{Z_{La}}{Z_{La} + Z_a}\left(V_a^{+} - Z_a I_{af}^{+}\right) \tag{2.6}$$

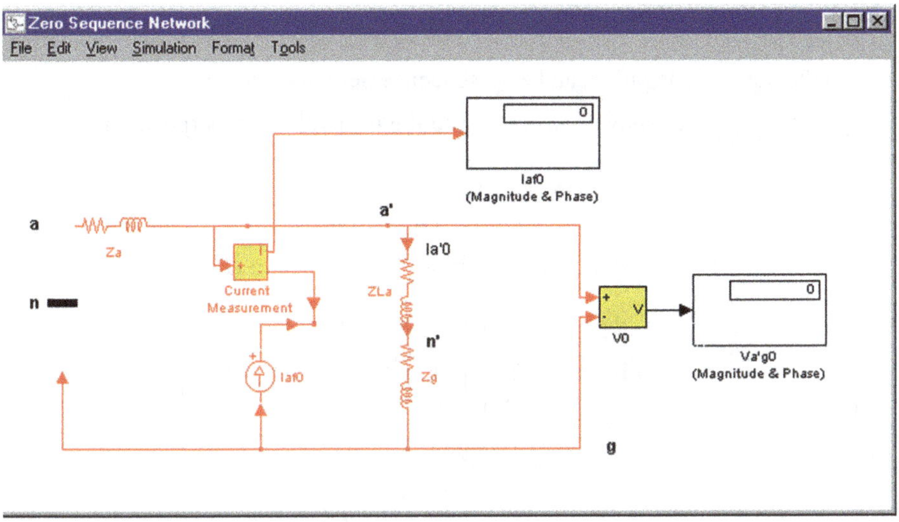

Fig. 2.17 Zero-sequence network

2.2.2.3 Negative Sequence Network

If only the negative sequence sources are kept in the model of Fig. 2.13, no source voltages will remain, but the system is still symmetrical and per-phase analysis can also be used to find the quantities in phase a. The corresponding system is shown in Fig. 2.16. As a result of analyzing the circuit, we get:

$$V_{a'g}^{-} = \frac{Z_{La}}{Z_{La} + Z_a}\left(-Z_a I_{af}^{+}\right) \tag{2.7}$$

2.2.2.4 Zero Sequence Network

As $I_a^0 = I_b^0 = I_c^0$ in Fig. 2.14, by applying KCL at node n ($I_a^0 + I_b^0 + I_c^0 = 0$), we get $I_a^0 = I_b^0 = I_c^0 = 0$. This means that in the zero-sequence network looking from the fault to the left side of the rest system, we see an open circuit. On the other hand $I_{a'}^0 = I_{b'}^0 = I_{c'}^0$, so looking to the right we get $V_{a'g}^0 = Z_{La} I_{a'}^0 + 3Z_g I_{a'}^0$ or $V_{a'g}^0 = -Z_{La}$ $I_{af}^0 - 3Z_g I_{af}^0$ with $V_{a'g}^0 = V_{b'g}^0 = V_{c'g}^0$.

From above equations, $V_{a'g}^0$ can be calculated from the single-phase zero-sequence network shown in Fig. 2.17. Note that in this circuit the value of the impedance connected from n' to g is $3Z_g$ in order to preserve the equivalence.

2.2.2.5 Connected Sequence Networks

From the positive, negative, and zero sequence networks shown above, we get $V^{+}_{a'g}$, $V^{-}_{a'g}$, and $V^{0}_{a'g}$ values shown. According to the principle of superposition, we have,

$$V_{a'g} = V^{+}_{a'g} + V^{-}_{a'g} + V^{0}_{a'g} \tag{2.8}$$

For a SLG fault, phase a to ground fault in this case using (2.4) we have,

$$\begin{bmatrix} I^{0}_{af} \\ I^{+}_{af} \\ I^{-}_{af} \end{bmatrix} = \frac{1}{3}\begin{bmatrix} 1 & 1 & 1 \\ 1 & a & a^2 \\ 1 & a^2 & a \end{bmatrix}\begin{bmatrix} I_{af} \\ I_{bf} \\ I_{cf} \end{bmatrix} = \frac{1}{3}\begin{bmatrix} 1 & 1 & 1 \\ 1 & a & a^2 \\ 1 & a^2 & a \end{bmatrix}\begin{bmatrix} I_{af} \\ 0 \\ 0 \end{bmatrix} \tag{2.9}$$

then

$$I^{0}_{af} = I^{+}_{af} = I^{-}_{af} = \frac{I_{af}}{3} \tag{2.10}$$

From expressions (2.8) and (2.10), it is clear that $V_{a'g}$ is the voltage across the series connection of the positive-, negative-, and zero-sequence networks when $I_{af}/3$ flows in the series connection. In addition, the terminal constraint

$$V_{a'g} = Z_f I_{af} = 3Z_f\frac{I_{af}}{3} \tag{2.11}$$

can be introduced by placing an impedance with value $3Z_f$ across the series connection. The sequence networks connected using the above-mentioned conclusions are shown in Fig. 2.18.

From the above description of the system, it can be seen that the fault current and voltage, as well as the voltages in the non-faulted phases, can be obtained if the sequence networks that represent the system are connected as in Fig. 2.14. Observe that the sequence networks are connected in series. From this model, the positive, negative, and zero sequence voltages and currents can be obtained and the corresponding phase voltage and currents are calculated using (2.3).

2.3 Short-Circuit Analysis

This section describes the use of symmetrical components and networks to represent asymmetrical (unbalanced) states of three-phase networks. A fault is placed in the power system model. The sequence networks are formed to correspond to a given system. The networks are connected to reflect the type of the fault placed in the original system. Once the two models (the original and the one connected from

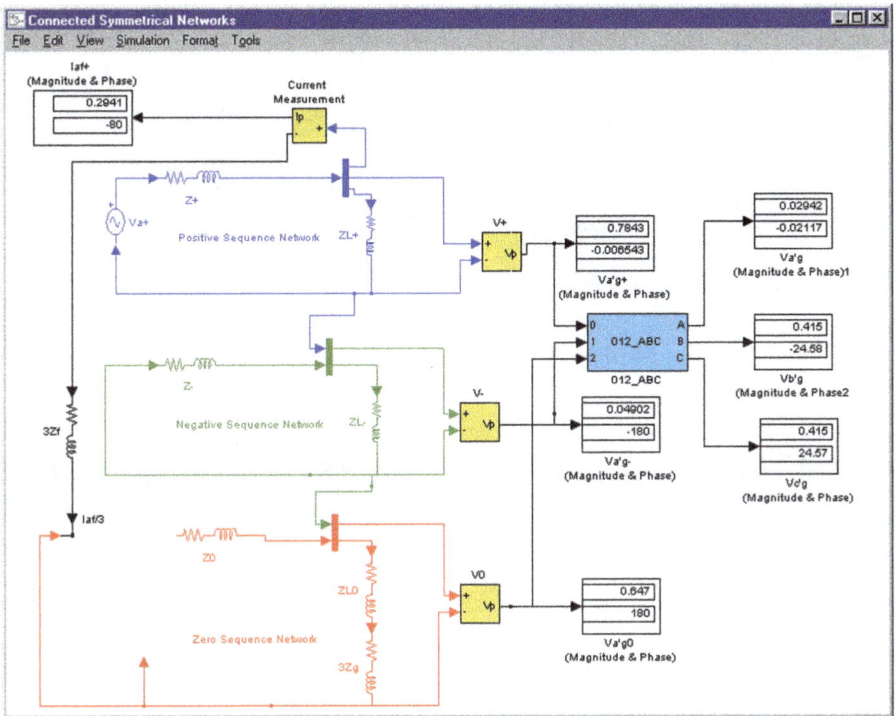

Fig. 2.18 Connected symmetrical networks

the sequence networks) are solved, the results should match. The phase voltages at the fault point and the fault current are to be compared for verification.

A Simulink module "SCo3.mdl" whose interface is shown in Fig. 2.19 is provided. It provides two options. The option on the left hand side is a simple three-phase one-machine power system model. Three ideal voltage sources supply the load via two three-phase impedances connected in series. A fault point is created between the two impedances. The voltage source works as an ungrounded Y, while the load works as a grounded Y. A SLG fault is placed in the system. The voltages in all three phases at the fault point are measured and displayed. The values of the voltages can be viewed in the time domain and in the phasor domain. In the latter case, the *abc* phasors are displayed in a figure at the end of the simulation; the corresponding symmetrical components are also displayed.

The default system parameters are:

Voltage source (phase *a*): 1.0 p.u./0°/60 Hz
Line impedances: $0.01 + j0.05$ p.u.
Load impedance: $0.72 + j0.54$ p.u.

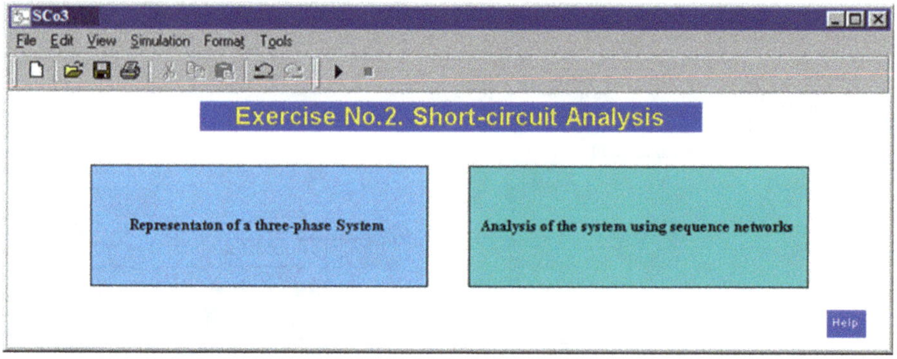

Fig. 2.19 Module interface for short-circuit analysis

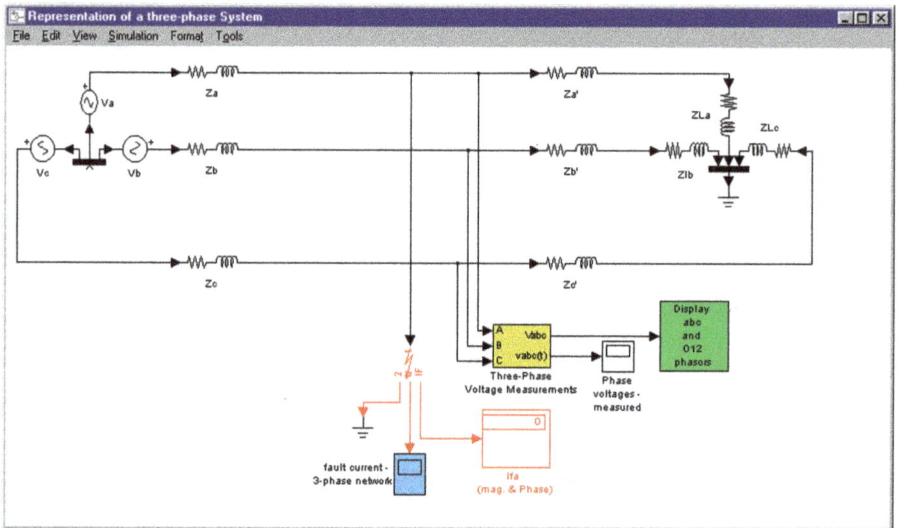

Fig. 2.20 Three-phase model of the system

If the option indicated in the left side of Fig. 2.19 is selected, the three-phase system is connected to a SLG fault. This is shown in Fig. 2.20. The symmetrical voltages at the fault location (V_1, V_2 and V_0) are supplied to a simple 012 → ABC converter and displayed for comparison with the voltages in the original three-phase model. Also, the fault currents from the original system and from the symmetrical networks are displayed. This selection is shown in Fig. 2.21. Note that there is a block named "Display abc and 012 phasors". This display shows, at the end of the simulation, the phasors in *abc* domain with both magnitudes and phases, as shown in Fig. 2.21. If the "continue" button of that display is pressed, the display will show the zero sequence phasor and how it is formed from the *abc* phasors. If the button is pressed again, the positive sequence phasor will be

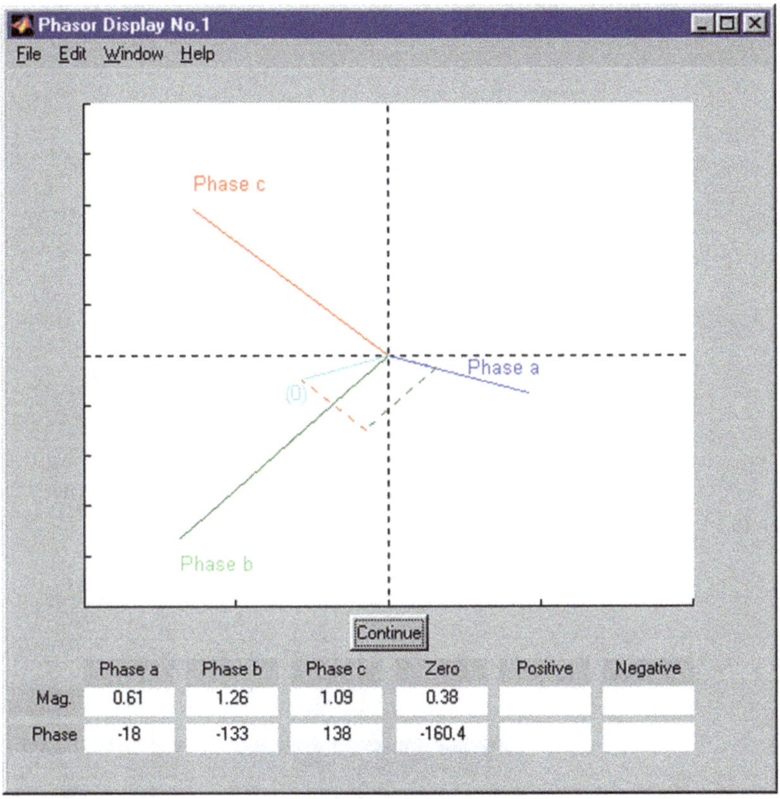

Fig. 2.21 Phasor display

displayed, and if pressed for a third time, the negative sequence phasor will be shown.

In the option shown on the right hand side of Fig. 2.19, three sequence networks for the three-phase system shown in Fig. 2.22 are included and connected to a SLG fault. The symmetrical voltages at the fault location (V_1, V_2, and V_0) are supplied to a simple 012 → ABC converter and displayed for comparison with the voltages in the original three-phase model. Also, the fault currents from the original system and from the symmetrical networks are displayed. The phasor representation of all these quantities is also available for comparison with the phasors obtained in the original three-phase system. Note that there is also a block that converts the symmetrical component phasors to the *abc* domain.

To execute the module for short-circuit analysis, simply run MATLAB and type SCo3. The models can be edited in a very easy and convenient way (*measurement element*) (Fig. 2.23).

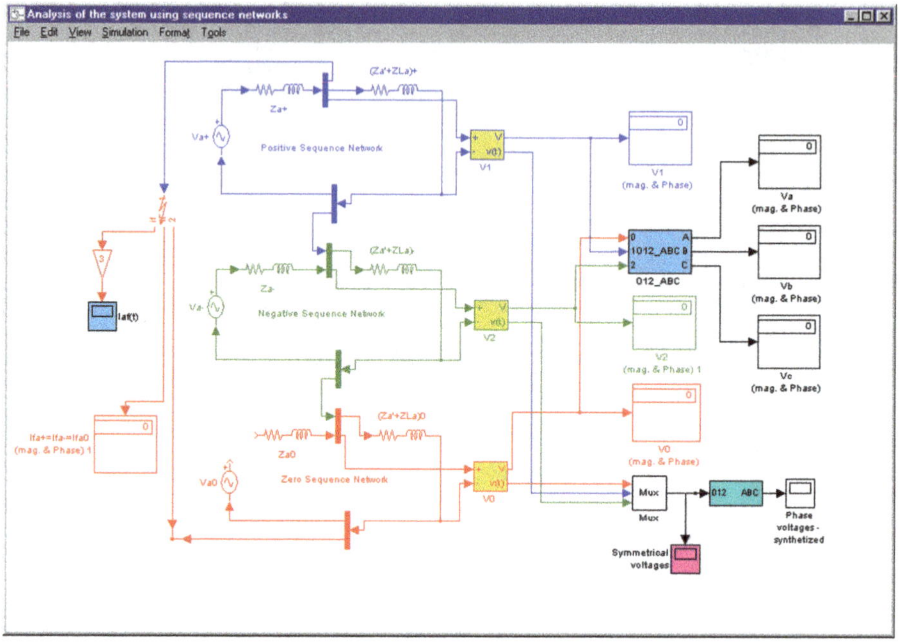

Fig. 2.22 Sequence networks for the original network

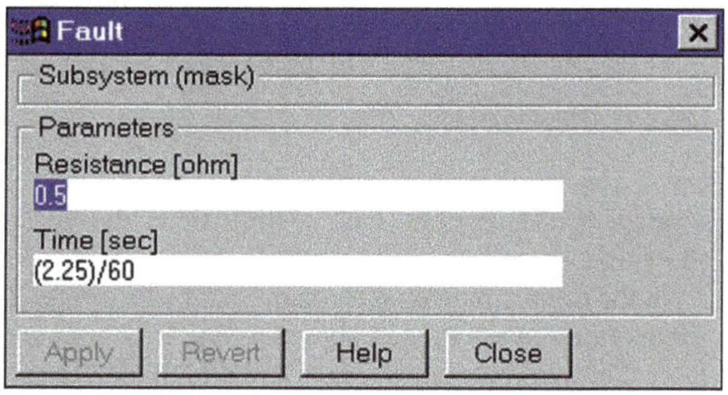

Fig. 2.23 The data entry for the fault model

2.4 Sequence Networks

2.4.1 Transmission Line

2.4.1.1 Transposed and Rotated Lines

Two models of the symmetric networks for a three-phase transmission line are shown in Figs. 2.24 and 2.25. The first of them shows the model of a transposed transmission line (two rotations) and can be used to model a fully symmetrical as

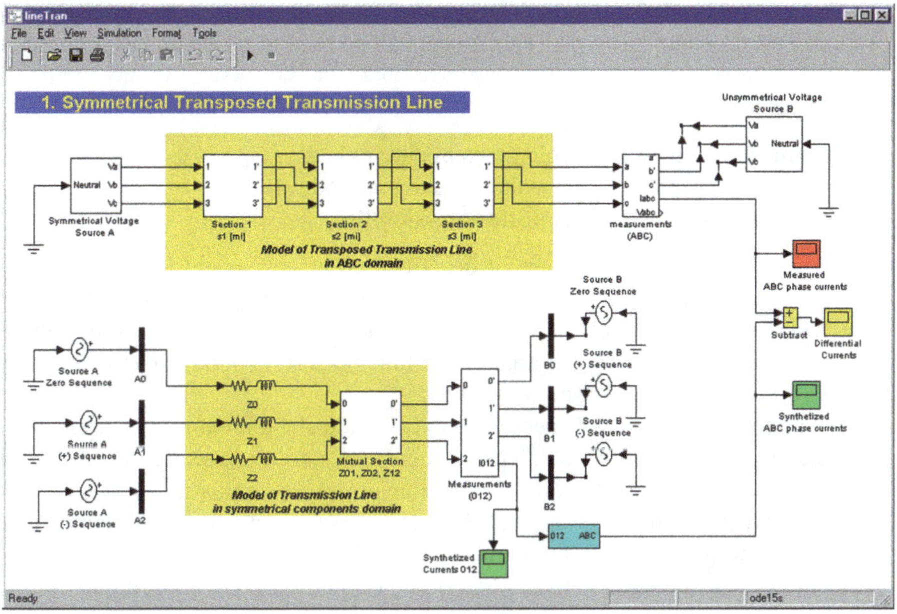

Fig. 2.24 Simulink model of transposed transmission line (two rotations)

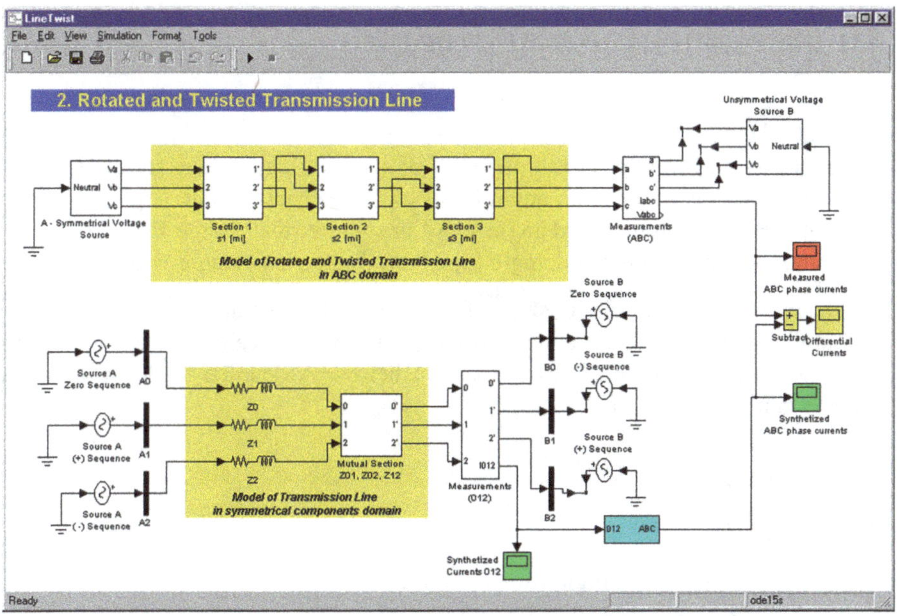

Fig. 2.25 Simulink model for unsymmetrical transmission line (rotation and twist)

well as an unsymmetrical line. The second figure describes a transmission line with one rotation and one twist. This model can be used to model an unsymmetrical line only.

Each model consists of:

- Symmetrical three-phase voltage source at node A
- Unsymmetrical three-phase voltage source at node B
- Three independent sections of transmission line

The parameters of every element described above are stored in the ASCII file *data.txt* (the format of the data is described in the MATLAB script files: "*lineTw.m*" and "*lineTr.m*"). The user can change the values of the parameters by simply modifying the file data.txt. Originally, the file "*data.txt*" contains parameters of a 69 kV, 40 miles length transmission line.

To run the simulation, it is necessary to calculate all parameters that are going to be used by the Simulink model. These calculations are done using the MATLAB scripts: "*lineTr.m*" and "*lineTw.m*", for transposed line and rotated/twisted line, respectively. After all necessary calculations, these scripts create the output files containing the results of the computations and the execution of the corresponding simulation. The files generated (ASCII) are outputtransp.txt and outputtwist.txt. The format for the input data is shown below:

Raa Xaa—self-impedances in Ohm per mile
Rbb Xbb
Rcc Xcc
Rab Xab—mutual impedances in Ohm per mile
Rbc Xbc
Rac Xac
s1 s2 s3—length of line sections in miles
EA angEA—magnitude [V] and angle [°] of symmetrical voltage source A
EBa angEBa—magnitude and angle [°] of unsymmetrical source B-phase *a*
EBb angEBb—magnitude and angle [°] of unsymmetrical source B-phase *b*
EBc angEBc—magnitude and angle [°] of unsymmetrical source B-phase *c*

The most important relationships of voltages and impedances in the *abc* domain and in the symmetrical components domain used for the calculations of the transposed and rotated/twisted lines are shown below. A detailed explanation of the process can be found in [1].

$Z_{abc} = A \times Z_{012} \times A^{-1}$
$Z_{012} = A^{-1} \times Z_{abc} \times A$
$E_{abc} = A \times E_{012}$
$E_{012} = A^{-1} \times E_{abc}$
$Z_{abc} = R^{-1} \times Z_{312} \times R$—single rotation 3-1-2
$Z_{abc} = T^{-1} \times Z_{132} \times T$—twist 1-3-2
$Z_{abc} = z_{123} \times s_i$, s_i—length of ith section of the line [mi]

where:

$$A = \begin{bmatrix} 1 & 1 & 1 \\ 1 & a^2 & a \\ 1 & a & a^2 \end{bmatrix}; \quad a = 1\angle 120^{\circ}$$

$$R = \begin{bmatrix} 0 & 1 & 0 \\ 0 & 0 & 1 \\ 1 & 0 & 0 \end{bmatrix} = \text{Rotation matrix}$$

$$T = \begin{bmatrix} 1 & 0 & 0 \\ 0 & 0 & 1 \\ 0 & 1 & 0 \end{bmatrix} = \text{Twist matrix}$$

$\boldsymbol{Z_{abc}}$ = mutual impedance of the line in [Ω]
z_{123} = mutual impedance of the line in [Ω/mi], when the conductors are in position 1, 2, and 3, respectively.

The matrix $\boldsymbol{Z_{012}}$ is not diagonal in general. One exception is when a transmission line is completely symmetrical. In that case, matrix $\boldsymbol{Z_{012}}$ is diagonal, having the nonzero elements: $\boldsymbol{Z_{00}}$, $\boldsymbol{Z_{11}}$, $\boldsymbol{Z_{22}}$.

2.4.1.2 Parallel Lines

The Simulink model used for demonstrating symmetrical components of parallel lines is shown in Fig. 2.26.

The model of the parallel lines in *abc* domain has three groups of ideal three-phase voltage sources connected via two parallel three-phase transmission lines. The three-phase voltage sources on the left hand side of the transmission lines are unbalanced, and the other two are balanced voltage sources. The phase currents of the two lines are measured and displayed.

The voltage sources have Y-grounded connection. Their parameters are:

Source 1 (Unbalanced)

Phase *a*: 10.0 V∠00/60 Hz
Phase *b*: 10.0 V∠1800/60 Hz
Phase *c*: 10.0 V∠900/60 Hz

Source 2 (Balanced)

Phase *a*: 8.0 V∠200/60 Hz

Source 3 (Balanced)

Phase *a*: 8.0 V∠-200/60 Hz

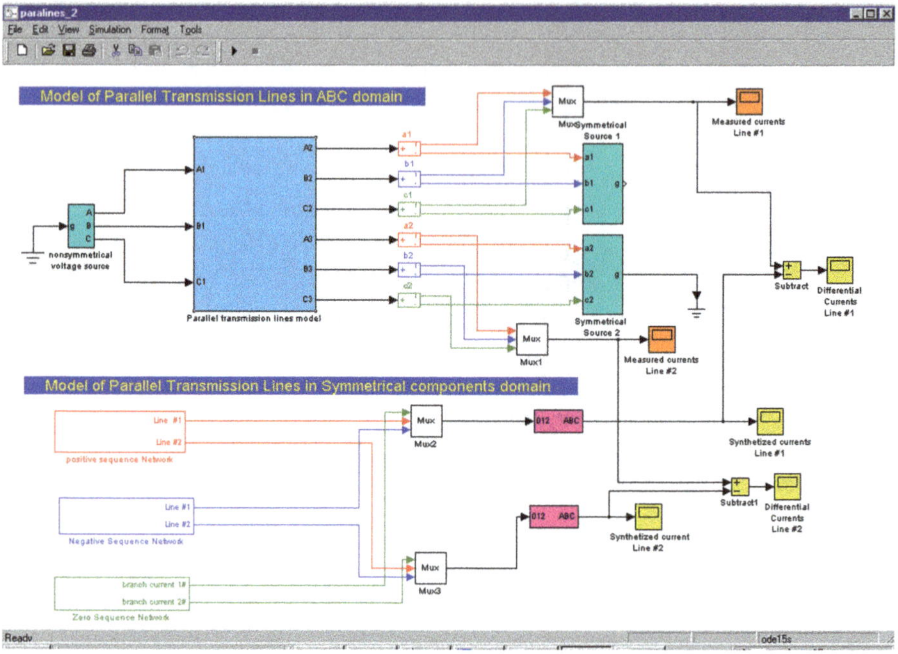

Fig. 2.26 Models of the parallel transmission lines in the *abc* and symmetrical components domains

The parameters for both parallel lines are:

$R = 1\mathrm{e} - 3\,\Omega$
$L = 0.5/(2 \times \pi \times 60)H$
$M_1 = 0.1/(2 \times \pi \times 60)H$
$M_2 = 0.01/(2 \times \pi \times 60)H$
M_1: mutual coupling between the phases in the same line
M_2: mutual coupling between the phase in one line and the phase in another line

The model of the symmetrical components domain consists of three independent networks (as there is no fault). The voltage sources of the positive, negative, and zero sequence sources that correspond to the voltage source to the left of the transmission lines are nonzero, as the source is unbalanced. As the other two sources are balanced, they are nonzero only for the positive network. The symmetrical branch currents ($I+$, I, I_0) are measured and converted to the phase currents (I_a, I_b, I_c) to display, in order to compare with the currents measured in the *abc* model.

The model of the parallel lines is shown in Fig. 2.27. It is divided into the impedance of each line and the mutual coupling between the two lines. Both lines are symmetrical.

The self-impedances of the lines are modeled with three-phase RL, positive, zero-sequence impedance block. In this case the positive and zero sequence

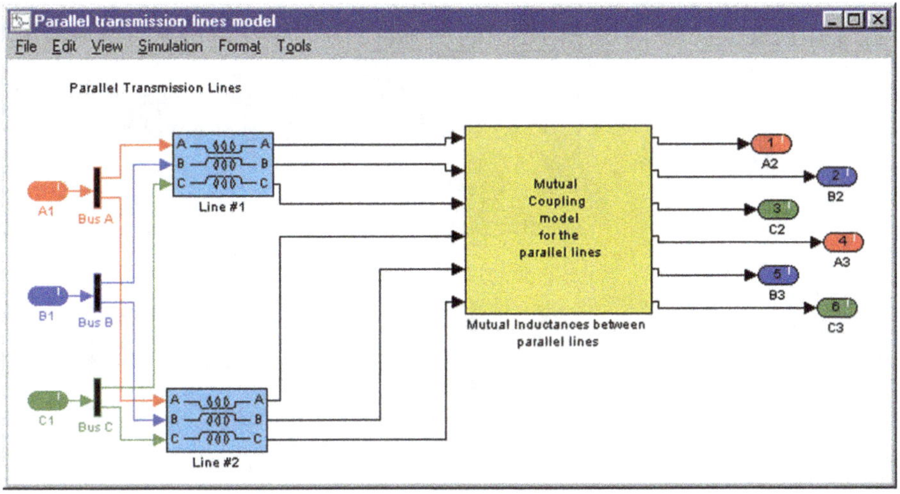

Fig. 2.27 Model of the parallel transmission lines

Fig. 2.28 Dialog box of the transmission line block

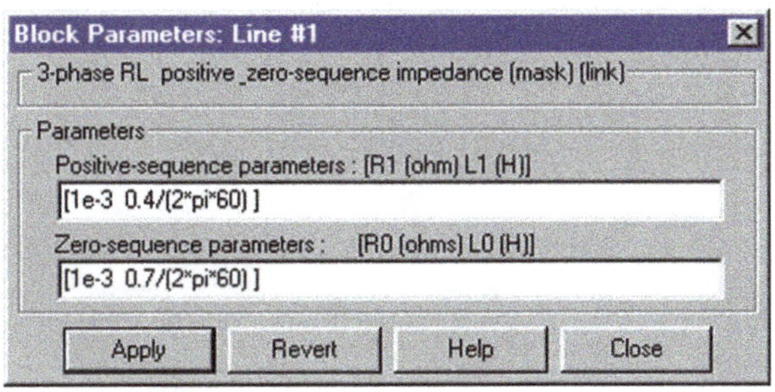

impedances must be entered. The self-impedances of the lines are in the Z1–Z0 model (named "line #1" and "line #2"). For this model two parameters are needed (see Fig. 2.28). One is the "Positive sequence parameter" and the other is the "Zero sequence parameter". For the model, the self-inductance of one phase is $L = 0.5/(2\pi f)$, and the mutual inductance between phases of the same line is $M_1 = 0.1/(2\pi f)$. As it is assumed that both lines are symmetrical, we have

$$L_{\text{positive}} = L_{\text{negative}} = L - M_1 = 0.4/(2\pi f)\text{H}$$

and

$$L_{\text{zero}} = L + 2M_1 = 0.7/(2\pi f)\text{H}$$

The resistances of the phases are the same for the positive and zero sequences.

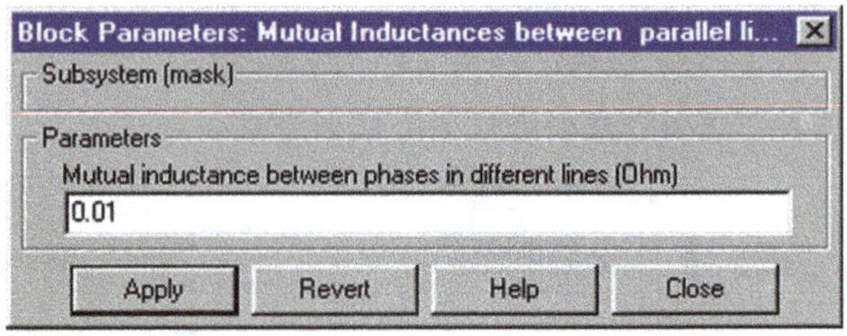

Fig. 2.29 Dialog box for the mutual coupling between the lines

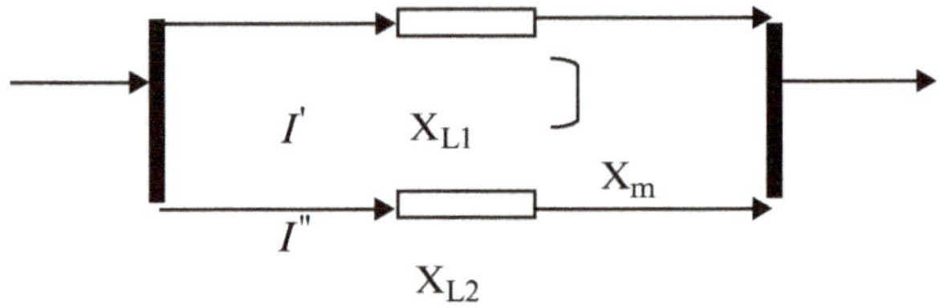

Fig. 2.30 Simplified model of parallel lines

$$R_{\text{positive}} = R_{\text{zero}} = R_l = (1\text{e} - 3)\ \Omega$$

The mutual inductances between the phases of the two lines are considered to be equal. Its value must be given in ohms (Fig. 2.29).

The calculations needed to obtain the impedances for the positive, negative, and zero sequence networks are shown below. Consider the following model (Fig. 2.30):

When considering the positive network, the voltage drop of the phase *a* of the first line is

$$V'_{a1} = I'_{a1}X'_{la1} + \left(I''_{a1} + I''_{b1} + I''_{c1}\right)X_{mp}$$

X_{mp} is the mutual reactance between a given phase in one line and other phase in the other line. As in the positive sequence network, we know that

$$I''_{a1} + I''_{b1} + I''_{c1} = 0$$

Thus, we can conclude that there is no mutual coupling between the two lines in the positive sequence network. For the negative sequence network, we can get the same conclusion similarly.

When considering the zero sequence network, the voltage drop of the phase *a* of the first line is:

$$V'_{a0} = I'_{a0}X'_{la0} + \left(I''_{a0} + I''_{b0} + I''_{c0}\right)X_{mp} = I'_{a0}X'_{la0} + 3I'_{a0}X_{mp}$$

Similarly, we have the voltage drop of the phase *a* of the second line

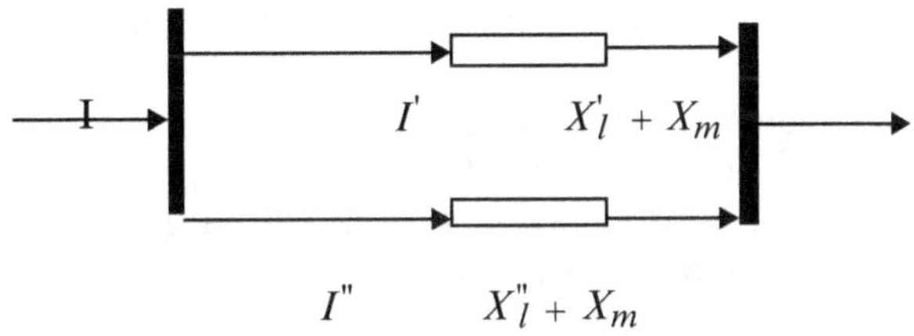

Fig. 2.31 Zero sequence equivalent model (general case)

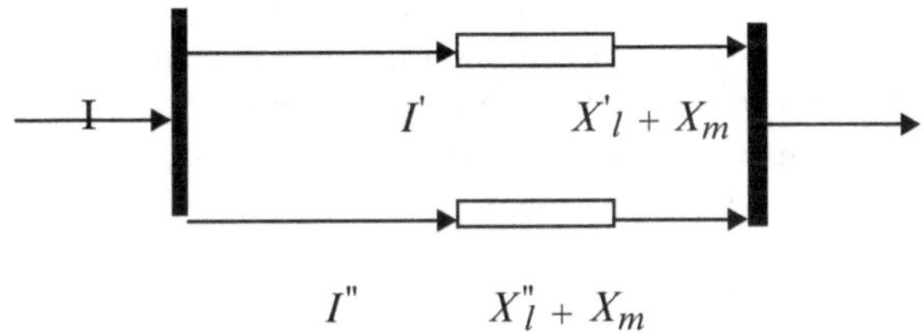

Fig. 2.32 Zero sequence equivalent model (identical lines)

$$V''_{a0} = I''_{a0}X''_{la0} + 3I'_{a0}X_{mp}$$

Assuming that

$$I_{a0} = I'_{a0} + I''_{a0} \quad \text{and} \quad X_m = 3X_{mp}$$

Solving these equations, we have:

$$V'_{a0} = I'_{a0}\left(X'_{la0} - X_m\right) + I_{a0}X_m$$
$$V''_{a0} = I''_{a0}\left(X''_{la0} - X_m\right) + I_{a0}X_m$$
$$V'_{a0} = V''_{a0}$$

From above equations, we can get the equivalent circuit of parallel lines of the zero sequence (Fig. 2.31):

X_l is the parameter that does not consider the mutual coupling between the two lines $X_m = 3X_{mp} = 3(\omega M2) = 0.03\ \Omega$. If the parameters of the two lines are identical, we can simplify the zero sequence model in the following form (Fig. 2.32).

The model illustrates the use of sequence networks to solve the asymmetrical states of three-phase networks. The sequence networks are formed to correspond to the original system as far as the structure and parameters are concerned. Because the line parameters are symmetrical and there are no faults, the sequence networks are decoupled.

2.4.1.3 Symmetrical Line with Tap Loads

To represent the symmetrical components model of a line with tap loads of delta and wye types, four models were developed. These Simulink models for wye and delta load are shown in Figs. 2.33 and 2.34, respectively. The two models are divided into two parts: the model in *abc* domain and the model in symmetrical components domain.

The first part of the both figures shows a simple three-phase transmission line. The line is symmetrical with a tap and a wye and delta load connected, respectively. One end of the transmission line is connected to a grounded Y voltage source, and the other end is short-circuited. Currents supplied to the transmission line and voltages at the tap point for all three phases are measured and displayed.

The system parameters are:

- Voltage sources (phase *a*): 1.0 p.u./0°/60 Hz
- Line impedance
  - Positive sequence: $0.01 + j0.05$ p.u.
  - Zero sequence: $0.02 + j0.1$ p.u.
- Load impedance: $0.72 + j0.54$ p.u.
- Ground impedance (wye connection only): 1 p.u. resistance

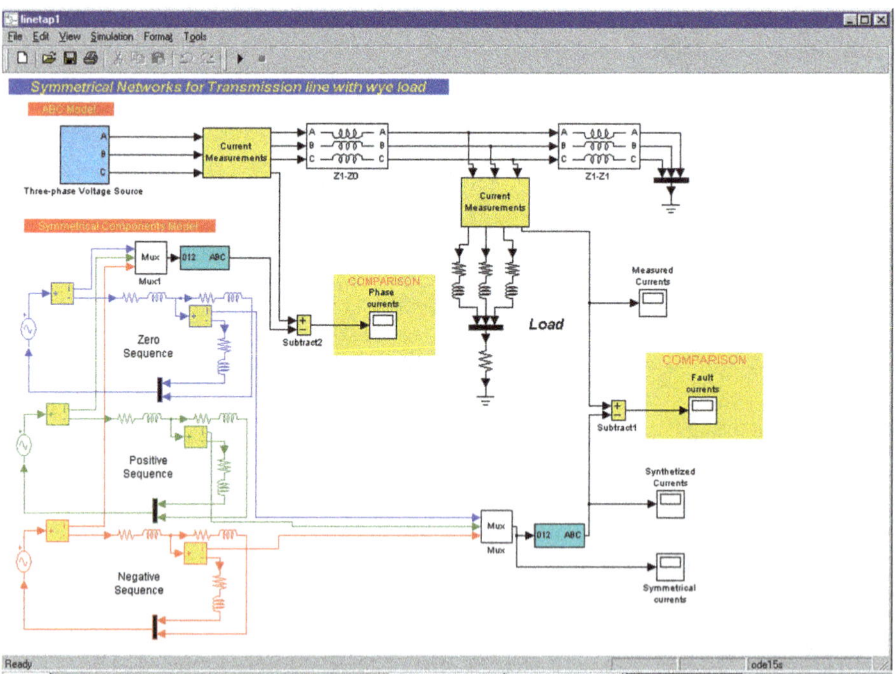

Fig. 2.33 Transmission line with wye-connected load

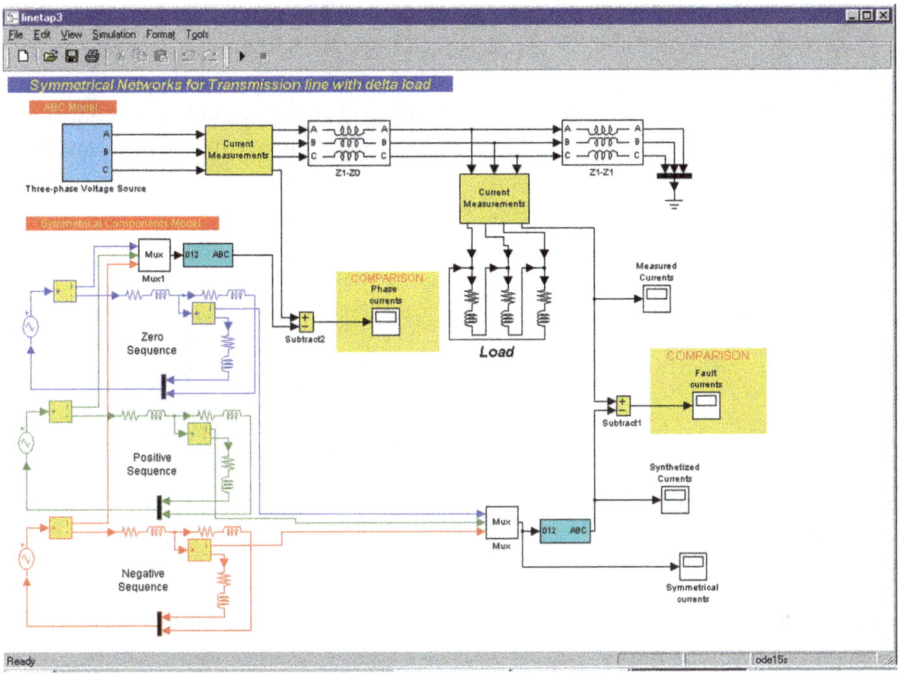

Fig. 2.34 Transmission line with delta-connected load

Since the transmission lines are tapped, we divide the line impedance corresponding to the location of the tap. For this model, it is assumed that the tap is located at 0.3 times the length of the transmission line from the generator. Thus, the line impedance on the left hand side of the tap becomes $0.3 \times (0.01 + j0.05) = 0.003 + j0.015$ p.u. for the positive sequence impedance and $0.3 \times (0.02 + j0.1) = 0.06 + j0.03$ p.u. for the zero sequence. And, to the right hand side of the tap point, it becomes $0.7 \times (0.01 + j0.05) = 0.007 + j0.035$ p.u. and $0.7 \times (0.02 + j0.1) = 0.014 + j0.07$ p.u. for the positive and zero sequence, respectively.

The second part consists of three symmetrical networks for the system shown in the first part. Currents coming out of the voltage sources and voltages at the tap locations are measured and supplied to a $012 \rightarrow abc$ converter and displayed for comparison with the corresponding currents and voltages in the original system.

2.4.2 *Load Model*

Four different models of load are presented in this section. They are:

- Wye-ungrounded
- Wye-grounded
- Wye-grounded with an impedance to ground
- Delta

Also, symmetrical and unsymmetrical loads are modeled.

In each model, a simple three-phase network is included along with its three symmetrical component networks. The method of symmetrical components is a tool for dealing with unsymmetrical problems that may occur in a multi-phase system. In the case of a three-phase system, there are three symmetrical components involved.

They are:

- Positive-sequence component
- Negative-sequence component
- Zero-sequence component

The sequence voltages are able to be found from these following equations:

$$V_0 = (1/3)(V_{\mathrm{A}} + V_{\mathrm{B}} + V_{\mathrm{C}})$$
$$V_1 = (1/3)(V_{\mathrm{A}} + aV_{\mathrm{B}} + a^2V_{\mathrm{C}})$$
$$V_2 = (1/3)(V_{\mathrm{A}} + a^2V_{\mathrm{B}} + aV_{\mathrm{C}})$$

where

$a = (-0.5 + j0.866)$
V_{A}, V_{B}, V_{C} are phase voltages,
V_0, V_1, V_2 are voltages of zero-sequence, positive-sequence, and negative-sequence, respectively.

2.4.3 Two-Winding Transformer

In this part, symmetrical component networks for three-phase two-winding transformers will be described, including various types: $\mathrm{Y_g}$–Δ, Δ–$\mathrm{Y_g}$, and $\mathrm{Y_g}$–Y. The Simulink models for the transformers with these connections are shown in Fig. 2.35. Note that all these transformers are composed of three independent single-phase transformers.

Regardless of the connection method, the positive and negative sequence networks of a three-phase transformer have the same impedance as a single-phase transformer that composes the three-phase transformer. However, the zero sequence impedance depends on the connectivity. Generally speaking, the delta connection side provides a loop for the zero sequence current that flows inside the transformer windings. The grounded-wye connection side provides a path for the zero sequence current to the ground. The ungrounded-wye side does not provide a path for the zero sequence current. Another point to be noted is that there is a phase shift for both the positive and negative sequence components for transformers of the

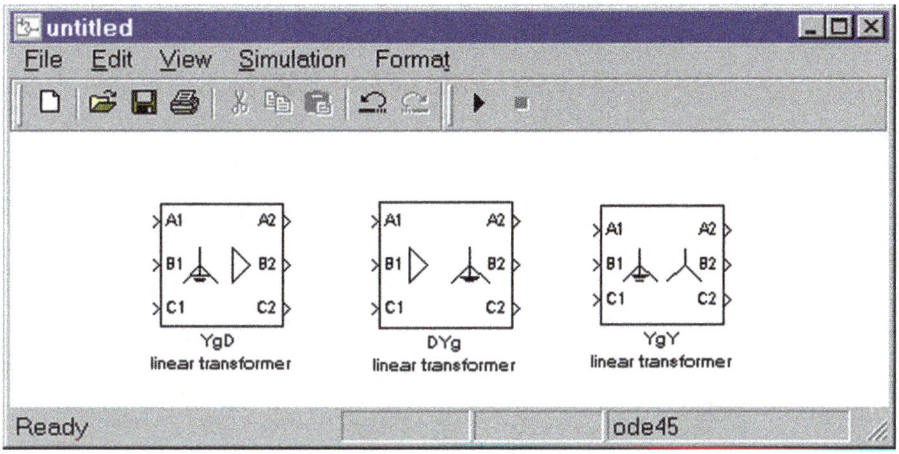

Fig. 2.35 Y_g–Δ, Δ–Y_g, and Y_g–Y transformers

types wye-delta and delta-wye, no matter if the wye side is grounded or not. So appropriate phase shifts need to be considered in building the sequence networks.

2.4.4 *Synchronous Machine*

Three Simulink models were built to demonstrate the symmetrical components of the synchronous machine. The first model, shown in Fig. 2.36, demonstrates the subtransient, transient, and steady state positive sequence impedance for the Synchronous machine. In this case it is used to represent a generator. In the second model (Fig. 2.37), the negative sequence impedance of the synchronous machine in steady state is demonstrated. In this case the synchronous machine is functioning as a motor. Finally, in the third case (Fig. 2.38), the modeling of the three sequence networks is evaluated in a system where an unbalanced source feeds the synchronous machine.

2.4.5 *Positive Sequence Network Model*

A three-phase short-circuit fault is placed at the machine terminals. The positive sequence current is measured and compared with the currents of symmetrical networks. Because the fault occurs at $t = 0.64$ s, transport delay blocks are used. There is also a delay in Clock.

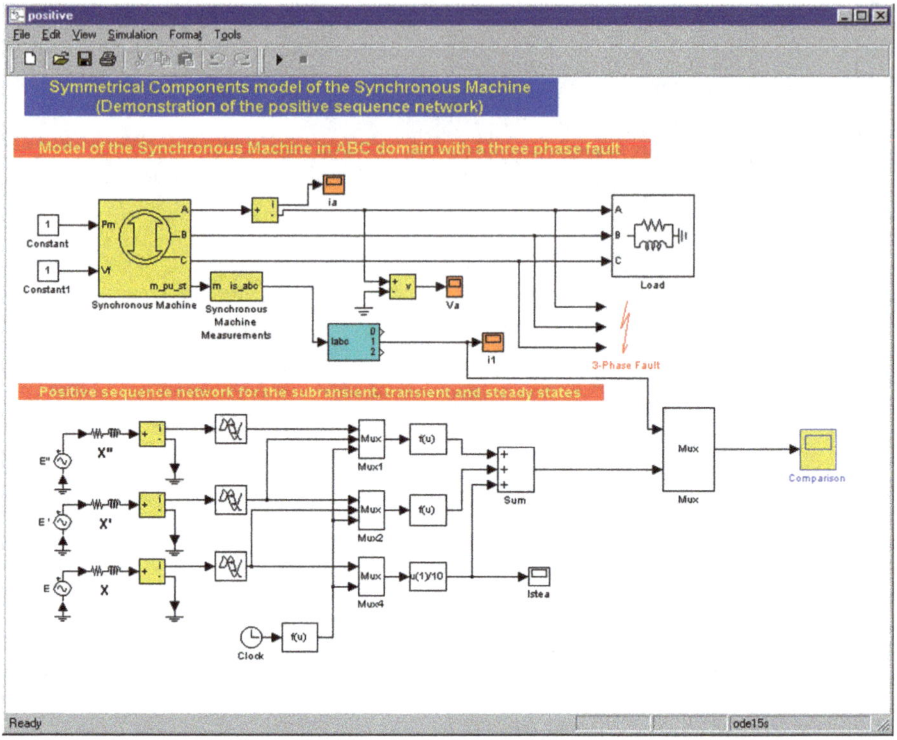

Fig. 2.36 Positive sequence network

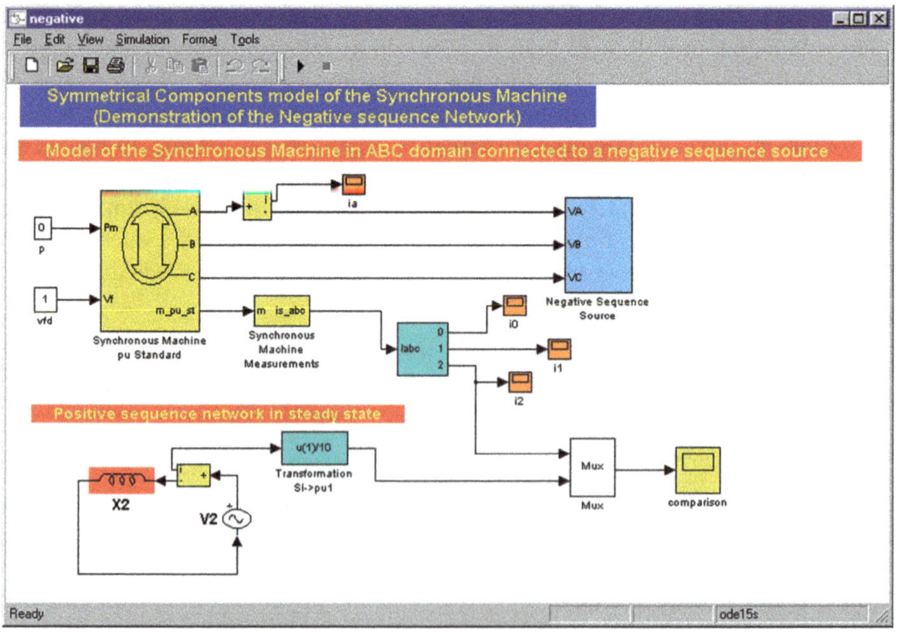

Fig. 2.37 Negative sequence network

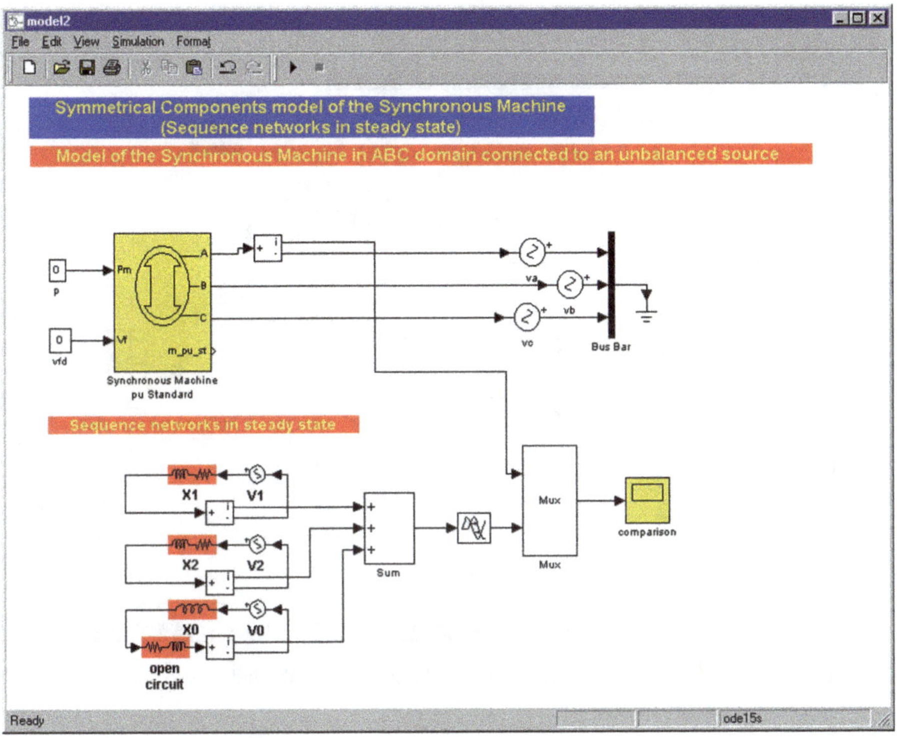

Fig. 2.38 Steady state sequence networks for the synchronous machine

The pre-fault values of E'', E', and E are calculated by the following formulas:

$$E'' = Va + jIa \times Xd'' = \sqrt{Ua^2 + \left(Ia \times Xd''\right)^2}$$

$$E' = Va + jIa \times Xd' = \sqrt{Ua^2 + \left(Ia \times Xd'\right)^2}$$

$$E = Va + jIa \times Xd = \sqrt{Ua^2 + (Ia \times Xd)^2}$$

2.4.6 Negative Sequence Network

A synchronous machine is connected with a negative sequence voltage source. This model is used to demonstrate the negative sequence reactance.

2.4.7 Sequence Networks in Steady State

This model shows the sequence parameters of the synchronous machine in steady state.

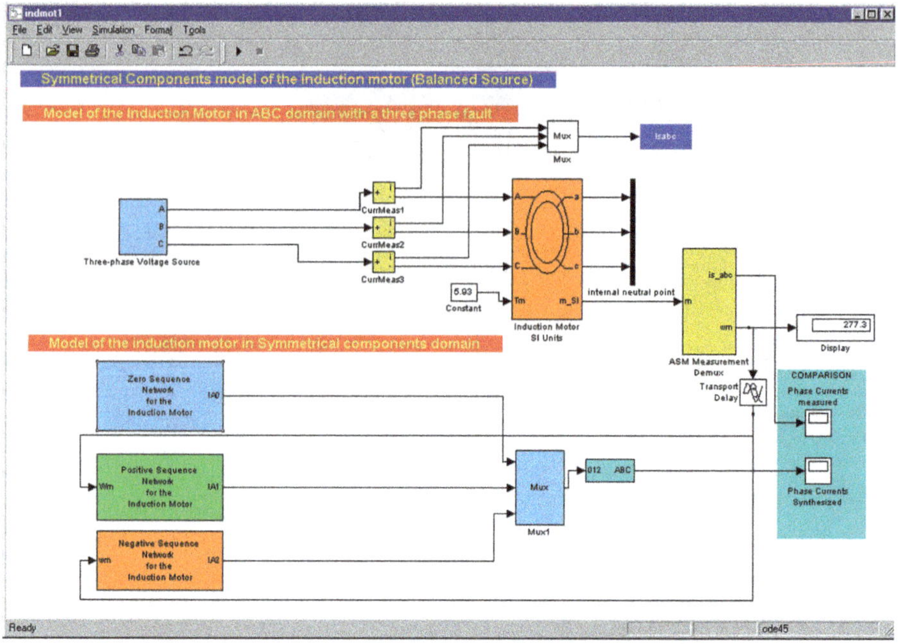

Fig. 2.39 Model in *abc* domain and symmetrical networks for the induction motor

2.4.8 *Induction Motor*

A Simulink model illustrating the symmetrical networks for the three-phase induction machine was created. The model consists of an ideal three-phase voltage source connected to the induction motor. The Simulink model is shown in Fig. 2.39.

The model is divided into two major parts. In the upper half of the Simulink diagram, a simple three-phase power system (including one induction machine) can be seen. The symmetrical networks for this simple three-phase power system are derived and implemented in the lower half of the block diagram.

As seen in Fig. 2.39, since the induction motors are usually wounded either for Δ or ungrounded Y connection, the zero sequence currents in the motor are always zero. The equivalent circuits for positive and negative sequences are directly taken from [1], and they can be seen in Fig. 2.40.

It should be noted that, the system shown in Fig. 2.39 contains a three-phase balanced source. Thus, the negative sequence equivalent circuit does not contain any voltage source. However, in the case of unbalanced situations, the negative sequence equivalent circuit has also a nonzero voltage source.

In the representation of these equivalent circuits in Simulink, the shunt resistances (iron losses) are neglected. With the rotor turning, however, the equivalent impedance of the rotor circuit is a function of the slip "*s*" as noted in Fig. 2.40. During the simulation, the slip value is calculated at each time step, by using the

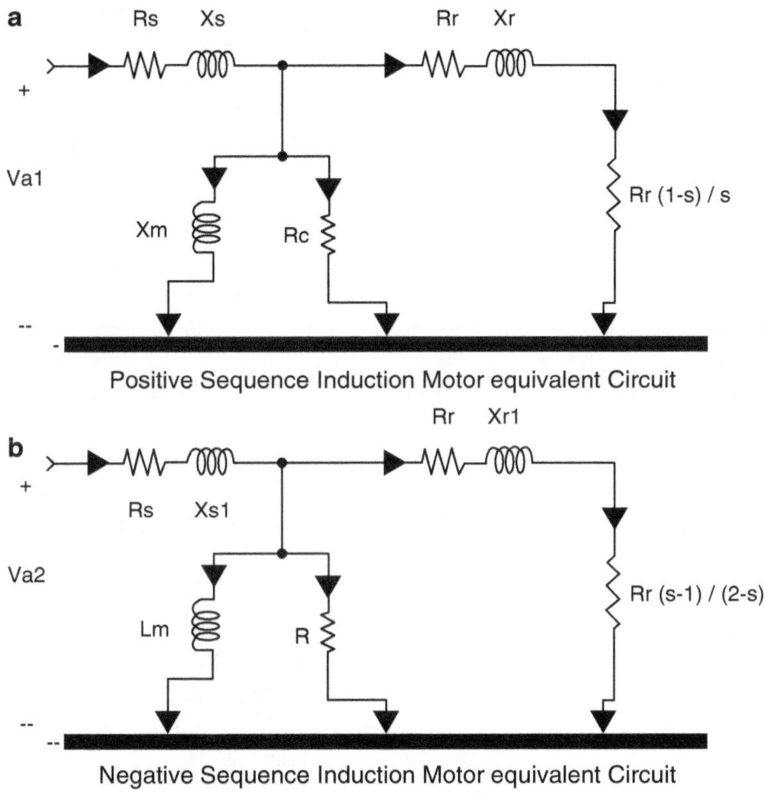

Fig. 2.40 Positive and negative sequence equivalent circuits

value of rotor speed, which is supplied by the induction motor model provided with the SimPowerSystems in Simulink. The calculation of the slip factor can be formulated simply as follows:

$$s = \frac{w_s - w}{w_s}$$

where

w_s = synchronous speed in rad/s
w = rotor speed in rad/s (obtained from the model)

The slip factor is calculated by using simple linear arithmetic Simulink blocks. The expressions:

$Rr\frac{(1-s)}{s}$ is for the positive sequence and $Rr\frac{(s-1)}{(2-s)}$ is for the negative sequence. The voltage drops are across the equivalent rotor impedance in positive and negative sequence. These voltage drops are calculated in Simulink using linear arithmetic blocks with the positive and negative sequence currents and the slip as inputs. The equivalent rotor impedance then is represented as a controlled voltage source.

Additionally, the nameplate values of the induction motor are directly obtained from the sample simulation in Simulink demos. Here, only the initial torque (i.e., the initial load on shaft) is calculated by using the following equation.

$$T_{\mathrm{m}} = P_{\mathrm{m}}/w \text{ (Nm / phase)}$$

During simulation, the currents obtained from symmetrical networks are transferred to the phase domain equivalents by using the block "012-to-*abc*" at each time step. At the end of simulation, the currents obtained from the induction motor model (i.e., measured currents) and the ones obtained from symmetrical networks (i.e., synthesized currents) should be compared in order to see if they match or not.

2.5 Matrix Method for Short-Circuit Calculation

The admittance matrix as well as the impedance matrix can be used to compute the fault currents in a power system. The characteristics of each of these matrices are described in exercises. The exercises illustrate admittance and impedance approaches for the short-circuit studies. After completion of this exercise, one will have an understanding of these approaches. The admittance and impedance approaches are explained in the following paragraphs.

2.5.1 Matrix Computation Approach

To obtain the short-circuit calculations, the Matlab programs are provided. The programs enable:

- Calculation of impedance matrices for both the positive sequence and zero sequence network of a given power system. This includes the step-by-step building of the matrices with explanation of each step.
- Calculation of the voltages at all the buses and the currents in all the lines for a given location of a fault. The program deals with SLG and three-phase faults.

The input data are stored in a separate file. The data of two systems are included in the following files:

- The 6-bus IEEE system with 100 MVA base (and6.dat)
- The 14-bus IEEE system with 100 MVA base (and14.dat)

The systems and related data may be found in [1] p. 480 for the 6-bus network, and p. 487 for the 14-bus network.

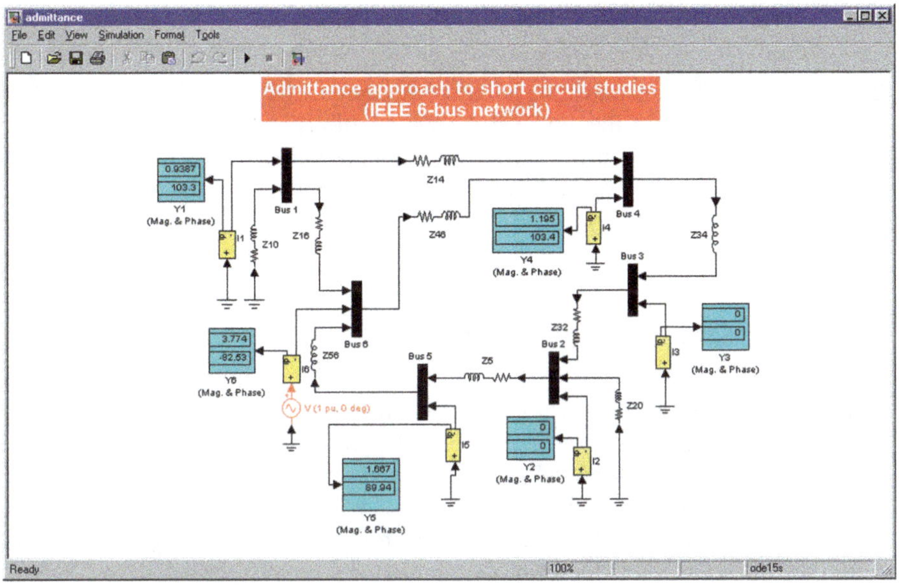

Fig. 2.41 Admittance approach to short-circuit studies

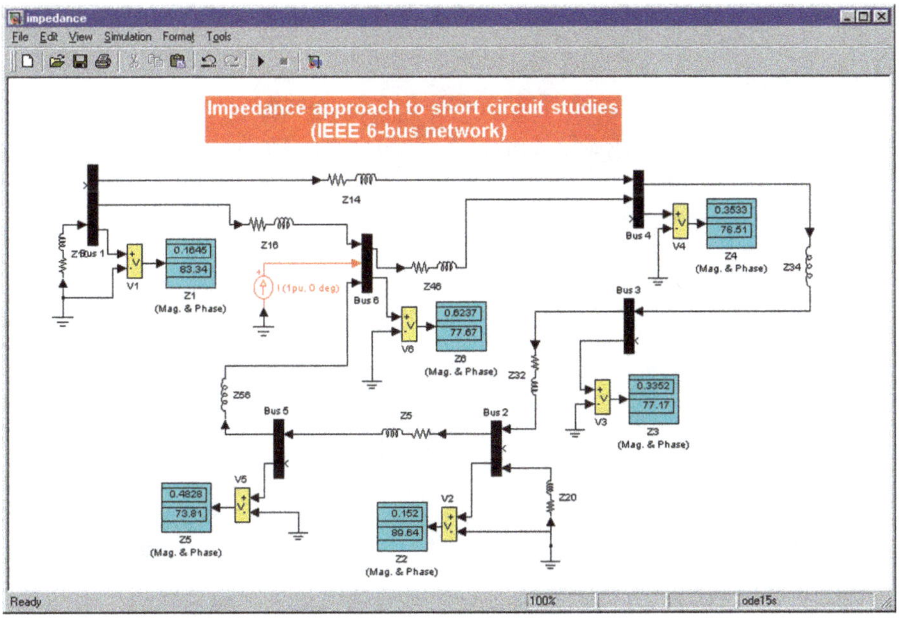

Fig. 2.42 Impedance approach to short-circuit studies

2.5.2 *Admittance and Impedance Approaches*

The model that demonstrates the admittance approach is shown in Fig. 2.41. Type in "admittance" to open the model. To run simulations, just press the "play" button.

The model that demonstrates the impedance approach is shown in Fig. 2.42. Type in "impedance" to open the model. To run simulations, just press the "play" button.

2.6 Summary

In this chapter, the materials for learning the basics of power system fault analysis and short-circuit calculation are described. The basic theory of symmetrical components and sequence networks is presented with examples implemented in MATLAB-based software. Then, the basics for creating sequence networks for common power system elements, transmission lines, loads, transformers, synchronous machines, and inductions motors are provided. Finally, the matrix methods for performing short-circuit analysis in large power systems are provided in both elementary and advanced learning levels.

Reference

1. Anderson PM (1995) Analysis of faulted power systems. IEEE Press, Piscataway

Chapter 3
Basics of Protective Relaying and Design Principles

3.1 Introduction

This chapter focuses on the basics of power system relaying with special attention paid to the overcurrent, impedance, and differential protection.

The MERIT software for those examples is a set of SIMULINK models in which:

- A single-phase model of a simple power system is developed using the Power System Blockset.
- Circuit Breakers (CBs), as well as Voltage and Current Transformers (VTs and CTs), are modeled as ideal elements.
- Appropriate relays are modeled using their generic description.
- The protective equipment (CBs, VTs, CTs, and relays) are connected together to enable closed-loop simulation, i.e., the trip signals of the relays are fed back to the CBs.

The configuration and parameters of the models may be changed by cut-and-paste and drag-and-drop manipulations with the power system elements' icons.

The sample exercises for this chapter include:

- Perform power system simulations of selected faults and observe how a given protection principle (overcurrent, impedance, and differential) works.
- Set the relays for a given power system.
- Verify by simulation that the relays operate as expected.
- Model malfunctioning of the protective equipment and verify operation of the back-up protection functions.

© Springer International Publishing Switzerland 2016

M. Kezunovic et al., *Design, Modeling and Evaluation of Protective Relays for Power Systems*, DOI 10.1007/978-3-319-20919-7_3

3.2 Overcurrent Relaying

3.2.1 Introduction

One of the basic strategies for protecting the power systems is overcurrent protection. When a fault happens in power systems, the current magnitude increases; the overcurrent relays measure fault current and compare it with the predefined thresholds (settings). If the current level increases more than the threshold value, after predefined time delay, trip command is issued and the corresponding circuit breaker operates and isolates the faulted area. In this section the principle of the overcurrent relay operation is discussed. The following issues are explained and covered by the MATLAB models and related simulations:

- Rules for protecting a network using overcurrent relays.
- Requirements for instrumentation (number and locations of instrument transformers) and switching apparatus (number and locations of circuit breakers).
- Analysis of the normal load conditions for selecting the instrument transformer ratio and setting the relays.
- Analysis of the fault conditions for selecting instrument transformer ratio and setting the relays.
- Setting and coordinating the relays.
- Simulation of the radial network protected with overcurrent relays. Checking the relay operation and coordination including protective equipment failures (relay and circuit breaker maloperation).

3.2.2 Relaying Basics

Figure 3.1 shows a simple radial network consisting of an equivalent system source (it typically represents a transformer connected to the transmission network) and two line segments (often called feeder segments) connected between three busbars. A number of loads are supplied from the busbars. This network is of the "radial" type because it is supplied from one source and does not contain any loops.

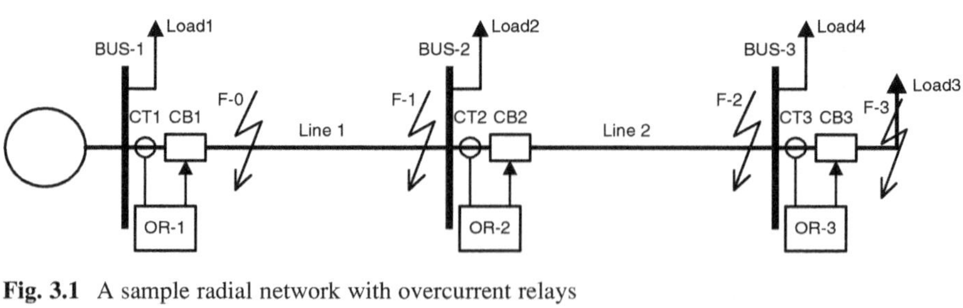

Fig. 3.1 A sample radial network with overcurrent relays

Radial networks can be protected against faults using the overcurrent relaying principle. Consider a power system element of a radial network that is located at the longest distance from the supplying source (loads 3 and 4 in the system of Fig. 3.1). If a fault occurs in this element, the fault current that is significantly larger than the normal load current flows. An instantaneous Overcurrent Relay (OR) is a device that measures the magnitude of a current and compares it against a threshold. If the current is higher than the threshold (i.e., a fault current is detected), then the relay operates by sending a signal to the Circuit Breaker (CB) to open the circuit (trip) and disconnect the faulted element from the rest of the system. For example, if the fault F-3 occurs in the system of Fig. 3.1, OR-3 operates and trips CB that removes the load 3 from the rest of the system. In this way the rest of the network keeps operating providing the power supply to the loads 1, 2, and 4.

Note:

- In order to protect a given element, one needs a Current Transformer (CT) to measure the current. The CTs should be installed at the element's terminal that is closer to the supplying source.
- One needs also a CB (installed at the terminal closer to the supplying source) to be able to disconnect the element from the supplying source in the case of a fault.
- An overcurrent relay must sense the current magnitude and compare it against the threshold.
- A fault in a given element will be cleared after the time equal to the operating time of the relay and the operating time of the corresponding CB.

Extending the above observations, one places the CB-2, CT-2, and OR-2 at the left terminal of the line 2 in order to protect this line. However, the OR-2 cannot distinguish between the faults F-3 (OR-2 must not operate) and F-2 (OR-2 must operate). Therefore, OR-2 must wait (certain time delay is applied) for the slowest relay protecting the lines and loads connected to the busbar 3 to operate. The ORs with fixed delay are called definite-time overcurrent relays. If the fault prolongs beyond the longest clearing time of the elements connected to the busbar 3, then the relay OR-2 should operate (the fault is not downstream from the line 2, thus it must be on the line 2 or the fault is downstream from the line 2 and OR-3 failed to operate).

Therefore:

- The ORs constituting the protection system must apply certain time delays in order to ensure selectivity of operation.

Similarly, to protect the line 1, one installs the CB-1, CT-1, and OR-1 between the line and the supplying source. The OR-1 must be coordinated with the relay downstream (i.e., OR-2) to ensure selectivity of operation.

The definite-time overcurrent principle is illustrated in Fig. 3.2. The figure shows the time delays of the relays in the protection system of the network.

The protection system of Fig. 3.2 operates as follows:

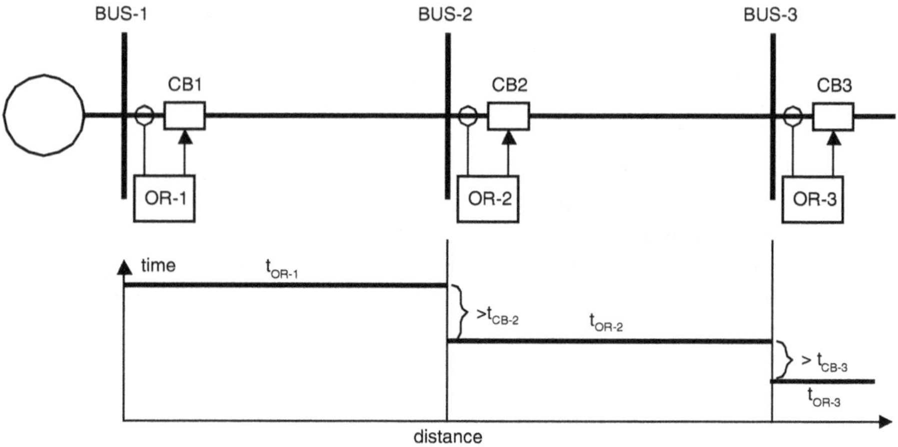

Fig. 3.2 Illustration of the definite-time overcurrent protection principle

- If there is a fault on one of the feeder elements, many or all relays upstream (i.e., between the fault and the supplying system) will pick-up (activate) and start timing-out. Since only the relay closest to the fault is supposed to trip, it has the shortest delay. After the relay trips, the remaining upstream relays will re-set and no more trips will take place. The relays installed downstream from the fault will not pick-up because there is no fault current flowing there. The part of the network downstream from the fault will be cut-off by the relay that operates. The part of the network upstream will operate normally after the fault is cleared.
- If the relay that is supposed to operate or the corresponding breaker fails to operate, the fault will be cleared by the next relay upstream. The upstream relay will operate in such a case as back-up protection. If it fails as well, the next relay upstream should operate, etc. In such a case, more network elements will get tripped. However, bearing in mind both the fault and equipment maloperation, the action is still optimal.

The following may be noted:

- To protect a radial network using ORs, one needs the CBs, CTs, and relays installed between each power system element and the supply system.
- The closer a fault to the source, the higher the fault current, but also, the longer the clearing time. The combination of these facts is the major disadvantage of this protection method. To improve the protection and shorten the average clearing time, one may use relays with the delay depending on the fault current magnitude, thus on fault location.

- According to IEC standard [1], the characteristic of inverse-time overcurrent relays (excluding induction type) can be described by the following expression:

$$T = \frac{C}{\left(\frac{I}{I_s}\right)^{\alpha} - 1}$$

where:

T is the relay operation time.
C is the constant for relay characteristic.
I_s is the current setting threshold.
I is the current detected by relay (normally the effective value) $I > I_s$.
α is the constant representing inverse-time type, $\alpha > 0$.

Depending on the value of α, different type of overcurrent relays are available: definite time, moderately inverse, inverse, very inverse, and extremely inverse. They are selected based on the location of the relays and coordination strategy for the relays [2].

3.2.2.1 Analysis of Load and Fault Conditions

- The load and fault conditions must be analyzed in order to select the CTs and CBs as well as to set the relays.
- The power flow and short-circuit calculations are very straight forward in radial networks and will not be discussed here.
- The fault locations that need to be considered are those producing the minimum and maximum fault currents for each feeder in the system (at the ends of each feeder segment, respectively).
- The maximum fault current is used to select the CBs and the CTs. The minimum fault current is used to check the sensitivity of the protection system.

3.2.2.2 Selecting the CTs and CBs

The following basic parameters of a CT should be considered from the protective relaying standpoint:

- The rated primary current should be higher than the load current during normal load conditions or acceptable overloads.
- The rated secondary current should match the rated input current of used relays (typically 5 A or 1 A).
- A CT should not saturate under the maximum fault current flowing through the CT in a given network (i.e., for resistance-less faults just downstream from the CT).

The following basic parameters of a CB should be considered:

- The maximum current that can be interrupted by a CB must be higher than the maximum fault current flowing through this CB in a given network (i.e., for faults just downstream from the CB).
- The maximum operating time of a CB must be known for proper setting of the corresponding overcurrent relay.

3.2.2.3 Selecting and Setting the Relays

The following parameters of an overcurrent relay should be considered:

- The rated relay input current should be matched with the rated secondary current of the corresponding CT.
- The setting ranges (pick-up current and time delay) should match the needs of a given relay application (location).

There are two settings of the definite-time overcurrent relay: the pick-up current and the time delay. The setting coordination process starts from the relay that is most distant from the source and progresses upstream relay by relay.

Pick-up Current

The pick-up current should be set as low as possible, but high enough to avoid picking up during overloading, transient, and switching conditions. Typically, the pick-up current is set at 120–150 % of the maximum load current. One must remember that the relay is set for the secondary amperes, and therefore, the load current must be first re-calculated into the secondary values by dividing it by the CT ratio.

Time Delay

The time delay is set according to the following rule:

$$\begin{aligned}\text{time delay} = &\ \text{maximum time delay of the relays downstream}\\ &+ \text{maximum CB operating time} + \text{security margin}\end{aligned}$$

The security margin depends on the accuracy of the installed relays and the variations of the operating time of the CB. Typically, it is in the range of 0.3–0.5 s.

Both the pick-up current and time delay are set in certain steps. When coordinating the relays, one should consider the actual time dials available on the relay (resulting from the existing steps), not the values calculated prior to setting the physical relays.

3.2.2.4 Sensitivity Check

After setting the relays, one should consider faults at the end of each line (feeder segment) and check if the relay protecting the line (primary protection) and at least one relay upstream (back-up protection) will pick-up for such minimum fault current.

3.2.3 *Software Models*

After completion of these simulation examples, the readers will learn how to protect a radial network using definite-time overcurrent relays. Particularly, the following issues are re-enforced: load flow and short-circuit calculations, selecting the protective equipment, setting and coordinating overcurrent relays, relay sensitivity check, analysis of the network operation under variety of conditions including faults and equipment maloperations.

The system shown in Fig. 3.1 has been implemented as a MATLAB model with the use of the Power System Blockset for modeling the network and SIMULINK for modeling the relays. In addition to the elements shown in Fig. 3.1, the Voltage Transformers (VTs) are added to facilitate voltage measurement. Also a number of measurement elements are connected to display the most critical signals in the model such as currents, voltages, and trip signals.

For simplicity, the model is a single-phase model. Therefore, the only fault type considered at this time is the single-line-to-ground fault.

The following models are used.

Voltage Source

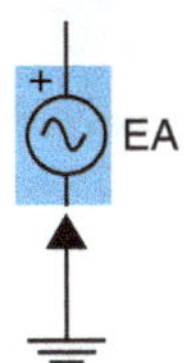

Double click on the icon to change the magnitude, phase, and frequency.

Lines and System Impedance

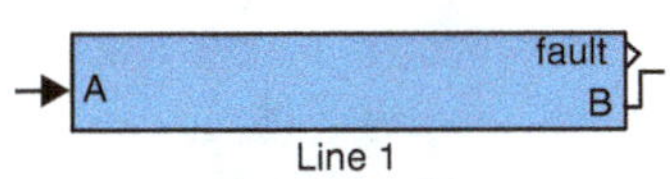

Double click on the icon to change the resistance, reactance, and the fault location. Third terminal is created that divides the line. The fault distance is counted from the terminal A of the line. To place a fault connecting the fault icon to the third terminal.

Fault

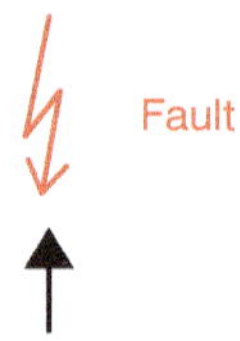

Double click on the icon to change the fault resistance and inception time. Connect the icon's only terminal to the place where the fault should occur.

Load

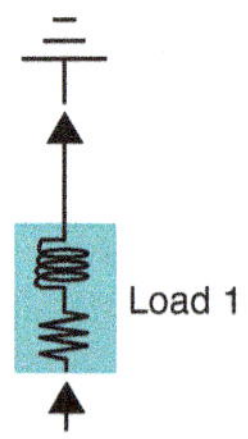

Double click on the icon to change the rated voltage, active and reactive power, and the frequency.

Current Transformer

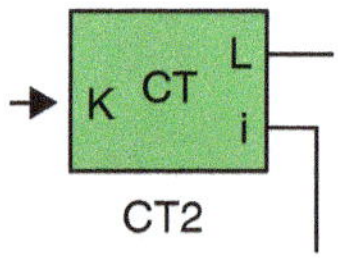

The transformer should be connected between the K and L terminals. The current signal is available at the "i" terminal. Double click on the icon to change the rated primary and secondary current. The transformation ratio is defined as the ratio between the primary and secondary current.

Voltage Transformer

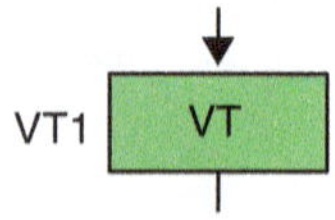

The transformer should be connected to a busbar. The voltage signal is available at the only output terminal. Double click on the icon to change the rated primary and secondary voltages. The transformation ratio is defined as the ratio between the primary and secondary voltage.

Circuit Breaker

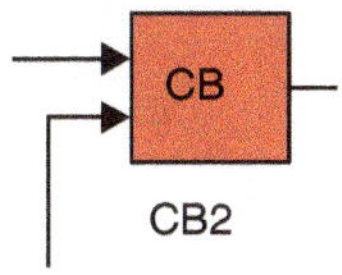

Double click on the icon to change the CBs operating (delay) time. The second input is the input for the trip signal (if the control signal is high, the CB opens after its operating time expires and when the current waveform goes through zero).

Scope (Measurement Element)

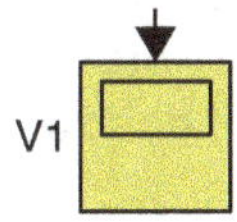

This element is used for displaying signal waveforms. Double-click to open a display window. Use the toolbar of the window to zoom in and out fragments of the waveform of your interest.

Display

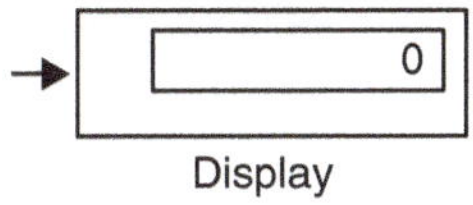

This element is used to display a numerical value of the signal at a given time instant. The block does not need to be opened; it just displays the numerical values on its icon.

Definite-Time Overcurrent Relay

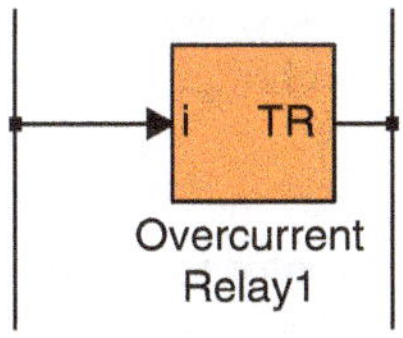

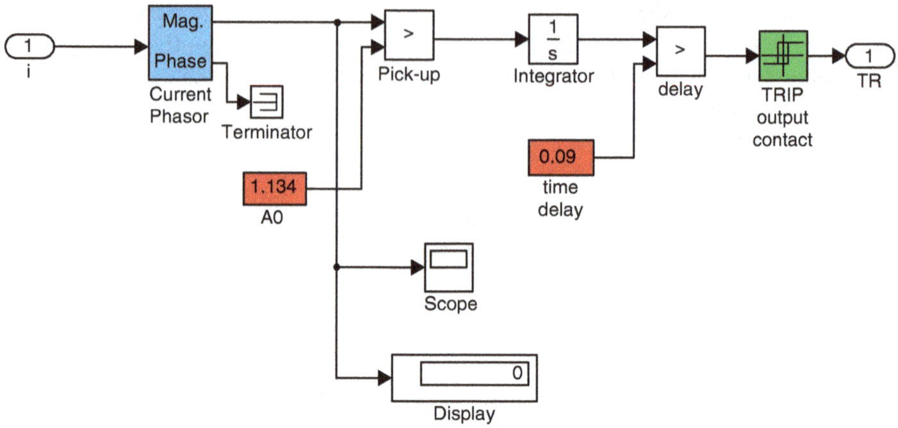

Fig. 3.3 The functional model of a definite-time overcurrent relay

The input terminal "i" should be fed with the current and the output terminal "TR" (for trip) should be fed-back to the CB.

Double-click to open a new window with the functional model of the relay (Fig. 3.3). In this model:

- The waveform of the current is processed into a phasor.
- The phasor's magnitude is utilized and compared with the threshold A0 (double click on the A0 block to change the pick-up current).
- The result comparison is integrated to reflect the time delay. The integrator's output is compared with the time threshold (double click on the block to change the delay).
- The hysteresis is used to model the relay operation (once tripped, the relay does not reset by itself).

3.2.3.1 Model Activation

The model is available as OCRe.mdl file. To use the model, put the file into the default working directory of MATLAB (typically MATLAB/BIN), run MATLAB, and type "OCRe."

The following are useful hints on how to use the model:

- Double-click on an element to change its parameters or see what is inside the element.
- Double-click on a scope (measurement element) to open it and see the plot.
- Use the zooming buttons of the plot to zoom in or out.
- Highlight (click-on) and press DEL to delete a given connection between two blocks.
- Draw a line between the terminals to connect the elements.

Default Data:

The model has been developed with the following default data:

- System voltage: 158 kV (114 % of 138 kV)
- System impedance: $1 + j10\ \Omega$
- Line 1 impedance: $2 + j20\ \Omega$, fault location 0.5
- Line 2 impedance: $2 + j20\ \Omega$, fault location 0.5
- Load 1: 138 kV, 100 MW, 30 MVAr
- Load 2: 138 kV, 100 MW, 30 MVAr
- Load 3: 138 kV, 50 MW, 15 MVAr
- Load 4: 138 kV, 50 MW, 15 MVAr
- Fault: 0 Ω, inception time 20 ms
- CB-1 operating time: 15 ms
- CB-2 operating time: 20 ms
- CB-3 operating time: 25 ms
- CT-1: 1600/1 A
- CT-2: 1000/1 A
- CT-3: 500/1 A
- OR-1: pick-up 1.134 A, delay 90 ms
- OR-2: pick-up 0.907 A, delay 50 ms
- OR-3: pick-up 0.907 A, delay 5 ms

Note: The time data are re-scaled to speed-up the simulation. The actual values would be approximately ten times higher.

3.2.3.2 Numerical Example

This subsection presents calculations for selecting the instrument transformer ratio, setting the relay, and checking the relay's sensitivity.

Load Flow Calculations

The load current is calculated as (apparent power)/voltage. Using the data from previous section one obtains:

- Load 4 draws 0.378 kA.
- Load 3 draws 0.378 kA.
- Load 2 draws 0.756 kA.
- Load 1 draws 0.756 kA.
- The load current through CT-3 is 0.378 kA (Load 3).
- The load current through CT-2 is 0.756 kA (Loads 4 and 3).
- The load current through CT-1 is 1.512 kA (Loads 4, 3, and 2).

Selecting the CTs

Assume we can use CTs with the 1 A secondary current and the primary current of the following choices: 500, 1000, 1500, 1600, 2000, 2500 A.

- CT-3: we select 500 A (the closest value higher than 378 A). Thus, the secondary load current at CT-3 is 378/500 × 1 = 0.756 A.
- CT-2: we select 1000 A (the closest value higher than 756 A). Thus, the secondary load current at CT-2 is 756/1000 × 1 = 0.756 A.
- CT-1: we select 1600 A (the closest value higher than 1512 A). Thus, the secondary load current at CT-1 is 1512/1600 × 1 = 0.945 A.

Short-Circuit Calculations

- The fault current for the faults F-3 and F-2 (Fig. 3.1) is equal to (system voltage)/(system impedance + line 1 impedance + line 2 impedance).
- The fault current for the fault F-1 (Fig. 3.1) is equal to (system voltage)/(system impedance + line 1 impedance).
- The fault current for the fault F0 (Fig. 3.1) is equal to (system voltage)/(system impedance).
- Using the data from Sect. 3 one obtains:
- Faults F-3 and F-2: 3.14 kA (6.28 A secondary at CT-3, 3.14 A secondary at CT-2 and 1.963 A secondary at CT-1)
- Fault F-1: 5.24 kA.
- Fault F-0: 15.72 kA

Setting the Relays

The relay OR-3 is set to operate without any delay. We assume that the used relay operates in 5 ms. We also assume that this is the slowest relay among those protecting the elements branching from the bus 3. The pick-up current of OR-3 should be set at 1.2 × load current = 1.2 × 0.756 A = 0.907 A.

The pick-up current of OR-2 should be set at 1.2 × load current = 1.2 × 0.756 A = 0.907 A (note that the load current is higher than for OR-3, but the secondary values are the same due to different ratios of CT-2 and CT-3). Assuming 20 ms security margin, the time delay of OR-2 should be 5 ms (delay of OR-3) + 25 ms (operating time of CB-3) + 20 ms (security margin) = 50 ms.

The pick-up current of OR-1 should be set at 1.2 × load current = 1.2 × 0.945 A = 1.134 A. Assuming 20 ms as the security margin and OR-2 to be the slowest relay at bus 2, the time delay of OR-1 should be 50 ms (delay of OR-2) + 20 ms (operating time of CB-2) + 20 ms (security margin) = 90 ms.

Sensitivity Check

The fault F-3 is the most distant fault in the system, thus producing the smallest current. We need to check if the relays will pick-up during this fault.

- OR-3 will pick-up because 6.28 A (F-3 fault current at CT-3) > 0.907 A (setting of OR-3).
- OR-2 will pick-up because 3.14 A (F-3 fault current at CT-2) > 0.907 A (setting of OR-2).
- OR-1 will pick-up because 1.963 A (F-3 fault current at CT-1) > 1.134 A (setting of OR-1).

The designed protection system will work correctly for the given system. OR-2 provides back-up for OR-3. OR-1 provides back-up for both OR-2 and OR-3.

Prediction of the Average Clearing Times

- Faults on load 3 (5 ms (relay) + 25 ms (breaker) = 30 ms.
- Faults on line 2 (50 ms (relay) + 20 ms (breaker) = 70 ms.
- Faults on line 1 (90 ms (relay) + 15 ms (breaker) = 105 ms.

3.3 Impedance Relaying

3.3.1 Introduction

Impedance relays compare the power system voltage and current and operate when the ratio is less than its preset value. They monitor the impedance between the relay location and the fault called apparent impedance. If the apparent impedance falls within the relay setting, the relay will operate [3]. Impedance relays are used whenever overcurrent relays do not provide adequate protection. This section provides exercises about how to use impedance (distance) relays to protect a power network. The following issues are explained in the introduction and covered by the MATLAB models and related simulations as well as by the examples and problem assignments:

- Rules for protecting a network using impedance relays.
- Requirements for instrumentation (number and location of the instrument transformers) and switching apparatus (number and location of the circuit breakers).
- Analysis of the normal load conditions for selecting the instrument transformer and setting the relay.
- Analysis of the fault conditions for selecting the instrument transformer and setting the relay.
- Setting and coordinating the relays.

- Simple communication schemes.
- Simulation of a simple transmission network protected with impedance relays. Checking the relays' operation and coordination including protective equipment failures (relay and circuit breaker maloperation).

3.3.2 Relaying Basics

Figure 3.4 shows a simple transmission network consisting of two equivalent sources (A and B) and two lines connected between three busbars. A number of loads are supplied from each busbar. This network is of the "transmission" type because both the load and fault currents are supplied from both sources.

Transmission lines can be and typically are protected against faults using the impedance relaying principle.

Consider the line 1 and its two impedance relays ZR-1 and ZR-2. Because the fault current is supplied from both ends of the faulted line, transmission lines must be equipped with CBs at all terminals. Therefore, the CB-1 and CB-2 are installed on the line 1 and both of them must trip in the case of a fault on the line 1.

An impedance relay is designed to sense the positive sequence impedance resulting from the voltages and the currents measured by the relay. The measured impedance includes the sign, therefore one may say that a given relay "looks" in a given direction. The relay ZR-1 looks at the line 1 in the direction from the bus 1 towards the bus 2. The relay ZR-2 looks at the line 1 in the direction from the bus 2 towards the bus 1.

The impedance measured by an impedance relay is proportional (at least in an ideal situation) to the distance between the relay location and the fault. Therefore, the value of the impedance indicates the fault that manifests itself as the impedance shorter than the impedance of the line, which enables the relay to trip when needed.

However, if the fault with significant fault resistance occurs, then the impedance "seen" by the relay is not an accurate enough measure of the distance. Figure 3.5 illustrates this by showing the line impedance vector and the "added" fault

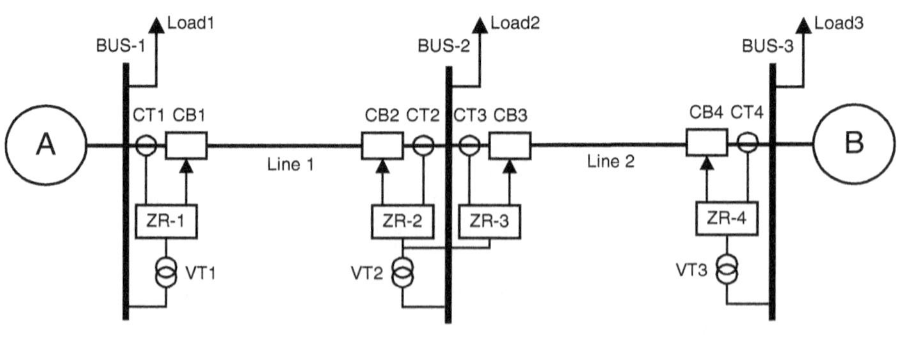

Fig. 3.4 Simple transmission network with impedance relays

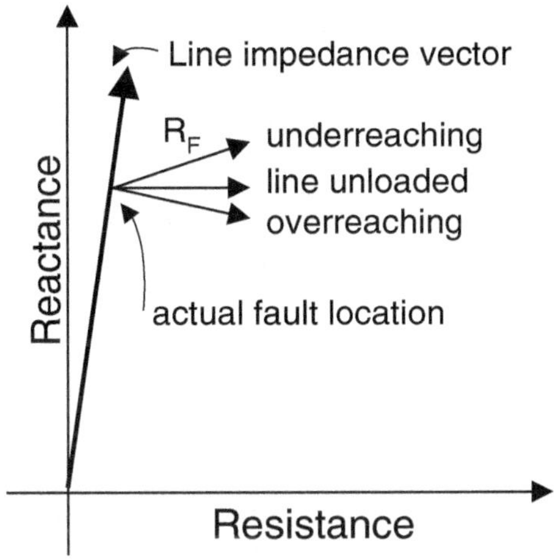

Fig. 3.5 Impedance measured during a fault with significant fault resistance

resistance, R_F, in three different cases: the unload line, the line loaded in one direction, and then in the opposite direction. If the line is not loaded then the fault resistance just adds with the portion of the line impedance vector proportional to the fault location. If the line is significantly loaded, then the fault resistance gets rotated by certain angle either up or down depending on the power flow direction. This causes the fault to appear closer than actually located (overreaching) or farther than actually located (underreaching).

Because of the possible overreaching and accuracy of the voltage and current transformers supplying a given relay, the relay cannot be set to trip if it senses the fault located within the 100 % line length. Typically, impedance relay protecting transmission lines is set to cover 75–80 % of the line length.

Since the fault resistance moves the measured impedance to the right hand side on the impedance plane (Fig. 3.5), impedance relays are designed to operate if the measured impedance is within a given region on the impedance plane. This region is called an operating characteristic and should contain 75–80 % of the line impedance vector and certain area right from that vector. Sample operating characteristic is shown in Fig. 3.6 (this is the shape used in the MATLAB example, practical shapes are different and depend on relay design, application, etc.). In this characteristic two parameters are shown: the reactance reach and the resistance reach.

The resistance reach should be as high as possible to enable the relay to operate under high fault resistance, but at the same time it should be as low as necessary to avoid false trips under normal load conditions.

The reactance reach is set at about 75–80 % of the line length to avoid false trips for faults located at the far end busbar or adjacent lines.

An impedance relay cannot distinguish between the faults on the protected line located at 80–100 % of the line length and external faults on the adjacent elements. However, those parts of the line are protected by the relay located on the other side of the transmission line and looking into the same line. Thus, as in the case of the

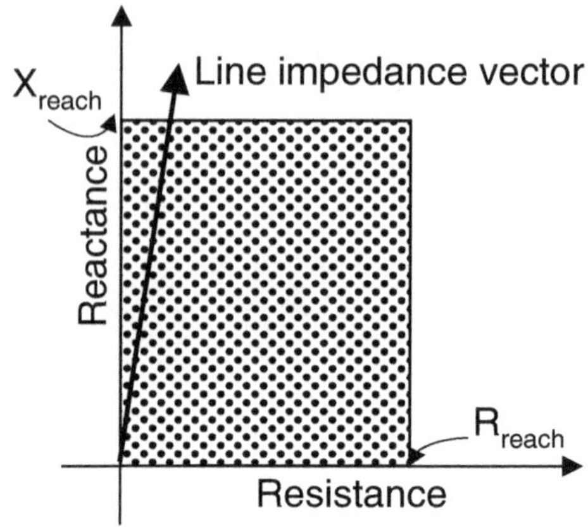

Fig. 3.6 Sample operating characteristic of an impedance relay

overcurrent protection, the impedance relay should trip the faults located farther than 80 % after certain time delay. Therefore, a typical relay incorporates not one, but several operating characteristics extending farther and farther. Those characteristics are called zones and are numbered consecutively: first zone covers 75–80 % of the line and operates without any delay; second zone typically covers 120–150 % of the line and operates with the delay longer than the slowest relay for the far end bus elements, etc. The second zone provides primary protection for remaining 20–25 % of the line and back-up protection for the elements adjacent to the far end busbar. The third and fourth zones provide back-up protection for certain distant elements in the system. Very often an impedance relay is equipped with an extra zone for the negative direction (so called reverse zone). This zone is meant basically to provide back-up protection for the local busbar where the line starts.

Figure 3.7 illustrates the protection system developed using the impedance relaying principle.

The protection system of Fig. 3.7 operates as follows:

- The primary protection for the line 1 is provided by ZR-1 and ZR-2. For faults located 0–20 % from ZR-1, ZR-1 operates immediately and ZR-2 opens CB-2 with its second zone time delay. For faults located 20–80 % of the line, both ZR-1 and ZR-2 see a fault in their first zones and operate immediately. For faults located at 80–100 %, ZR-2 operates immediately (a fault is in its first zone), while ZR-1 operates with the second zone time delay.
- The back-up protection for the line 1 is provided by the second and third zones of ZR-4 (and the reverse zone of ZR-3—not shown in the figure).
- The primary protection for the line 2 is provided by ZR-3 and ZR-4. For faults located 0–20 % from ZR-3, ZR-3 operates immediately and ZR-4 opens CB-4 with its second zone time delay. For faults located 20–80 % of the line, both ZR-3 and ZR-4 see a fault in their first zones and operate immediately. For faults located at 80–100 %, ZR-4 operates immediately (a fault is in its first zone), while ZR-3 operates with the second zone time delay.

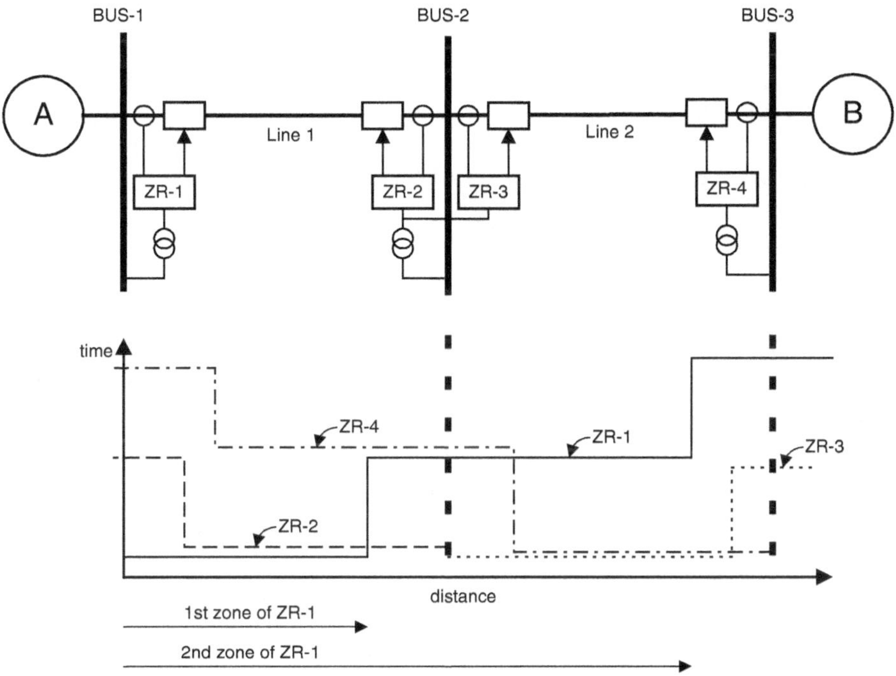

Fig. 3.7 Illustration of the impedance protection principle

- The back-up protection for the line 2 is provided by the second and third zones of ZR-1 (and the reverse zone of ZR-2—not shown in the figure).
- The back-up protection for the bus 2 is provided by the second zones of ZR-1 and ZR-4 (and the reverse zones of ZR-2 and ZR-3—not shown in the figure).
- The back-up protection for the bus 1 is provided by the second zone of ZR-2 and the third zone of ZR-4.
- The back-up protection for the bus 3 is provided by the second zone of ZR-3 and the third zone of ZR-1.

Note that not all of the faults on a line are cleared with the first zone time delay. For some of the faults, one of the CBs is opened after the second zone time delay. One may, however, shorten the fault clearing time by connecting the impedance relays on the opposite ends with a communication channel. The connected relays are called a "scheme," which may be accomplished in a number of ways. One of the simplest is called a "transfer trip".

The transfer trip scheme works as follows: if a given relay operates with the first zone time delay, it sends the signal to the relay at the opposite end of the line to trip the CB there. The relay at the opposite end either applies some extra checking or just sends the incoming signal to the local CB. A simple transfer trip scheme is implemented in the MATLAB model for the relays ZR-3 and ZR-4.

3.3.2.1 Analysis of the Load and Fault Conditions

The load and fault conditions must be analyzed in order to select the CTs and CBs as well as to set the relays.

The fault locations that need to be considered are those producing the maximum fault current for each line in the system. The maximum fault current is used to select the CBs and the CTs.

3.3.2.2 Selecting the CTs and VTs

The following basic parameters of instrument transformers should be considered from the protective relaying standpoint:

- The rated primary current should be higher than the load current during normal load conditions or acceptable overloads.
- The rated secondary current should match the rated input current of relays (typically 5 A or 1 A).
- A CT should not saturate under the maximum fault current flowing through the CT in a given network.
- The rated primary voltage of a VT should match the network rated voltage.
- The rated secondary voltage of a VT should match the rated input voltage of used relays (typically 100 V).

3.3.2.3 Selecting the CBs

The following basic parameters of a CB should be considered:

- The maximum current that can be interrupted by a CB must be higher than the maximum fault current flowing through this CB in a given network.
- The maximum operating time of a CB must be known for proper setting the corresponding impedance relay.

3.3.2.4 Selecting and Setting the Relays

The following parameters of an impedance relay should be considered:

- The rated input current and voltage should be matched with the rated secondary current of the corresponding CT and rated secondary voltage of the VT, respectively.
- The setting range (zones and time delays) should match the needs of a given relay location.

There are two settings of the impedance relay: the zones' reach and shape and the zones' time delay. The setting process starts from the first zone of all relays, continues with the second zone, etc. as explained below.

First Zone Reach

The shape of the first zone relay operating characteristic should be selected depending on the specific application. Sometimes, impedance relays (particularly digital relays) offer a number of the characteristic shapes. Typically, different shapes are applied for SLG (Single Line to Ground) and LL (Line to Line) faults. The reactance reach is set at about 75–80 % of the line length. The resistance reach is set to avoid false operation during load conditions and depends on the maximum load of a given line.

First Zone Time Delay

Typically there is no extra delay for the first zone operation (some communication schemes may require certain short delay). However, the relay does not operate instantaneously for the faults in this zone due to an inherent time delay associated with the relay decision-making process.

Second Zone Reach

The reactance reach must be securely shorter than the shortest reach of the first zones in the positive direction adjacent to the far-end busbar. Figure 3.7 explains this. For example, the second zone of ZR-1 must not overlap with the second zone of ZR-3.

Second Zone Time Delay

The zone 2 time delay is calculated as follows:

Time delay = maximum time delay of the relays downstream + maximum CB time + security margin.

The security margin depends on the accuracy of the installed relays and the variations of the operating time of the relays and CB. Typically, it is in the range of 0.1–0.3 s.

For example, if the first zone of ZR-3 operates in 20 ms, CB-3 operates in 100 ms, and the security margin is 100 ms, then the second zone delay of ZR-1 should be 20 + 100 + 100 = 220 ms (not less than that). If the first zone of ZR-3 is not the slowest protection adjacent to the bus 2, and, for example, the busbar protection operates in 40 ms, then for selectivity the second zone of ZR-1 should be delayed at 40 + 100 + 100 = 240 ms.

The third and fourth zones are set according to the same general rules as illustrated in Fig. 3.7.

3.3.3 Software Models

After completion of the simulation examples, the students will learn how to protect a transmission network using impedance relays. Particularly, the following issues are re-enforced: setting and coordinating multi-zone impedance relays, and analysis of the network operation under variety of conditions including faults and protective equipment maloperations.

The system shown in Fig. 3.4 has been implemented as a MATLAB model with the use of SimPowerSystems for modeling the network and the relays. In addition, a number of scopes (measuring elements) are connected to display the most critical signals in the model such as currents, voltages, and trip signals.

For simplicity, the model is a single phase model. Therefore, the only fault type considered at this time is the single-line-to-ground fault. The modules of voltage source, line and system impedance, fault, load, current and voltage transformers, circuit breaker, scope (*measurement element*), and display are presented in previous section.

Two-Zone Impedance Relay

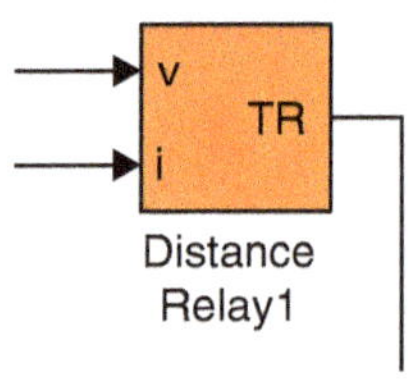

The input terminals "i" and "v" should be fed with the current and voltage, respectively, while the output terminal "TR" (trip signal) should be fed-back to the CB.

Double-click on the relay's icon to open a new window with the functional model of the relay (Fig. 3.8).

In this model:

- The waveforms of the current and voltage are processed into phasors.
- The phasors are divided to obtain the impedance. The impedance is transformed from the polar to rectangular coordinates and fed to the zone models.
- The zones are implemented as shown in Fig. 3.6. Double-click on the zone block to change the reach of the reactance and resistance.
- The hysteresis is used to model the relay operation (once tripped, the relay does not reset by itself).

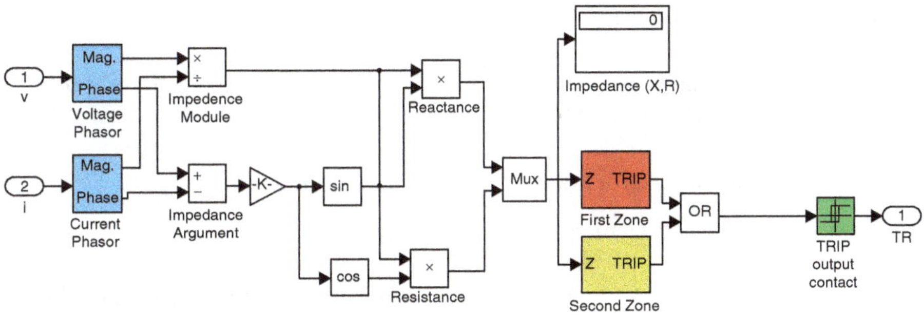

Fig. 3.8 The functional model of an impedance relay

The "relay" in the computer exercise is a functional model only capable of simulating the basic function of an impedance relay, i.e., measure the impedance, compare it to the zones, and time-out certain time delays.

The icons in the MATLAB model are consistently colored to indicate different functions of the blocks. For example, the displays are yellow and the instrument transformers are green, etc.

Default Data

The model has been developed with the following default data:

Source A voltage: 158 kV (114 % of 138 kV), 0°
Source A impedance: $1 + j10\ \Omega$
Source B voltage: 138 kV, −15°
Source B impedance: $1 + j10\ \Omega$
Line 1 impedance: $2 + j20\ \Omega$, fault location 0.5
Line 2 impedance: $2 + j20\ \Omega$, fault location 0.5
Load 1: 138 kV, 100 MW, 30 MVAr
Load 2: 138 kV, 100 MW, 30 MVAr
Load 3: 138 kV, 100 MW, 30MVAr
Fault: 0 Ω, inception time 20 ms
CB-1 operating time: 15 ms
CB-2 operating time: 17 ms
CB-3 operating time: 20 ms
CB-4 operating time: 20 ms
CT-1, CT-2, CT-3, CT-4: 2000/5 A
VT-1, VT-2, VT-3: 138 kV/100 V
For all the relays: first zone reactance reach of 4.35 Ω and resistance reach of 3 Ω (secondary ohms) and time delay 8 ms; second zone reactance reach of 8.85 Ω and resistance reach of 4 Ω (secondary ohms) and time delay 50 ms.

Note: The time data are re-scaled to speed-up the simulation. The actual values would be approximately ten times higher.

3.3.3.1 Model Activation

The model is available as ImRe.mdl file. To use the model, put the file into the default working directory of MATLAB (typically MATLAB/BIN), run MATLAB, and type “ImRe”.

The following are useful hints on how to use the model:

- Double-click on an element to change its parameters or see what is inside the element.
- Double-click on a scope (measurement element) to open it and see the plot.
- Use the zooming buttons of the plot to zoom in or out.
- Highlight (click-on) and press DEL to kill a given connection between two blocks.
- Draw a line between the terminals to connect the elements.

3.3.3.2 Numerical Examples

This subsection presents calculations for selecting the protective equipment, setting the relays, and checking the relays’ sensitivity.

Load Flow Calculations

The maximum load current needs to be calculated for selecting the primary current of the CTs. The maximum transferred power must sometimes be known to calculate the impedance locus under load conditions and select the resistance reach of the relays. Those calculations are skipped.

Selecting the CTs and VTs

Assume CT with the following ratio is used for all the relays: 2000 A/5 A.

Assume VT with the following ratio is used: 138,000 V/100 V.

Short-Circuit Calculations

The short-circuit calculations are needed for selecting the CTs and CBs. Those calculations are skipped here, but appear in the software manual.

Setting the Relays

For each relay, the transformation ratio of the impedance due to instrument transformers must be calculated first. Since the currents and voltages are altered by the ratios of the CTs and VTs, the impedance resulting from the secondary current and voltage is not the actual impedance from the system, but is proportional to the actual impedance.

The impedance ratio = voltage ratio of the VT divided by current ratio of the CT.

For the assumed data we obtain: impedance ratio = 138,000:100/2000:5 = 3.45. This means, for example, that the reactance of the line 1 is seen by the relay ZR-1 as 20/3.45 = 5.8 Ω instead of 20 Ω.

Assume, next we set the first zones of all the relays at 75 % of the line length. Thus:

- ZR-1: first zone reactance reach = 20/3.45 × 0.75 = 4.35 Ω.
- ZR-2: first zone reactance reach = 20/3.45 × 0.75 = 4.35 Ω.
- ZR-3: first zone reactance reach = 20/3.45 × 0.75 = 4.35 Ω.
- ZR-4: first zone reactance reach = 20/3.45 × 0.75 = 4.35 Ω.

Assume that there is a 30 % security margin between the two overlapping zones (for example, second zone of ZR-1 and first zone of ZR-3). Thus (see Fig. 3.7 for explanation):

- ZR-1: second zone reactance reach = 20/3.45 (line 1) + 4.35 × 0.7 (70 % of ZR-3's first zone) = 8.85 Ω.
- ZR-2: second zone reactance reach assumed also 8.85 Ω (since there is no information about the protective relays adjacent to the bus 1).
- ZR-4: second zone reactance reach = 20/3.45 (line 2) + 4.35 × 0.7 (70 % of ZR-2's first zone) = 8.85 Ω.
- ZR-3: second zone reactance reach assumed also 8.85 Ω (since there is no information about the protective relays adjacent to the bus 4).

Assume next, the load analysis enables one to assume that the resistance reaches 3 Ω for the first zone and 4 Ω for the second zone.

The second zone time delays should be set as follows:

- ZR-1: 20 ms (CB-3) + 8 ms (ZR-3) + 20 ms (security margin) = 48 ms.
- ZR-2: Assume 50 ms as for ZR-1 since there is no data on the protection system adjacent to the bus 1.
- ZR-4: 17 ms (CB-2) + 8 ms (ZR-2) + 20 ms (security margin) = 45 ms.
- ZR-3: Assume 50 ms as for ZR-3 since there is no data on the protection system adjacent to the bus 3.

Assume 50 ms for the second zone of all the relays in the system.

3.4 Differential Relaying

3.4.1 Introduction

This exercise shows how to use the current differential principle to protect power networks. The following issues are explained in the introduction and covered by the MATLAB model and related simulation as well as by the examples and problem assignments:

- Rules for protecting a network using differential relays.
- Requirements for instrumentation (number and locations of the instrument transformers) and switching apparatus (number and locations of the circuit breakers).
- Analysis of the normal load conditions for selecting instrument transformers and setting the relays.
- Analysis of the fault conditions for selecting instrument transformers and setting the relays.
- Setting the relays.
- Simulation of a simple transmission network protected with differential relays. Checking the relay operation.

3.4.2 Relaying Basics

Figure 3.9 shows a simple transmission network consisting of two equivalent sources (A and B) and two lines connected between three busbars. A number of loads are supplied from the busbars. This network is of the "transmission" type because the load and fault currents are supplied from both sources.

Two differential relays (DR-1 and DR-2) are installed in the system to protect the lines. The current differential relays are considered. It is assumed that the currents at both ends of a protected line are measured via the Current Transformers (CTs) by the relay whose trip signal is sent to the Circuit Breakers (CBs) at both

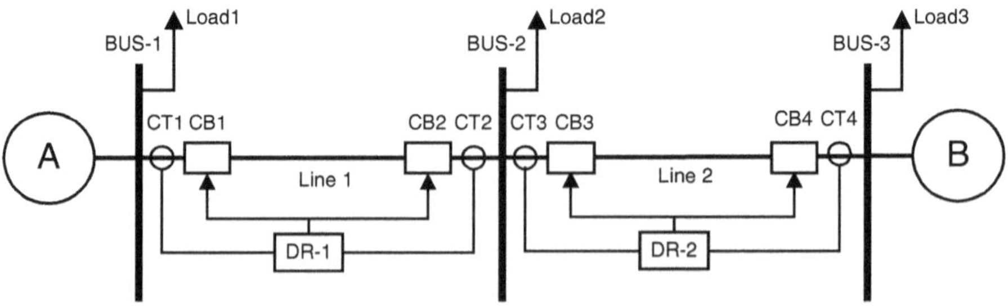

Fig. 3.9 A simple transmission network with impedance relays

ends. The differential rule is a powerful relaying principle and can be applied to virtually any power system element, i.e., lines, transformers, busbars, motors, etc., if the currents at all the terminals of a given element are available for measurement.

Consider the line 1 and its relay DR-1. Because the fault current is supplied from both ends of the faulted line, transmission lines must be equipped with CBs at all terminals. Therefore, the CB-1 and CB-2 are installed on the line 1 and both of them must trip in the case of a fault on the line 1.

A differential relay is designed to measure the currents at all the terminals of a protected element. The relay generates internally so-called "differential current", i.e., the current that is a geometrical sum of the currents phasors at all the terminals of a protected element. Applying the Kirchhoff's Current Law (KCL) to the protected element that is sound (does not have a fault), one concludes that such differential current is ideally zero or assumes very small values (due to measuring inaccuracies). If an internal fault occurs, the differential current is simply the fault current and assumes very large values. Thus, one obtains a very powerful discriminator between external and internal faults.

An ideal differential relay would thus measure the differential current and compare it to a threshold (unbiased differential relay).

In some cases, however, such as when the CTs saturate, or their ratios are not matched perfectly, the differential signal may assume considerably large values even though there is no internal fault. In order to prevent false trips in such cases, one would have to set the threshold high. But this would limit the relay sensitivity. Other practical solution is to use the second signal (so-called "restraining current" or "bias current") and to apply so-called "percentage" or "biased" characteristic.

The restraining signal is created by the relay as the sum of the currents (magnitudes) at all the terminals of a protected element. The restraining signal is much higher during external faults than during internal faults.

Table 3.1 compares the values of the differential and restraining currents in the following three key situations.

Bearing in mind the relations from the table, one defines the operating characteristic of a so-called "biased" differential relay as follows:

$$\text{TRIP} = (I_{\text{DIFF}} > A0) \quad \text{and} \quad (I_{\text{DIFF}} > K\ I_{\text{RES}})(1)$$

where, I_{DIFF} and I_{RES} are the magnitudes of the differential and restraining currents, respectively.

$A0$ and K are the pick-up threshold and bias of the relay, respectively.

The pick-up value $A0$ is typically set at 10–50 % of the rated current. The bias K is typically set at 0.2–0.4 depending on the quality of the CTs.

Table 3.1 Differential and restraining currents

Current	Load	External fault	Internal fault
Differential	Very small	Small	Very large
Restraining	Average load current	Large	Large

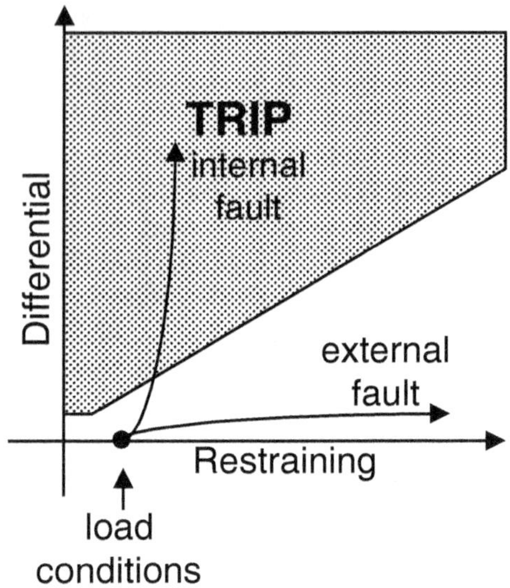

Fig. 3.10 The operating characteristic of a biased differential relay

As shown in Fig. 3.10, the load and external fault operating points fall into the NO-TRIP region of the characteristic, while the internal faults fall into the TRIP region.

Differential relays are built in many variants depending on the application. They may compare instantaneous values of the current phasors; they may apply many extra trip conditions such as the specific harmonic content (protection of power transformers or busbars).

The protection system of Fig. 3.9 operates as follows:

- The primary protection for the line 1 is provided by DR-1. The relay trips for faults are located anywhere along the line. Thus, the relay reach is 100 % of the line length. Generally, the operating zone of any differential relay is defined by the location of the CTs used to measure the currents.
- The primary protection for the line 2 is provided by DR-2.
- Note that differential relays do not provide any back-up protection. In the system of Fig. 3.9, the DR-1 will not operate for faults at the bus 2 nor on the line 2. The DR-2 will not operate for faults on the line 1 nor the bus 2. Often impedance relays are installed to provide back-up protection for differential relays.

3.4.2.1 Analysis of the Load and Fault Conditions

The load and fault conditions must be analyzed in order to select the CTs and CBs.

The fault locations that need to be considered are those producing the maximum fault currents for each line in the system. The maximum fault current is used to select the CBs and the CTs.

The maximum external fault current must also be known to approximate the maximum unbalance current during external faults in order to set the bias of the differential relay.

3.4.2.2 Selecting the CTs and CBs

The following basic parameters of instrument transformers should be considered from the protective relaying standpoint:

- The rated primary current should be higher than the load current during normal load conditions or acceptable overloads.
- The rated secondary current should match the rated input current of the relays (typically 5 A or 1 A).
- A CT should not saturate under the maximum fault current flowing through the CT in a given network.
- If any of the CTs that feed a differential relay can saturate, then the relay should be set appropriately to ensure stability of relay operation during external faults.

The following basic parameters of a CB should be considered:

- The maximum current that can be interrupted by a CB must be higher than the maximum fault current flowing through this CB in a given network.
- The maximum operating time of a CB must be known for proper setting of the impedance back-up relays (differential relays are NOT coordinated with any other relays and they operate independently).

3.4.2.3 Selecting and Setting the Relays

The following parameters of a differential relay should be considered:

- The rated input currents should be matched with the rated secondary currents of the connected CTs.
- The setting ranges (pick-up current and bias) should match the needs of a given relay location.

There are three basic settings of the differential relay: the current matching ratio, the pick-up current, and the bias.

Matching Ratio

A differential relay compares the currents of a protected element by forming the differential signal. Prior to the comparison, however, the currents must be brought to a common base. In general, this operation includes compensation for the phase shift (when power transformers are considered) and magnitudes (almost for any application). In electromechanical and static relays, this operation was accomplished by a special matching transformer. In digital relays, this operation is done by calculations.

The matching ratios for the currents fed to a differential relay are simple functions of the assumed base (typically one of the secondary currents) and the ratios of the installed CTs.

Pick-Up Current

The pick-up current must be set above the maximum differential current during maximum load conditions. In this way, the relay does not pick-up during load conditions. It will pick-up during fault conditions, both internal and external. Typically, the pick-up current is set at about 10–50 % of the maximum load current.

Bias

The bias must be high enough to prevent false trips during external faults and saturation of the CTs. Depending on the accuracy of the CTs installed; the bias is set at 20–40 %.

3.4.2.4 Software Models

After completion of these simulation examples, the readers will learn how to protect a transmission network using differential relays. Particularly, the following issues are re-enforced: setting differential relays and analysis of the network and relay operation under variety of loading and fault conditions.

The system shown in Fig. 3.9 has been implemented as a MATLAB model with the use of the Power System Blockset for modeling the network and SIMULINK for modeling the relays. In addition, a number of scopes (measurement elements) and displays are connected to view the most critical signals in the model such as currents, voltages, and trip signals.

For simplicity, the model is a single phase model. Therefore, the only fault type considered at this time is the single-line-to-ground fault. The modules of voltage source, line and system impedance, fault, load, current and voltage transformers, circuit breaker, scope (*measurement element*), and display are presented in previous section.

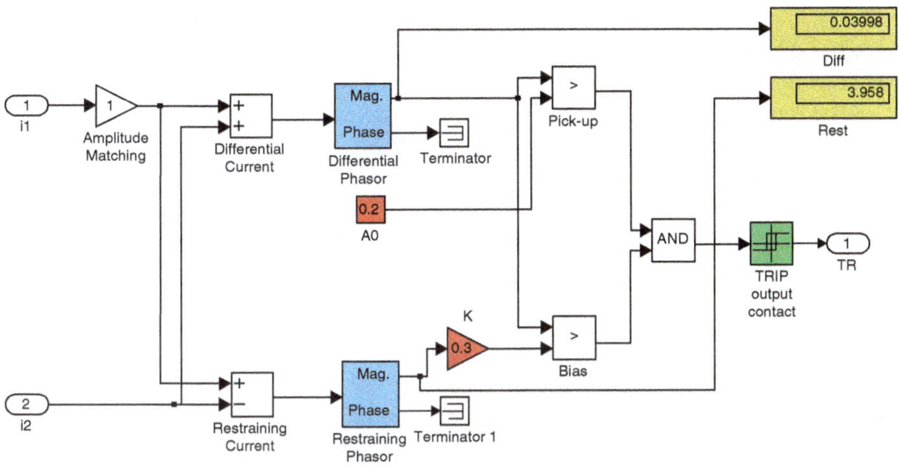

Fig. 3.11 The functional model of a single-slope differential relay

Single-Slope Differential Relay

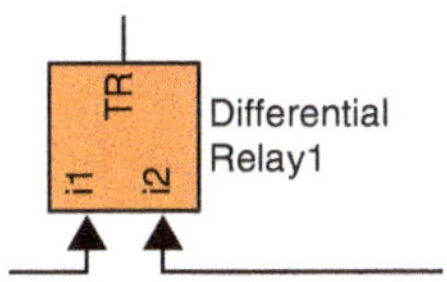

The input terminals "i1" and "i2" should be fed with the currents from both ends of the line, respectively, while the output terminal "TR" should be fed-back to the CBs.

Double-click on the relay's icon to open a new window with the functional model of the relay (Fig. 3.11).

In this model:

- The differential current is formed as a sum of the terminals' currents (depending on the way the CTs are connected).
- The restraining current is formed as a difference of the currents (depending on the way the CTs are connected).
- A gain element is present in the i1-channel to match the ratios of the CTs.
- The waveforms of the differential and restraining currents are processed into phasors.
- The magnitude of the differential current is compared against the threshold $A0$ (pick-up). Double-click on the block to change the pick-up value.
- The magnitude of the differential current is compared against a given portion of the restraining current (bias). Double-click on the block denoted as K to change the bias.
- The trip signal is formed according to Equation (1).
- The hysteresis is used to model the relay operation (once tripped, the relay does not reset by itself).

The "relay" in the computer exercise is a functional model capable of simulating the basic function of a differential relay only, i.e., ratio matching, pick-up, and bias characteristic.

The icons in the MATLAB model are consistently colored to indicate different functions of the blocks. For example, the displays are in yellow, the instrument transformers are green, etc.

Default Data

The model has been developed with the following default data:

- Source A voltage: 158 kV (114 % of 138 kV), 0°.
- Source A impedance: $1+j10\ \Omega$
- Source B voltage: 138 kV, −15°.
- Source B impedance: $1+j10\ \Omega$
- Line 1 impedance: $2+j20\ \Omega$, fault location 0.5
- Line 2 impedance: $2+j20\ \Omega$, fault location 0.5
- Load 1: 138 kV, 100 MW, 30 MVAr
- Load 2: 138 kV, 100 MW, 30 MVAr
- Load 3: 138 kV, 100 MW, 30 MVAr
- Fault: 0 Ω, inception time 20 ms.
- CB-1 operating time: 15 ms
- CB-2 operating time: 25 ms
- CB-3 operating time: 18 ms
- CB-4 operating time: 21 ms
- CT-2: 2000/1 A
- CT-1, CT-3, CT-4: 2000/5 A
- DR-1: matching gain = 0.2 (2000:5/2000:1 = 0.2), pick-up = 0.2, bias = 0.3
- DR-2: matching gain = 1 (2000:5/2000:5 = 1), pick-up = 0.2, bias = 0.3

Note: The time data are re-scaled to speed-up the simulation. The actual values would be approximately ten times higher.

3.4.2.5 Model Activation

The model is available as DiRe.mdl file. To use the model, put the file into the default working directory of MATLAB (typically MATLAB/BIN), run MATLAB, and type "DiRe".

The following are useful hints on how to use the model:

- Double-click on an element to change its parameters or see what is inside the element.
- Double-click on a scope (measurement element) to open it and see the plot.
- Use the zooming buttons of the plot to zoom in or out.
- Highlight (click-on) and press DEL to kill a given connection between two blocks.
- Draw a line between the terminals to connect the elements.

3.4.2.6 Simulation Examples

This subsection presents calculations for selecting the protective equipment, setting the relays, and checking the relays' operation.

Load Flow Calculations

The maximum load current needs to be calculated for selecting the primary current of the CTs. Those calculations are given in the Exercise 3.

Selecting the CTs

Assume the following CT-1, CT-3, and CT-4 are selected as: 2000 A/5 A, while CT-2 has the ratio of 2000 A/1 A.

Short-Circuit Calculations

The short-circuit calculations are needed for selecting the CTs and CBs. Those calculations are given in the Exercise 3 in the Manual.

Setting the Relays

Assume the pick-up values are selected at 0.2, while the bias is set at 0.3. The DR-1 requires setting the matching gain at 0.2 in the channel 1 (for the CT-1 current).

3.5 Summary

This section introduces the basics of power system relaying including overcurrent, impedance, and differential protections. The protection principles are described using several simple power system network models. The software models for each type of protection and examples are provided to further explain the critical issues that must be considered such as the analysis of load and fault conditions, selection of current and voltage transformers, settings, and coordination of relays. After this section, the readers will learn the relaying fundamentals and the principles to select and configure the overcurrent relays, impedance relays, and differential relays.

References

1. IEC 60255-3 (1989-06) electrical relays—Part 3: Single input energizing quantity measuring relays with dependent or independent time
2. Elmore WA (2003) Protective relaying theory and applications. CRC Press, Boca Raton
3. Lewis Blackburn J, Domin TJ (2006) Protective relaying: principles and applications, 3rd edn. CRC Press, Boca Raton

Chapter 4
Modeling of Digital Relay and Power System Signals

4.1 Introduction

This chapter contains documentation of the developed MATLAB software used for development of the models as a part of the learning process. The developed blocks are gathered in three libraries: Relay Elements, Input Signals, and Tools, which will be presented in the following subsections.

4.2 Major Elements of a Digital Relay

A digital relay design may be simplified by three major blocks, as shown in Fig. 4.1:

- The Data Acquisition block which constitutes the front-end of the relay and links the digital processing segment of the relay with its analog inputs.
- The Measuring block which estimates certain parameters of the input signals such as magnitude, phase angle, resistance and reactance, active and reactive power, etc. Those parameters are used for the tripping decision.
- The Decision Making block which uses selected relaying principles by comparing the signal parameters from the measuring unit with given settings (thresholds). It also uses time delays and logic functions in order to issue the trip and alarm signals.

In this section, we address those blocks in detail.

The Data Acquisition block functioning requires understanding of the sampling theorem, analysis of the error of signal representation as a function of the sampling frequency, explanation of the concept of the aliasing frequency, an introduction to analog filtering together with common approximations of Analog Filters (AFs), and representation of the horizontal resolution of the Analog to Digital (A/D) converter.

© Springer International Publishing Switzerland 2016

M. Kezunovic et al., *Design, Modeling and Evaluation of Protective Relays for Power Systems*, DOI 10.1007/978-3-319-20919-7_4

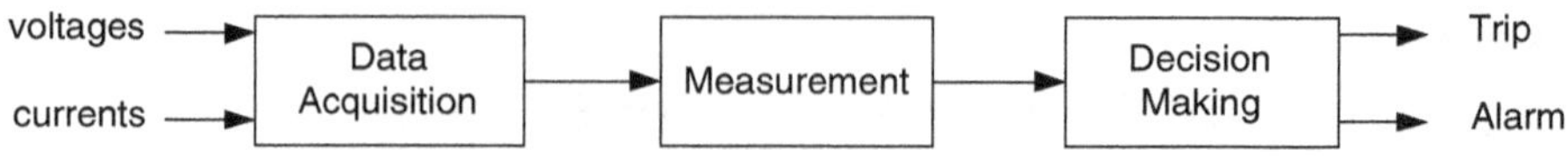

Fig. 4.1 Major components of a digital relay

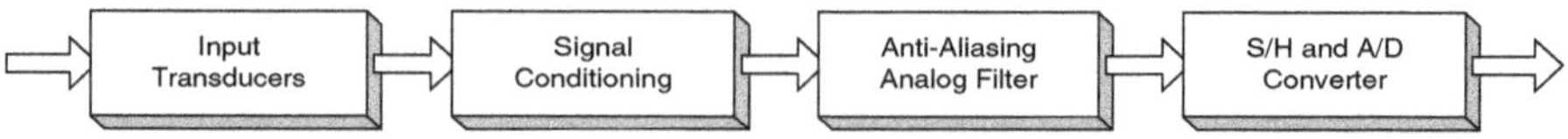

Fig. 4.2 The front-end part of a digital relay

The Measuring block focuses on phasor estimation using signal samples. The issues such as the data window, accuracy of estimation, as well as time and frequency responses are addressed. As an example, we focus on the full- and half-cycle Fourier algorithm as the means of reconstructing the signal from its samples.

The Decision-Making block is used to analyze an overcurrent relay, one of the most common relaying principles. We present instantaneous, definite-time, and time-dependent modes of operation.

4.2.1 Data Acquisition Block

The front-end of a digital relay consists of four elements (Fig. 4.2):

- Input transducer
- Signal conditioner
- Analog anti-aliasing Filter (AF)
- Analog to Digital (A/D) converter

The input transducer matches the standard secondary voltage and current with the input signal level appropriate for a numerical protective relays. It converts current into voltage and isolates the relay input circuit from the secondary wiring of the substation.

The Signal Conditioner scales the signals down to match the input range of the subsequent signal processing elements.

The Analog Filter (AF) provides necessary, usually low-pass, anti-aliasing filtering. It passes all the signal components that are used by the relaying algorithm, and it stops all the remaining components assumed to be a noise. Depending on the operating principle of a given relay, different signal components may be considered as the information and noise.

The Sample and Hold (S/H) element samples its input signal usually at regular time intervals and the Analog to Digital (A/D) converter converts the samples into their numerical representation.

4.2.1.1 Sampling

The sampling circuit takes samples of the analog input signal usually at regular time intervals and converts them into the digital (numerical) representation. This operation is accomplished by a Sample and Hold (S/H) element, which can be a part of the A/D converter controlled by a sampling clock. Usually, not only the present sample of the input signal, but both the present sample and some number of "historical" samples are required for performing protection functions. This calls for a first-in last-out buffer which stores the needed number of samples.

Several design options for the A/D converter are possible: (a) One A/D converter may serve a number of channels, or (b) each channel may have its own A/D converter. In the first case either an S/H circuit is used to take the samples in separate channels synchronously, or the S/H circuit is not used and the samples are shifted in time. However, we do not focus on those hardware aspects, but on functional specifications only.

Three parameters of the S/H and A/D converter are important:

- Sampling frequency (horizontal resolution)
- Number of bits (vertical resolution)
- Conversion time

The sampling frequency must be correlated with the assumed frequency spectrum of the input signal to meet the requirements of the sampling theorem. The number of bits of the A/D output word must correlate the rate of change and dynamic change of the signal with the required accuracy. The conversion time must be short enough to enable the A/D converter to sample all the channels and to leave certain time for the online computations related to the protection functions performed by a given relay, before the next set of samples is taken.

4.2.1.2 Sampling Frequency

The sampling theorem states: A signal to be reconstructed from its digital (sampled) form must be sampled at least twice in its period.

A given relaying principle uses selected frequency components from its input signals. The highest frequency component determines the minimum sampling frequency.

The higher the sampling frequency, the better the representation of the analog signal.

Let us consider two simple examples to illustrate how the sampling frequency must be correlated with the relaying principle used by a given relay.

1. Assume that an overcurrent relay uses the amplitude of the 60 Hz component of the current signal. Ultimately, the sampling frequency may be set at any value higher than 120 Hz. Practically, we may specify the sampling rate as a number of samples per cycle of the fundamental (60 Hz) frequency. Thus, in this case, more

than two samples per cycle are theoretically sufficient (some early digital relays worked on four samples per cycle).

2. Assume a transformer differential relay uses the fifth harmonic for restraining during overexcitation conditions. If so, the minimum sampling frequency assumes $2 \times 5 \times 60$ Hz = 600 Hz (ten samples per cycle). However, much higher sampling rates are practically used. For certain optimization purposes, it is convenient if the number of samples per cycle is a multiplier of four. Thus, the typical sampling rates assume 16, 20, 32, 64 samples per cycle. The higher the sampling frequency, the better representation of the signal. The costs considerations of the signal processing circuits make today's relays to operate typically on 20 or 32 samples per cycle.

4.2.1.3 Aliasing

As we may check using the MERIT 2000 software, the analog signal of the frequency $n \times f_s \pm f_1$ sampled at the frequency f_s appears in its sampled form as a signal of the phantom frequency f_1 (n is any integer number, f_s is a sampling frequency). For example, signals of the frequencies 540, 660, 1140, 1260 Hz, etc. sampled at 600 Hz will be seen in their sampled form as a false 60 Hz component (Fig. 4.3a). This phenomenon is called *aliasing* and should be avoided in digital signal processing. This explains the sampling theorem—*A signal to be reconstructed from its digital (sampled) form must be sampled at least twice in its period.*

This phenomenon may lead to maloperation of a digital relay. Assume the relay uses the 60 Hz component for its tripping decision. The aliasing frequencies will be seen as a phantom 60 Hz component and introduce a false information to the relay.

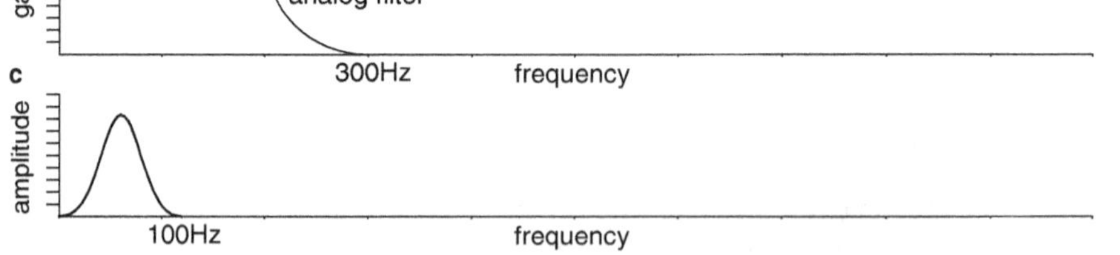

Fig. 4.3 Illustration of the aliasing frequencies (**a**) and the application of an anti-aliasing analog filter (**b**). The resulting frequency spectrum (**c**) does not contain any aliasing frequencies

Analog filtering applied prior to the A/D conversion provides a solution to this problem. The AF should reject all the aliasing frequencies (Fig. 4.3b) so that only the basic spectrum of the signal goes through (Fig. 4.3c).

4.2.1.4 Analog Filter

The AF has basically two functions:

First, it stops the aliasing frequencies and prevents violation of the sampling theorem. Second, it rejects majority of components not used by a given relay. Depending on the operating principle of a given relay, different components in the input signals are treated as the information and the noise. The noise filtering is shared between the analog anti-aliasing and digital filters used for parameter estimation.

Usually, low-pass type of filtering is used. If the d.c. components are to be rejected, it is more convenient to filter them out using digital filters. In order to provide good filtering properties, active filters are often used. This also enables one to combine analog filtering with signal conditioning. The typical order of the filter varies from 2 up to 4. Often, a standard approximation of an analog filter is used. There are several standard filters such as Buttherworth, Chebyshev, Bessel, etc. They provide optimal time and frequency responses for a given order of a filter.

The more the extensive filtering (the closer the cut-off frequency to the operating frequency), the larger the phase shift introduced by the filter between its output and input signals. This phase shift corresponds to certain time delay in the flow of information. Therefore, under some circumstances, it may slow the operation of a relay by fraction of a millisecond up to a few milliseconds.

4.2.1.5 Word Length of an A/D Converter

If the A/D converter covers the range of the signal from $-X_{\max}$ to $+X_{\max}$ with N bits (including the sign bit), it uses only 2^N values to represent the signal in that range. Consequently, the distance between the neighboring values equals ΔX. Given by the following equation:

$$\Delta X = 2X_{\max}/2^N$$

If the signal is between the two "steps", it will be represented by one of the closest values. Statistically, the error of the half of ΔX will be made. This relative error is small if the amplitude of the signal is close to the signal range, but it will be considerable if small signals are processed. Considering the maximum percentage error for the smallest utilized signal magnitudes, one may compute the required number of bits of the A/D converter.

Since the voltage signals mostly drop during faults, 12 bits are considered sufficient for the voltage channels. The currents, in turn, may build up to hundred times the nominal value during short-circuits, and therefore, the 16-bit resolution is used in the current channels.

Due to the cost considerations, in some designs the cheaper but fast 12-bit converter is used with a circuit for dynamic scale changing. In some cases, conversion is performed twice, but the resulting vertical resolution corresponds to a 16-bit converter.

4.2.1.6 Conclusions

- The AF must be used to eliminate the aliasing frequencies; it may also eliminate some noise
- The sampling frequency must meet the needs of the applied relaying principle
- The word length of the A/D converter must meet the expected range of input signals
- Considering the AF and A/D converter only, a number of design options are possible
- Digital simulation is an efficient tool for analyzing variety of those design options

The AF is a mandatory part of any digital relay. Even if the relay is based on the information contained in the high frequency spectrum (such as the travelling wave principle), the AF must be used. In order to minimize the delay introduced by the filter, the cut-off frequency may be set high (close but below half of the sampling frequency). If the AF is used to filter out certain amount of noise, the cut-off frequency would be set lower. The sampling frequency depends on the operating principle, assumed accuracy of the measurement, and available hardware. The vertical resolution of the A/D converter depends on the assumed accuracy and on the nature of the measured signals. Voltages show smaller dynamic range and may be treated by 12-bit converter. Currents, in turn, typically call for 16-bit converter because of the larger dynamic range.

A number of design options are possible considering the front-end of a digital relay. Some of the options are interrelated. For example, if very extensive analog filtering is applied (the cut-off frequency is close to the operating frequency), there is no need for the high sampling frequency.

Since the sampling is a nonlinear operation, precise analysis and comparisons of various design options are possible by means of digital simulations.

4.2.2 *Phasor Estimation*

4.2.2.1 Definition

A pure sine signal of a known frequency may be mathematically represented by two numbers: the magnitude and the phase. The pair of the magnitude and phase is called a phasor. Mathematically, a phasor is often denoted as a complex number: magnitude/*phase*.

Knowing its magnitude and initial phase, a sine waveform may be reconstructed from those two parameters. Depending whether sine or cosine function is used as the base for reconstruction, two formally different phasor definitions are possible. The phasor represents a waveform in the steady state conditions. There is no definition of a transient phasor (although the concept of wavelets is close to that).

The phasor represents adequately the waveform only if the third parameter (the frequency) is known and constant. If this condition is violated, the phasor is not a perfect representation of the waveform. In such a case, a triple (not a pair) of numbers (amplitude, phase, and frequency) should be used. Protective relays for generators are an example of such an extended approach. However, in phasor considerations the frequency is typically assumed to be the nominal one (60 Hz).

4.2.2.2 Application

Most of the contemporary digital relays are based on the classical relaying principles using phasors. Those principles use phasors of the input signals and the relations between the phasors (such as impedance or power) to make the trip decision. Usually, the phasors of the fundamental frequency components are used for protection.

In digital protective relaying, phasors of the relay input signals are used to measure (calculate) the quantities utilized in the trip criteria. Those quantities include the voltage and current phasors themselves or their combinations such as impedance and power. Those signals suit as the basis for the trip decision.

Typically, phasors of the fundamental frequency components are used (60 Hz). In some applications, such as transformer protection, other frequency components may be represented by their phasors.

4.2.2.3 Requirements

Since the relay makes its decision using the information contained in the phasors, the phasors should be estimated as fast as possible. When, due to a fault, the phasor changes rapidly, its estimate provided by a digital-measuring algorithm should reflect this change immediately. In addition, the dynamic measuring errors should be minimized. Particularly, for the overcurrent principles the amplitude should not be overestimated, while for the undervoltage principle the amplitude should not be underestimated.

On the other hand, the estimate of the phasor should not be affected by the frequency components other than the nominal one. A perfect estimator would have the frequency response passing the nominal frequency and blocking completely all the other frequencies.

There is no perfect algorithm that meets the requirements of the speed and accuracy. The faster the algorithm, the more sensitive it is to the off-nominal frequencies. A number of different families of digital algorithms for phasor estimation have been developed. Here, we focus on the Fourier algorithms only.

4.2.2.4 Orthogonal Components

Majority of methods for digital phasor estimation may be presented using the concept of orthogonal components. In this approach the input signal x is split into a pair of its orthogonal components x_{d} and x_{q} using a pair of so-called orthogonal filters $\mathrm{F_d}$ and $\mathrm{F_q}$ as shown in Fig. 4.4. The magnitude and phase are calculated next using the equations shown below. The orthogonal filters actually define the algorithm.

In many applications, the filters are the Finite Impulse Response (FIR) filters. A FIR filter when exposed to the impulse input shows a non-zero output for a finite time. After certain time the output assumes zero. Mathematically, a FIR filter is represented by a non-recursive equation that links the filter output with the samples of its input.

Theoretically, the Infinite Impulse Response (IIR) filters may be also used. Their impulse response lasts for the infinite time because they have the recursive form. The IIR filters provide better filtering, but they may be a source of certain computational errors and are mostly avoided in digital power system relaying.

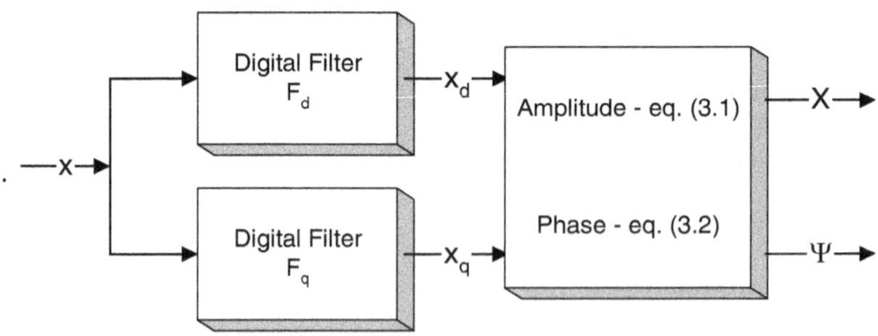

$$X = \sqrt{x_d^2 + x_q^2} \qquad \psi = \arctan\left(\frac{x_q}{x_d}\right)$$

Fig. 4.4 Phasor estimation using orthogonal components

Sample Orthogonal Filters

Using the FIR filters, either the direct or quadrature orthogonal component of the output signal is a weighted sum of the input signal samples. The collection of those samples constitute so-called data window.

The general equations are the following:

$$x_{\mathrm{d}(n)} = \sum_{k=0}^{N-1} a_k x_{(n-k)}, \quad x_{\mathrm{q}(n)} = \sum_{k=0}^{N-1} b_k x_{(n-k)}$$

Example of filter coefficients:

$$\begin{aligned}
a &= C_a[1 \quad 1 \quad 1 \quad 1] \\
b &= C_b[1 \quad 1 \quad -1 \quad -1] \\
a_k &= C_a \cos(\omega_0 k) \\
b_k &= C_b \sin(\omega_0 k) \\
a &= C_a[1 \quad 1 \quad 1 \quad 1 \quad -1 \quad -1 \quad -1 \quad -1] \\
b &= C_b[1 \quad 1 \quad -1 \quad -1 \quad 1 \quad 1 \quad -1 \quad -1]
\end{aligned}$$

In the examples, the numbers C_a and C_b are appropriate scaling factors (used in order to adjust the filter gain at the fundamental frequency to magnitude equal to 1). The 4-sample zero-order and first-order Walsh filters are the first example. The cosine-sine filters (Fourier filters) are the second example. The 8-sample first-order and second-order Walsh filters are the third example.

Data Window

Since there are two unknowns, ultimately two samples of the signal are needed to calculate the phasor. For example:

$$x_{\mathrm{d}(n)} = C_1\left(x_{(n)} + x_{(n-1)}\right) \quad x_{\mathrm{q}(n)} = C_2\left(x_{(n)} - x_{(n-1)}\right)$$

This algorithm shows that in the simplest case only two samples are needed to estimate the phasor. A number of samples needed for calculation of the quantity are called a data window. Practically, longer data windows are necessary (half a cycle or full cycle). The data window should not be confused with the sampling frequency—the signal may be sampled at comparatively high frequency but short data window may be used for the measurement. Vice versa, the sampling frequency may be low, but the data window consists of large number of samples.

Approximately, the reaction time of a given measuring algorithm (in terms of its settling time) equals the length of the data window. For example, if the signal is sampled at 20 samples per cycle and the data window consists of 10 samples (half cycle algorithm), the settling time for a pure sine waveform will be around half a cycle.

As may be checked using the attached software, algorithms with short data windows display very poor accuracy and are very sensitive to various waveform distortions. Therefore, practically longer data windows are used. Half a cycle or full cycle data windows are typical.

Fourier Algorithm

In the Fourier algorithm, the data window covers either a full cycle of the fundamental frequency (full-cycle algorithm) or half a cycle (half-cycle algorithm). The coefficients of the orthogonal filters are in the shape of the cosine (d) and sine (q) waveforms, respectively.

The Fourier algorithm is perhaps the most popular measuring algorithm used in today's digital relays. Using the orthogonal components approach, it may be seen as a pair of filters having the coefficients in the shape of sine and cosine waveforms. Using the Digital Fourier Transform (DFT) or Fast Fourier Transform (FFT), a phasor is estimated as the first harmonic.

The full cycle Fourier algorithm has the settling time around one cycle and is not sensitive to the harmonic frequencies (including the d.c. component). On the other hand, the half cycle Fourier algorithm is sensitive to the even harmonics (including the d.c. component), but it is faster (the reaction time is approximately half a cycle).

Frequency Response

Figure 4.5 shows the frequency response of the full cycle Fourier algorithm. The gain for the nominal frequency is one. The gain for other frequencies is lower. Particularly, for the harmonic frequencies it is zero, which means that the harmonic frequencies are damped perfectly. The non-harmonic frequencies affect the measurement to a certain extent given by the value of the gain. Under a given frequency, the estimated amplitude is not accurate, and in addition, it oscillates between certain lower and upper boundary. The shaded area indicates those boundaries. When considering an overcurrent relay, the upper boundary should be taken into account. When considering an undervoltage relay, the lower boundary counts. Frequency response is a very useful tool for analysis of the properties of the algorithms.

4.2.2.5 Conclusions

- There is a number of different measuring algorithms developed for power system protection
- Short-window algorithms are fast but inaccurate; long-window algorithms are more accurate but slower

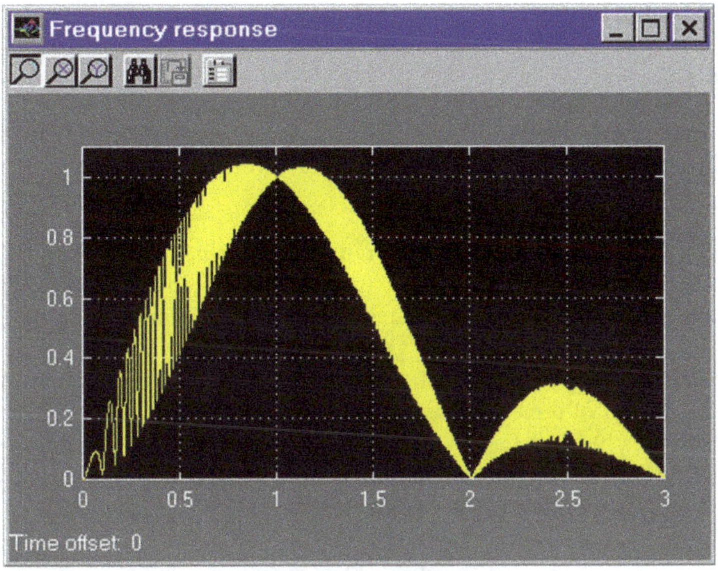

Fig. 4.5 Frequency response of the full cycle Fourier algorithm

- Always certain compromise between the speed and accuracy requirements must be made
- By digital simulation one is able to investigate behavior of different measuring algorithms in realistic conditions

4.3 Library of Modeling Elements

4.3.1 Bias Characteristic

4.3.1.1 Purpose

Compares its two input signals using a bias characteristic.

4.3.1.2 Input

Two inputs feed this block. The first is the operating and the second is the restraining signal. Both signals are given in instantaneous values.

The output of this block is the result of the comparison using the bias characteristic. Its value is one if the operating signal is greater than the bias characteristic at the given restraining, otherwise it is zero.

4.3.1.3 Dialog Box

See Fig. 4.6.

4.3.1.4 Description

This block determines whether the operating signal is in the operating region at the value of restraining signal when comparing to any bias characteristics. This means that if the operating signal is greater than the bias characteristic at the given restraining, the output signal is on. Otherwise, the output is zero.

The Bias characteristic must be entered as a two-column matrix. Each row represents a point of the bias characteristic.

The way this block determines if its output is one or zero is as follows:

Let (R_m, O_m) and (R_n, O_n) be the points on the plane.

The line connecting those two points has the following equation:

$$O = \frac{O_m - O_n}{R_m - R_n}(R - R_m) + O_m$$

Therefore, if we want to test whether point (O, R) is above the line connecting (R_m, O_m) and (R_n, O_n), the following equation is presented:

$$O > \frac{O_m - O_n}{R_m - R_n}(R - R_m) + O_m$$

If the point is above the line, the condition shown above is true.

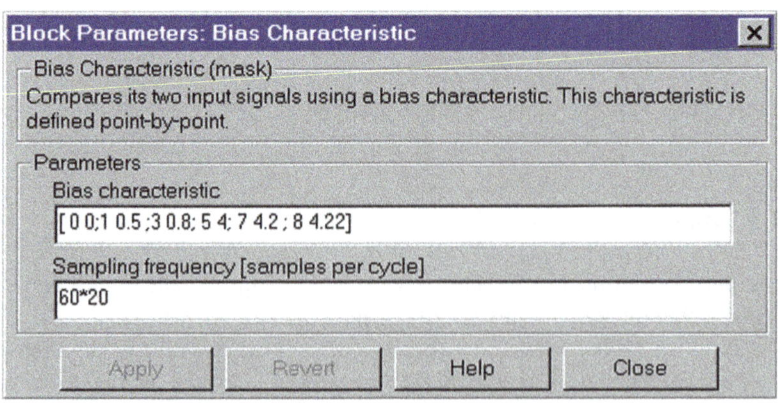

Fig. 4.6 Dialog box of the bias characteristic

4.3.1.5 Parameters

Bias characteristic: Array that contains the points that define the bias characteristic.

Sampling frequency: The sampling rate in samples per cycle.

4.3.1.6 Recommended Solver

Ode45 (Dormand-Prince) is recommended. There is no need for time step for this solver. The discrete solver can also be used.

4.3.1.7 Example

In the following example, the bias characteristic block has the following bias characteristic: [0 0; 15 15; 30 0]. The operating signal is a sinusoidal signal of amplitude 5 and a DC component of 5 (Figs. 4.7, 4.8, and 4.9).

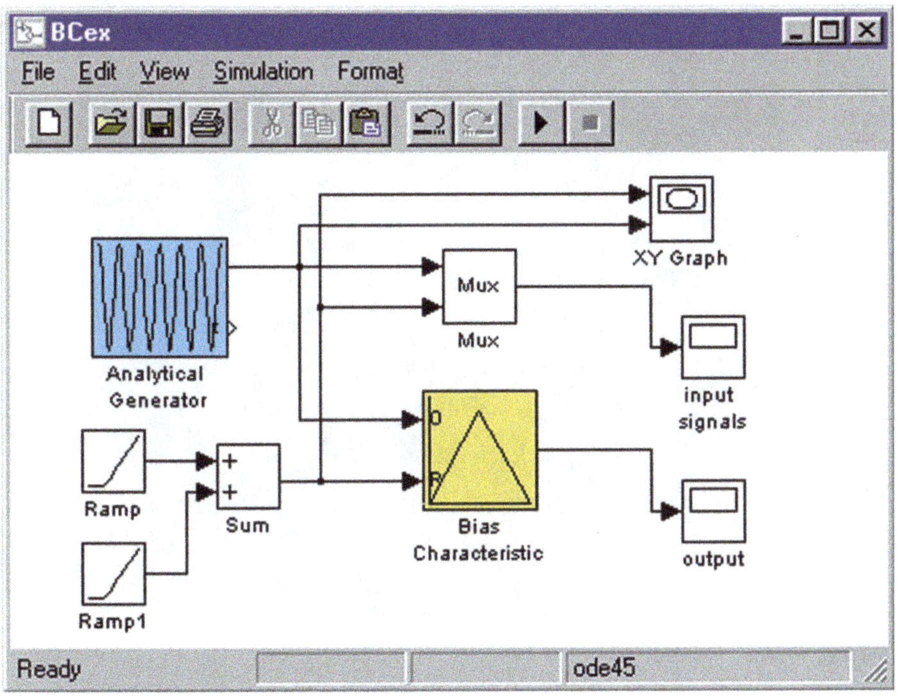

Fig. 4.7 Example for the bias characteristic block

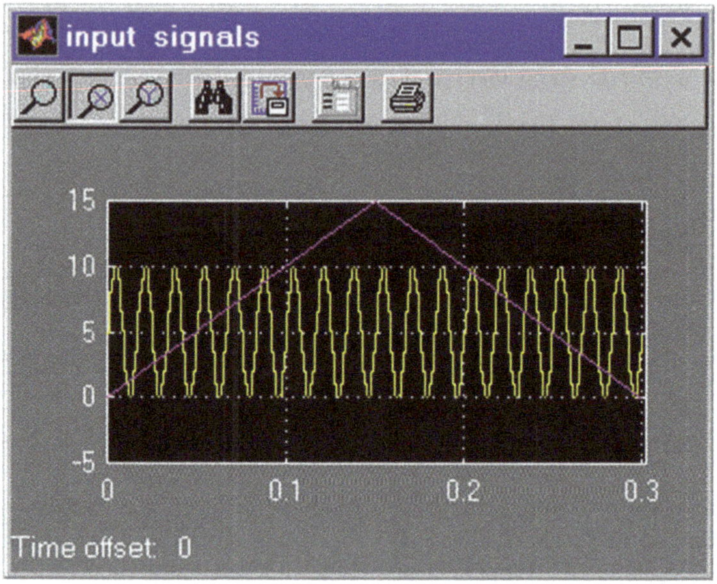

Fig. 4.8 Operating and restraining signals

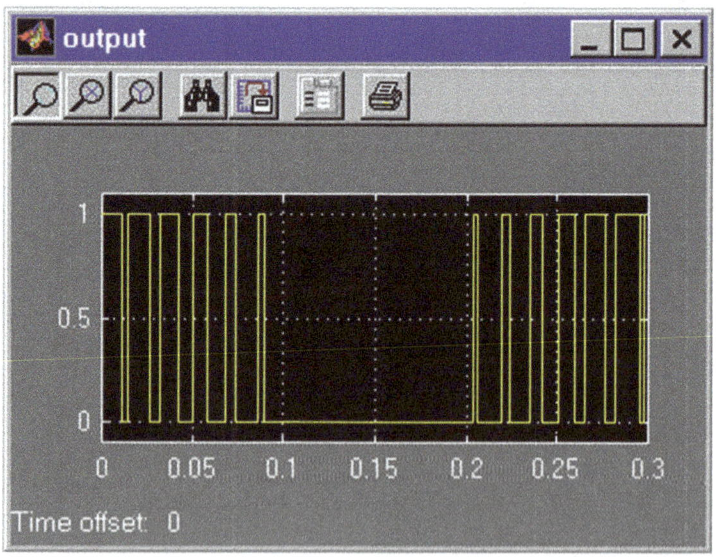

Fig. 4.9 Output of the block

4.3.2 Basic Measurements

4.3.2.1 Purpose

Given the current and voltage signals in the orthogonal form, computes their amplitudes, as well as the active and reactive powers and impedance components.

4.3.2.2 Dialog Box

See Fig. 4.10.

4.3.2.3 Description

This block of orthogonal components takes the voltage and current at the present sampling time and obtains the magnitude of the current and voltage as well as the

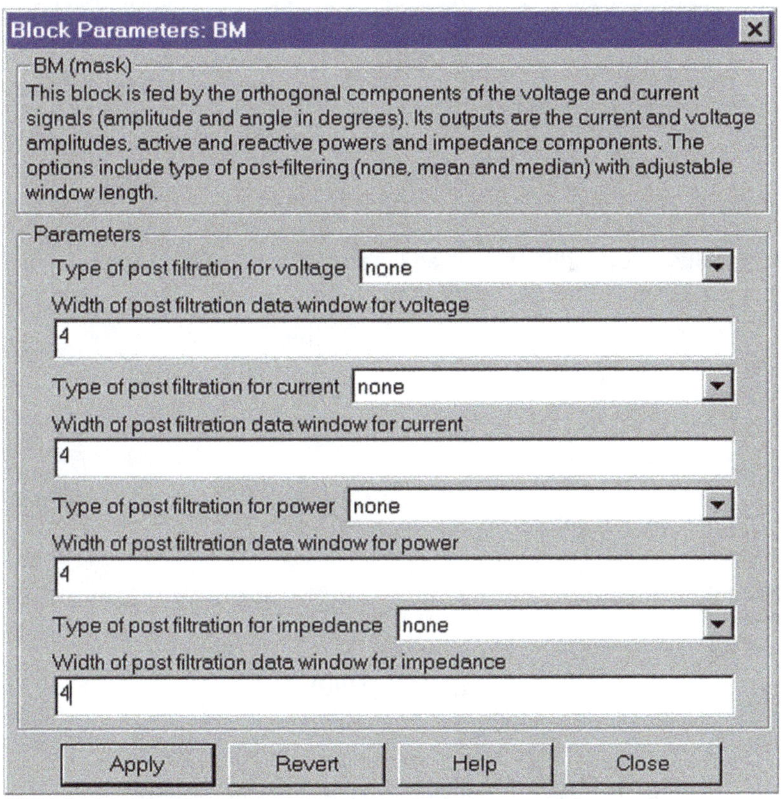

Fig. 4.10 Dialog box of bias characteristic

complex power and impedance. The outputs may be filtered. Two types of filters, mean and median, are implemented. All these filters may have different length of their data windows.

Basic Processing

Define the input orthogonal components of voltage and current as

$$\begin{aligned} v\_ort &= |V|\angle\Phi_v \\ i\_ort &= |I|\angle\Phi_I \end{aligned}$$

The output impedance and complex power are:

$$\begin{aligned} S &= v\_ort \times i\_ort^* = |V||I|\angle(\Phi_V - \Phi_I) = |V||I|(\cos(\Phi_V - \Phi_I) + i\cdot\sin(\Phi_V - \Phi_I)) \\ Z &= \frac{v\_ort}{i\_ort} = \frac{|V|\angle\Phi_V}{|I|\angle\Phi_I} = \frac{|V|}{|I|}\angle(\Phi_V - \Phi_I) = \frac{|V|}{|I|}(\cos(\Phi_V - \Phi_I) + i\cdot\sin(\Phi_V - \Phi_I)) \end{aligned}$$

Note that the current and voltage magnitudes can be obtained directly from the inputs.

Post-processing Filter

Because the noise and parameter mismatch, there are some oscillations in the results of the basic processing. Therefore, the post-processing filter should be used. In this block, the mean and the median post-processing filters are implemented. These filters are defined as follows:

Mean Filter

$$X_C^p = \frac{1}{p}\sum_{k=0}^{p-1} X_{Ck}$$

Median Filter

$$X_C^p = \mathrm{median}(X_{Ck})$$

where:

X_C^p is the post-filtered value for power, voltage, current, or impedance.
X_{Ck} is the data window of samples.

4.3.2.4 Parameters

Type of post-filtration for voltage/current/impedance: Mean, median, or no filter can be used for post-filtering of the measured voltage, current, or impedance.

Width of post-filtration data window for voltage/current/impedance: The size of the data window for post-filtration of the measured quantity.

4.3.2.5 Recommended Solver

This block has discrete states only, so the discrete solver can be used if there is no need to use the DAB (Data Acquisition Board). In this case the time step should be selected so that all the harmonics of the signal can be detected. If the model used uses the DAB, as in the example shown below, the ode-45 (Dormand-Prince) solver is recommended. For this solver, a time step does not need to be specified.

4.3.2.6 Example

An example with this block is shown in Fig. 4.11. The results are shown in Fig. 4.12. The voltage signal in this case goes from a magnitude of 7 to a magnitude of 4 when the fault occurs. Note that in block BM1 there is no post-filtering, while in BM2 median post-filtering was used. For that reason, BM2 filters out the high frequency noise signal, though there is a delay added to the signal. This delay is caused by the post-filtering method. The window size used for both blocks is 40.

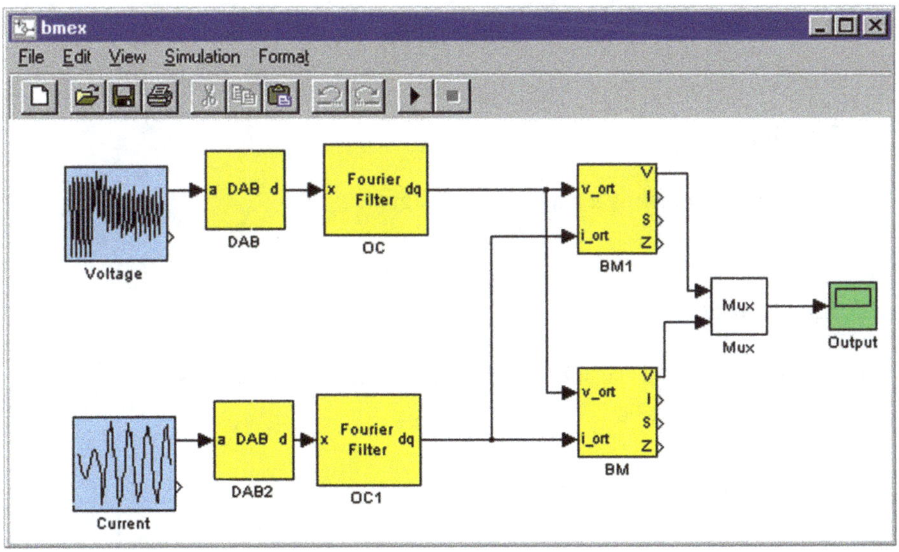

Fig. 4.11 Example for the BM block

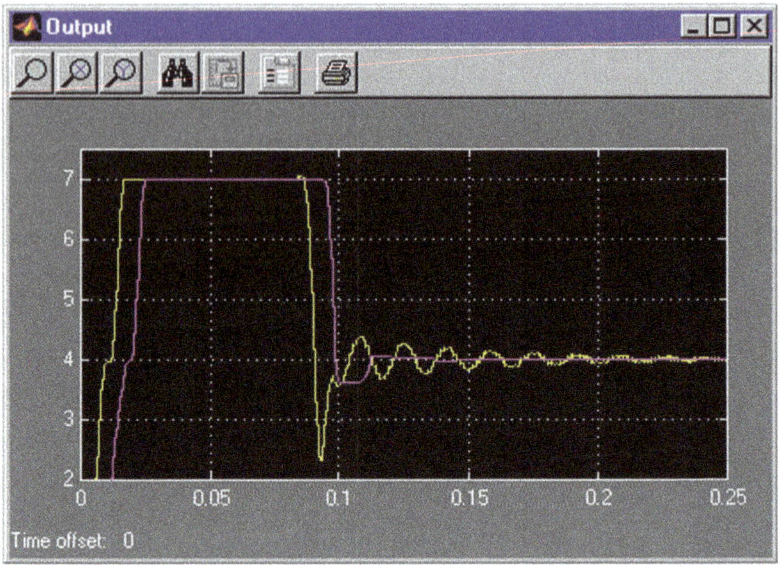

Fig. 4.12 Simulation results for BM block

4.3.3 Data Acquisition Board

4.3.3.1 Purpose

Performs analog filtering, signal conditioning, and sampling.

4.3.3.2 Inputs

The input of this block is a single analog signal.

4.3.3.3 Outputs

The output of this block is a buffer with the input signal samples.

4.3.3.4 Dialog Box

See Fig. 4.13.

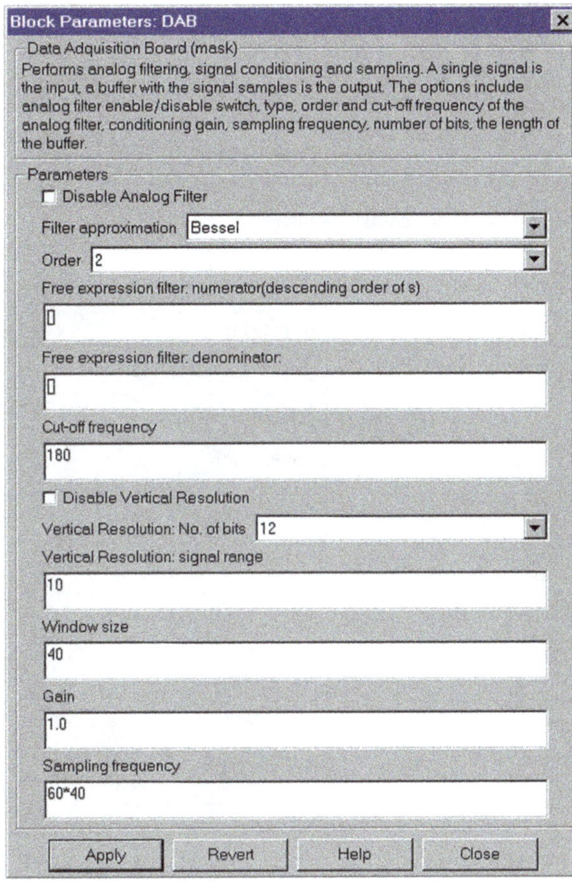

Fig. 4.13 Dialog box of DAB

4.3.3.5 Description

This block enables modeling and simulation of the analog anti-aliasing filter, analog signal conditioner, and Analog to Digital converter (A/D). The components of this block are shown in Fig. 4.14.

Analog Filter

To convert signals from analog to digital form, it is necessary to remove unwanted high frequencies before sampling. The sampling theorem requires that the analogue signals above certain frequency be attenuated to avoid errors in subsequent digital processing. The analog filter performs this function.

The cutoff frequency for the filter is chosen, so the sampling theorem is satisfied, i.e., $f_s \geq 2f_m$, where f_s is the sampling frequency and f_m is the maximum frequency of the signal, otherwise, an "aliasing error" will be introduced. The filter approximation

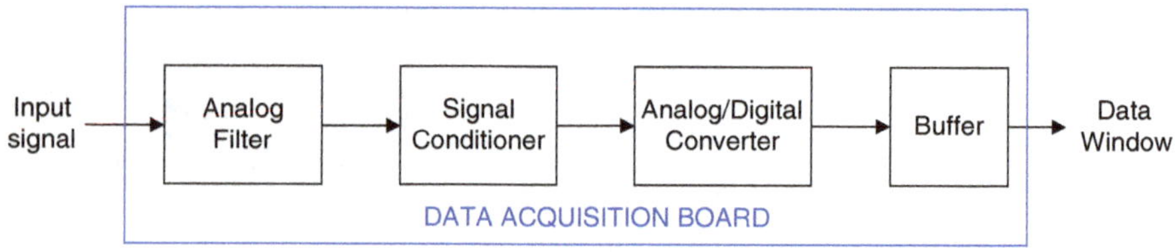

Fig. 4.14 Data acquisition board

Table 4.1 Filter approximations

Order	Approximation	Transfer function
Second	Bessel	$\frac{3\omega_c^2}{s^2+3\omega_c s+3\omega_c^2}$
	Tschebyscheff	$\frac{1.43\omega_c^2}{s^2+1.42\omega_c s+1.52\omega_c^2}$
	Butterworth	$\frac{\omega_c^2}{s^2+\sqrt{2}\omega_c s+\omega_c^2}$
Third	Bessel	$\frac{15\omega_c^3}{s^3+6\omega_c s^2+15\omega_c^2 s+15\omega_c^3}$
	Tschebyscheff	$\frac{0.716\omega_c^3}{s^3+1.25\omega_c s^2+1.53\omega_c^2 s+0.716\omega_c^3}$
	Butterworth	$\frac{\omega_c^3}{s^3+2\omega_c s^2+2\omega_c^2 s+\omega_c^3}$
Fourth	Bessel	$\frac{105\omega_c^4}{\left(s^2+5.79\omega_c s+9.14\omega_c^2\right)\left(s^2+4.22\omega_c s+11.14\omega_c^2\right)}$
	Tschebyscheff	$\frac{0.358\omega_c^4}{\left(s^2+0.35\omega_c s+1.06\omega_c^2\right)\left(s^2+0.84\omega_c s+0.36\omega_c^2\right)}$
	Butterworth	$\frac{\omega_c^4}{\left(s^2+0.76\omega_c s+\omega_c^2\right)\left(s^2+1.85\omega_c s+\omega_c^2\right)}$

implemented in the DAB and its transfer function is shown in Table 4.1 shown below. In the table, ω_c is the cut-off radian frequency. The user can also provide directly a transfer function of the filter. The analog filter can also be disabled.

Signal Conditioner

The signal conditioner is an ideal gain. This trivial operation of rescaling the signal is useful when modeling such phenomena as ratio mismatch with respect to differential relays or when considering relations between the primary and secondary currents and voltages.

A/D Converter

This part of the block performs the Analog to Digital conversion. It uses a zero-order sample and hold element. The vertical resolution or word length can be 8, 12, and 16 bits. The maximum signal range for the signal must be specified. It was implemented as a combination of a zero-order element and an S-function block with the following function used for the analog to digital conversion.

$$X_{\text{digital}} = \text{round}\left(\frac{x_{\text{analog}}}{\Delta x}\right) \cdot \Delta x$$

where:

$$\Delta x = \frac{x_{\max}}{2^N - 1}$$

N is the vertical resolution of the A/D converter (number of bits excluding the sign bit).

$x_{\max}$ is maximum input value.

Buffer

The input for this block is the present sample from the A/D converter. The output is the data window. The data window is a row vector that contains the current output of the A/D converter and $k - 1$ back samples of the input, i.e.,

$$\text{Data Window} = [y(n)y(n-1)y(n-2)\ldots y(n-k-1)]$$

where

k is the data window size.
n is the current sample index.
y is the output from the A/D converter.

4.3.3.6 Parameters

Disable Analog Filter: This option, if checked, disables the Analog low-pass filter.

Filter approximation: The type of approximation of the filter: Bessel, Tschebyscheff, Butterworth, or a transfer function introduced by the user.

Order: The order of the approximation. For the free expression option, this parameter is not needed.

Free expression—numerator: An array that contains the coefficients of the numerator (descending order of s). This array is needed only if the free expression option is selected.

Free expression—denominator: an array that contains the coefficients of the denominator (descending order of s). This array is needed only if the free expression option is selected.

Cut-off frequency: Contains the cut-off frequency for the analog filter.

Disable Vertical Resolution: If this option is checked, the vertical resolution is infinite, i.e., there are no quantization errors.

Vertical resolution—number of bits: Number of bits used to display the output of the A/D converter excluding the sign bit. It can be 8, 12, or 16 bits.

Vertical resolution—signal range: Maximum absolute value that the input signal is expected to reach.

Window size: Number of elements for the buffer of the DAB.

Gain: Ideal gain to condition the signal.

Sampling frequency: Rate at which the input signal is sampled.

4.3.3.7 Recommended Solver

Ode45 (Dormand-Prince). No time step needed for this solver.

4.3.3.8 Example

The following diagram shows an example of the use of the DAB. In this example, a fault signal is feed into the DAB. The sampling rate is 20 samples/cycle of the fundamental frequency of 60 Hz. Second-order Bessel approximation is used for the analog filter. The cut-off frequency is 200 Hz and the vertical resolution used is 12 bits. The analog input and digital output for this example are shown in Figs. 4.15 and 4.16.

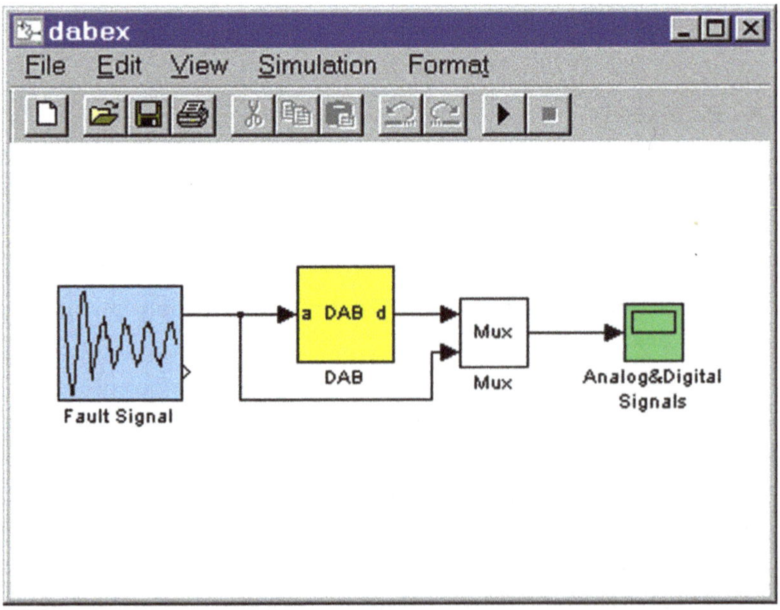

Fig. 4.15 Example for DAB

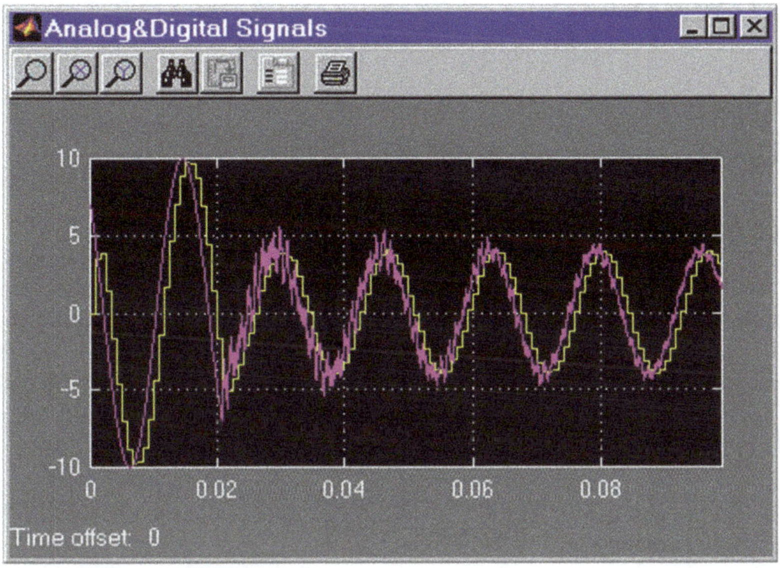

Fig. 4.16 Input and output signals for DAB

4.3.4 Directional Element

4.3.4.1 Purpose

Implementation of a directional element is discussed next. The operating characteristic is defined by two angles.

4.3.4.2 Inputs

This block has two phasor inputs: voltage and current.

4.3.4.3 Outputs

This block has one output. Its value is one when the current phasor falls into the operating characteristic and zero otherwise.

4.3.4.4 Dialog Box

See Fig. 4.17.

4.3.4.5 Description

The angles Alpha and Beta define the lower and upper limits of the operating characteristic, respectively. If the difference between the current and voltage phasor is greater than alpha and less than beta, then the output of the directional element is one, otherwise, the output of this block is zero. Angles Alpha and Beta are shown in Fig. 4.18.

Note that the icon of the block reflects the values of the angles Alpha and Beta.

The sampling rate for the signal must be provided. This frequency should be given in multiples of the fundamental frequency (60 Hz).

4.3.4.6 Parameters

Angle Alpha: This angle defines the lower limit of the operating characteristic. Its value is given in degrees.

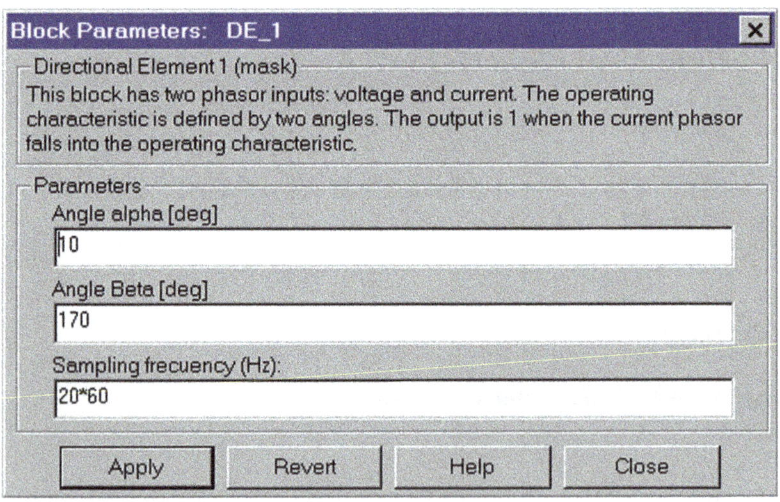

Fig. 4.17 Dialog box of directional element (DE)

Fig. 4.18 Operating characteristic

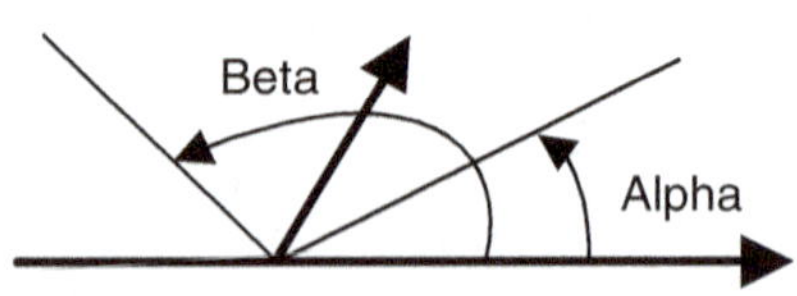

Angle Beta: This angle defines the upper limit of the operating characteristic. Its value is given in degrees.

Sampling frequency: The sampling rate for the input signal in Hz.

4.3.4.7 Recommended Solver

Ode45 (Dormand-Prince).

4.3.4.8 Example

The following example shows how the DE_1 block works. Figure 4.19 shows the block diagram for the example. In this case, $\alpha = 10°, \beta = 170°$, and the input voltage phasor is $1\angle 0°$. The current phasor, on the other hand, has a constant magnitude of one, but its phase increases. In Fig. 4.20, the results of the simulation are shown. Note that the output is zero, whenever the difference between the voltage and current phases is in the ranges 0–10° and 170–360°. The output is one whenever the relative angle is in the range 10–170°.

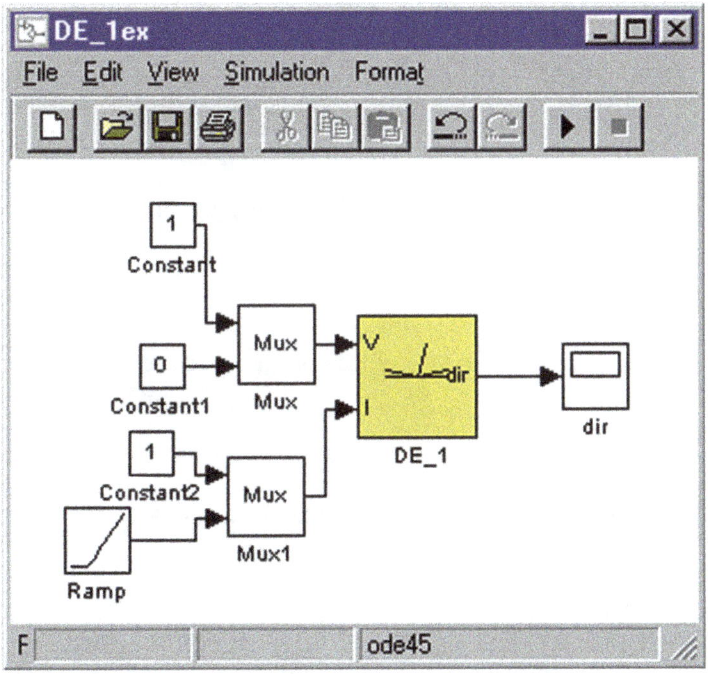

Fig. 4.19 Example for DE_1 block

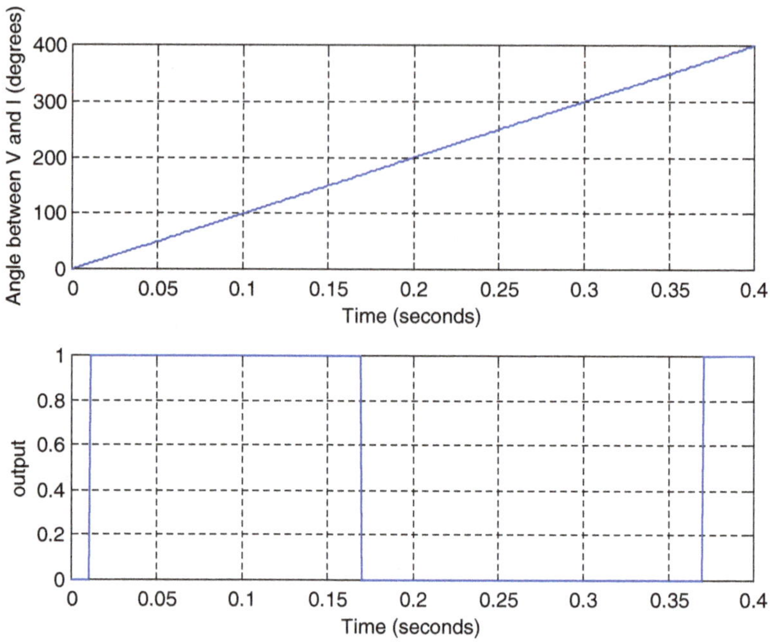

Fig. 4.20 Simulation results for DE

4.3.5 Differential Equation-Based Impedance Measurement

4.3.5.1 Purpose

Measures the impedance based on the differential equation approach.

4.3.5.2 Inputs

Samples of the instantaneous value of the voltage and current signals.

4.3.5.3 Outputs

Resistance and reactance signals calculated by the block (Fig. 4.21).

4.3.5.4 Description

The block of impedance measurement based on a differential equation model (DEIM) is composed of four parts:

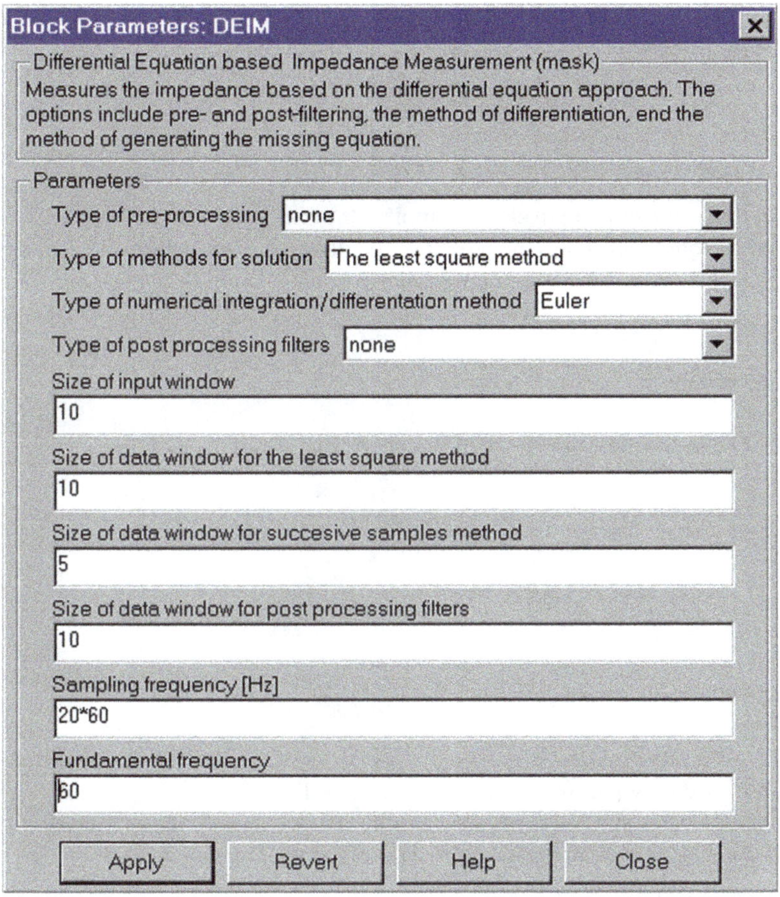

Fig. 4.21 Dialog box of differential equation-based impedance measurement (DEIM)

- Pre-processing
- Solution methods
- Numerical integration/differentiation method
- Post-processing

The inputs in this block are discrete voltage and current samples. The output is the impedance in complex form.

Pre-processing

For the pre-processing, half and quarter of cycle Fourier and Walsh filters were implemented. These filters belong to the bandpass type. The purpose of this process is to take out some unwanted frequencies from the signal. The algorithm for these filters has the form:

$$y(n) = \sum_{k=0}^{p} a_k \cdot x(n-k)$$

where $y(n)$ is the nth output signal sample, $x(n)$ is nth input signal sample, and a_k is a constant. Note that $p+1$ is the length of the data window, where p is the number of samples per cycle.

Fourier Filter

For this type of filter, the constant has the form:

$$a_k = \sin\left[\omega_0 T_s\left(\frac{p}{2} - k\right)\right]$$

where T_s is the sample time.

Walsh Filter

The Walsh filter implemented in this block is a zero order Walsh filter (it has rectangular data window), so in this case $a_k = 1$.

In the Fourier and Walsh filters, for the half of cycle data window $p = p/2$ and $\omega_0 = \frac{\pi}{T_W}$. For the quarter, a cycle data window $p = p/4$ and $\omega_0 = \frac{\pi}{2T_W}$, where $T_W = (p+1)T_S$.

To obtain the half or quarter of a cycle filters, the size of the input data window as well as the sampling frequency must be selected accordingly.

Solution Methods Combined with Numerical Integration/Differentiation Methods

Two solution methods are implemented in this block: two points in time dislocated by m samples and the least square method over the data window of d samples, and two numerical integration/differentiation methods: Euler and the trapezoidal rule.

Numerical Integration/Differentiation Methods

The trapezoidal rule calculates the square over two points of function $f(x1)$ and $f(x1)$. It is very dependable on $\Delta t = t_1 - t_2$ where t_1 and t_2 are times when the function samples are taken. This method in the most cases is very reliable, but under some circumstances can give poor estimate.

The Euler rule is very sensitive for small number of samples and can produce inaccurate results under some situations.

Two Points in Time Dislocated by m Samples Method

In this method R and L are calculated over a window of m samples. Typically, m is equal to 1.

We have the following equation: $Ri + L\frac{di}{dt} = v$

(a) Solution using Euler method

$$Ri(n) + \frac{L}{T_S}[i(n) - i(n-1)] - v(n) = 0$$
$$Ri(n-m) + \frac{L}{T_S}[i(n-m) - i(n-m-1)] - v(n-m) = 0$$

where $i(n)$, $i(n-1)$, $i(n-m)$, $i(n-m-1)$, and $v(n)$, $v(n-m)$ are current and voltage samples, respectively. The values of R and L can be obtained from the two equations shown above as there are two unknowns and two equations.

(b) Solution using the Trapezoidal rule

$$R\frac{\Delta t}{2}\left[i(n) + i(n-1\right] + L\left[i(n) - i(n-1) - \frac{\Delta t}{2}[v(n) + v(n-1)] = 0\right.$$
$$R\frac{\Delta t}{2}\left[i(n-m) + i(n-m-1\right] + L[i(n-m) - i(n-m-1)$$
$$- \frac{\Delta t}{2}[v(n-m) + v(n-m-1)] = 0$$

From these two equations, the values of R, L, and X can be obtained.

Least Square Method Over a Data Window of d Samples

The advantage of this method is that mean square approximation error is minimal.

The equation to solve is the following: $\int\limits_{n-d+1}^{n} \left(R + L\frac{di}{dt} - v\right)^2 = 0$

(a) Solution using Euler method
Applying Euler method and converting the integral to summation, we have:

$$S = \sum_{u=n-d+1}^{u=n} \left[Ri(u) + \frac{L}{T_S}(i(u) - i(u-1)) - v(u) \right]^2$$

From the conditions $\frac{\partial S}{\partial R} = 0$ and $\frac{\partial S}{\partial L} = 0$, the expressions for R and L can be obtained.

(b) Trapezoidal rule

Applying trapezoidal rule and converting the integral to summation, we have the following:

$$S = \sum_{u=n-d+1}^{u=n} \left[R\frac{T_S}{2}(i(u) + i(u-1)) + L(i(u) - i(u-1)) - \frac{T_S}{2}(v(u) + v(u-1)) \right]^2$$

From the conditions $\frac{\partial S}{\partial R} = 0$, and $\frac{\partial S}{\partial L} = 0$, the expressions for R and L can be obtained.

Post-Processing Methods

The post-processing methods implemented in this block are the mean and median filters of a data window of length p.

(1) Mean filter

The mean filter is a linear filter. One disadvantage of this filter is that it does not reject non-logical values (errors), which can occur as the result of the process. This type of filter is expressed by the following form:

$$y(n) = \frac{1}{p}\sum_{k=0}^{p} x(n-k)$$

where p is the window size, $x(n-k)$ are the input samples, and $y(n)$ is output at time n.

(2) Median filter

The median filter obtains the median value of the data window.

4.3.5.5 Parameters

Type of pre-processing: The types of pre-processing filter are: none, half and quarter of cycle Fourier, and half and quarter of cycle Walsh Filters.

Type of methods for solution: The method of solution used. There are two methods available: two points dislocated by m samples and least square.

Type of numerical integration/differentiation method: Two methods are available: Euler and the Trapezoidal rule.

Type of post-processing filter: Mean and median, as well as no filtering can be selected for post-filtering.

Size of input window: Number of samples of the voltage and current signals the block will need. This parameter has to be selected according to the pre-processing filtering selected.

Size of data window for the least square: Number of samples that the least square method will use.

Size of data window for successive sample method: Number of samples that the successive sample method will use.

Size of data window for post-processing filters: Number of samples needed for the post-filtering.

Sampling frequency: Sampling rate in Hz.

Fundamental frequency: Value of the fundamental frequency.

4.3.5.6 Recommended Solver

Ode 45 (Dormand-Prince). No time step needed for this method.

4.3.5.7 Example

The following example shows the operation of the DEIM block. The impedance components are shown for the cases with median filtering and with no post-filtering. Note that the transients in both signals are greatly reduced when post-filtering is used (Figs. 4.22, 4.23, and 4.24).

4.3.6 Digital Filter

4.3.6.1 Purpose

Implementation of a digital filter.

4.3.6.2 Inputs

Data window of a sampled signal.

4.3.6.3 Outputs

Data window which contains the filtered signal.

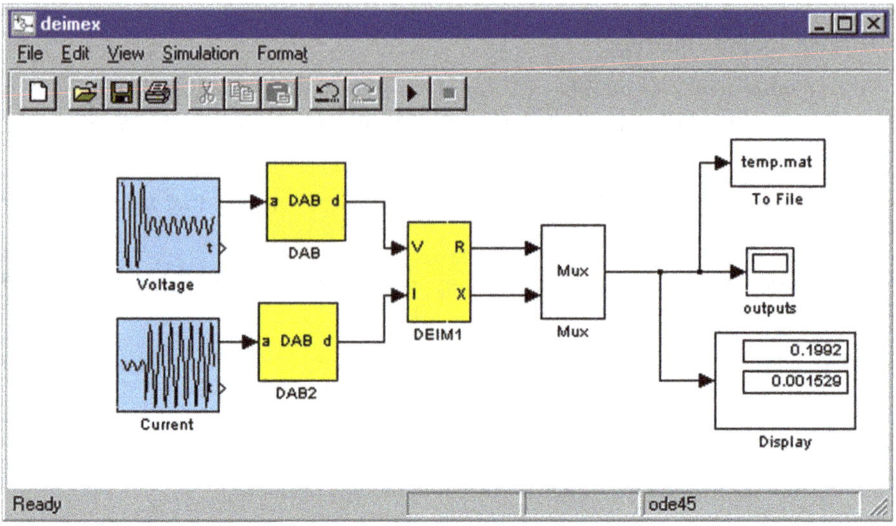

Fig. 4.22 Example for the DEIM block

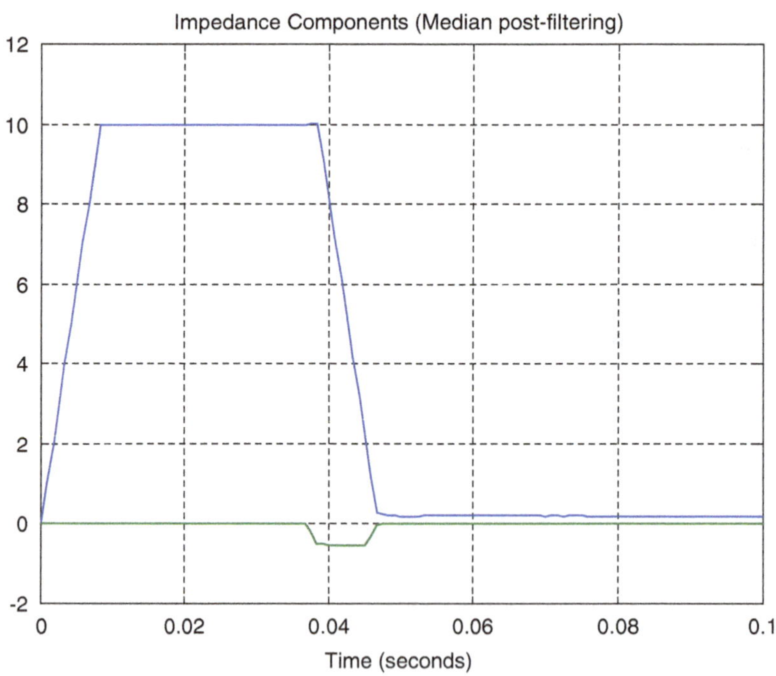

Fig. 4.23 Simulation results for DEIM (mean post-filtering)

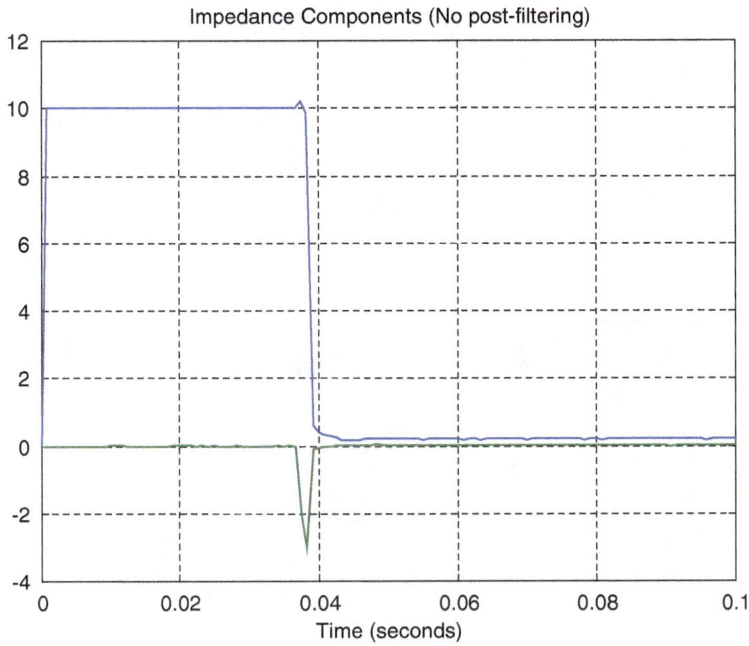

Fig. 4.24 Simulation results for DEIM (no post-filtering)

4.3.6.4 Dialog Box

See Fig. 4.25.

4.3.6.5 Description

Both finite impulse response (FIR) and infinite impulse response (IIR) filters are implemented in this block.

The FIR filter types available are Fourier Sine, Fourier Cosine, and Walsh. For this kind of filters, only the band-pass type is available. The length of the data window for these filters must be provided. This parameter together with the sampling frequency determines the frequency response for this type of filters.

The expression for the FIR filters is shown below:

Sine Data Window Band-Pass Filters

$$y(n) = \sum_{k=0}^{(p-1)/2} \left[x(n - p/2 + k + 0.5) - x(n - p/2 - k - 0.5)\right] \sin \omega_0 T_i \left(k + 0.5\right)$$

Cosine Data Window Band-pass Filters

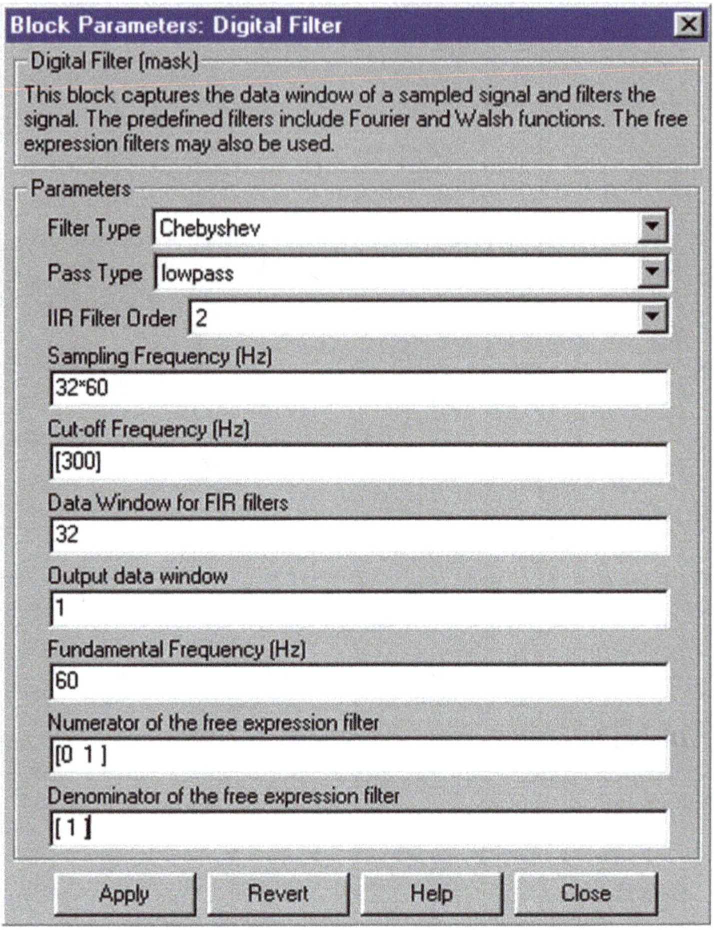

Fig. 4.25 Dialog box of digital filter (DF)

$$y(n) = \sum_{k=0}^{(p-1)/2} \left[x(n - p/2 + k + 0.5) + x(n - p/2 - k - 0.5)\right] \cos \omega_0 T_i \left(k + 0.5\right)$$

First-Order Walsh Filters

$$y(n) = \sum_{k=0}^{(p-1)/2} x(n-k) \; - \sum_{k=(p+1)/2}^{p} x(n-k)$$

Second-Order Walsh Filters

$$y(n) = -\sum_{k=0}^{(p+1)/4-1} x(n-k) + \sum_{k=(p+1)/4}^{3(p+1)/4-1} x(n-k) - \sum_{k=3(p+1)/4}^{p} x(n-k)$$

Third-Order Walsh Filters

$$y(n) = \sum_{k=0}^{(p+1)/4-1} x(n-k) - \sum_{k=(p+1)/4}^{(p-1)/2} x(n-k) + \sum_{k=(p+1)/2}^{3(p+1)/4-1} x(n-k) - \sum_{k=3(p+1)/4}^{p} x(n-k)$$

Fourth-Order Walsh Filters

$$y(n) = \sum_{k=0}^{(p+1)/8-1} x(n-k) - \sum_{k=(p+1)/8}^{3(p+1)/8-1} x(n-k) + \sum_{k=3(p+1)/8}^{5(p+1)/8-1} x(n-k) - \sum_{k=5(p+1)/8}^{7(p+1)/8-1} x(n-k) + \sum_{k=7(p+1)/8}^{p} x(n-k)$$

where:

x and y are inputs and outputs, respectively.
$p+1$ is the number of points of the filter window, i.e., $T_w = (p+1)T_i$.
T_w is the data window of the filter.
T_i is the sampling frequency of the input signal x.
ω_0 is the fundamental frequency of the input signal x.
T_0 is the fundamental period of the input signal x, with $\omega_0 = 2\pi/T_0$.

The IIR filter approximations available are Butterworth, Bessel, Chebyshev, and a free expression. The first three represent digital filters designed from their analog low-pass prototype filters. There are four types of filters available for each approximation: low-pass, high-pass, band-pass, and band-stop. The order for the IIR filters can be first, second, and third. The free expression represents any filter expressed in Z-transform. In this case, the numerator and denominator must be specified.

The cut-off frequency or frequencies for the Butterworth, Bessel, and Chebyshev filters must be provided. For low-pass and high-pass filter, one cut-off frequency is needed. For band-pass and band-stop, two cut-off frequencies need to be specified, with the first quantity beings the lower cut-off frequency, while the second the higher cut-off frequency.

This block contains a buffer like the one in the Data Acquisition Block. The user must provide the length of the output window desired.

4.3.6.6 Parameters

Filter type: The following filters are available: Fourier sine, Fourier cosine, Walsh 1, Walsh 2, Walsh 3, Walsh 4, Butterworth, Bessel, Chebyshev, and free expression.

Pass type: The filter can be: low-pass, high-pass, band-pass, and band-stop.

IIR filter order: The order of the filter, if the filter type selected, is an IIR filter. The order can be 1, 2, or 3.

Sampling frequency: Sampling rate of the signal in Hz.

Data window for FIR filters: Length of the input data window. This parameter is needed only if the filter selected is of type FIR.

Output data window: Number of samples that the output data window contains.
Fundamental frequency: The fundamental frequency of the system.

Numerator of the free expression filter: Array with the coefficients of numerator of the filter expression. It is needed only if the free expression option was selected.

Denominator of the free expression filter: Array with the coefficients of denominator of the filter expression. It is needed only if the free expression option was selected.

4.3.6.7 Recommended Solver

Ode45 (Dormand-Prince).

4.3.6.8 Example

In the following example, the discrete signal has frequency components at 60 and 120 Hz. The fundamental frequency is 60 Hz and the input data window is 20. The output data window is one (Figs. 4.26 and 4.27).

4.3.7 Digital Fourier Transform

4.3.7.1 Purpose

Computes the phasors of up to five selected harmonics of the input signal.

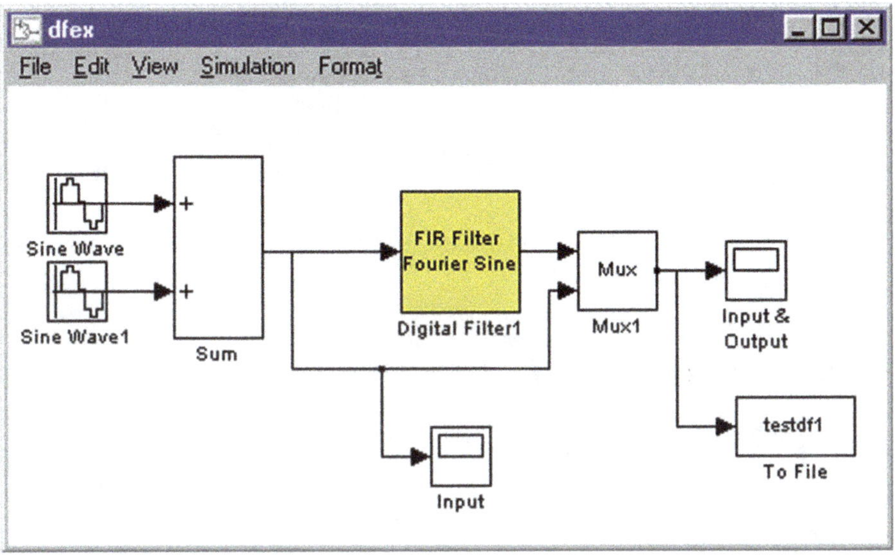

Fig. 4.26 Example for DF block

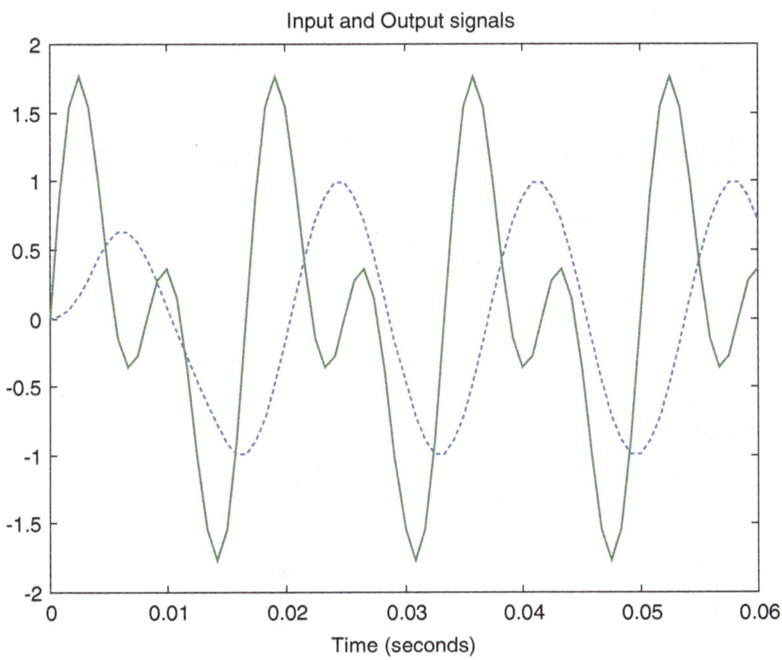

Fig. 4.27 Simulation results for DF

4.3.7.2 Input

A data window which contains samples of a continuous signal.

4.3.7.3 Output

The outputs are the magnitude and phase of the harmonics components of the input signal specified in the dialog box.

4.3.7.4 Dialog Box

See Fig. 4.28.

4.3.7.5 Description

The DFT block represents a filter that only allows signals that are multiples of the fundamental frequency to pass through. Up to five harmonics including the DC component can be selected.

To obtain the frequency component of a signal, the Digital Fourier Transform can be used. The DFT transforms a time series of samples to a series of frequency-domain samples.

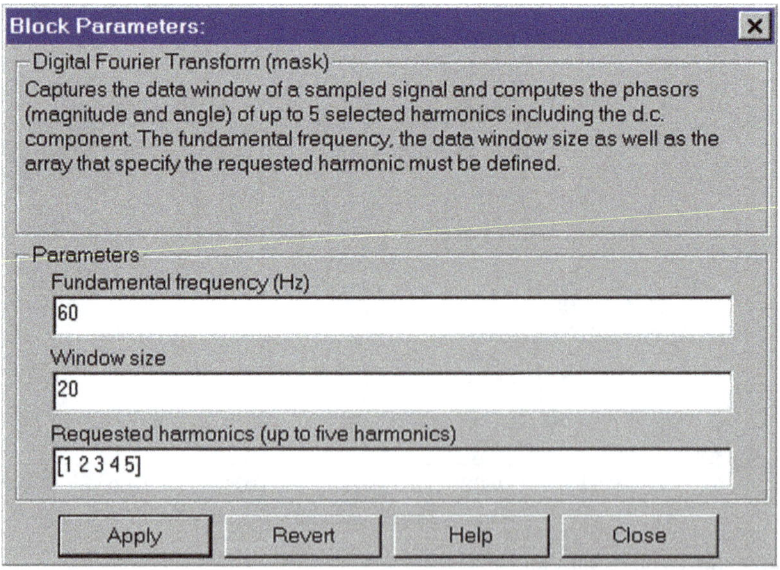

Fig. 4.28 Dialog box of Digital Fourier Transform (DFT)

The equation for the DFT is the following

$$X_{\mathrm{d}}(k)=\sum_{n=0}^{N-1} x(n)\mathrm{e}^{-j2\pi kn/N},\quad k=0,1,2,3,\ldots,N-1$$

where:

$x(n)$ is the discrete set of time samples.
$X_{\mathrm{d}}(k)$ is the set of frequency-domain samples obtained by the DFT of $x(n)$.
N is number of samples being considered.
n is time sample index.
k is the index for the computed set of discrete frequency components.

For an input vector of length N, the DFT is a vector of length N.

Computing the DFT is essentially a repetitive task. The major operations are the same over and over again. Therefore, the FFT is introduced to speed up the calculation. Practically, the FFT is an algorithm for computing the DFT. In this block the FFT, rather than the DFT, is used to calculate the signal harmonics.

The block obtains the discrete waveforms, u, from the Data Acquisition Board, and then uses the following MATLAB functions for computing magnitudes and phases.

$$\begin{array}{ll}\text{magnitude} = fft(u)/N & \text{for dc component}\\ \text{magnitude} = 2fft(u)/N & \text{for ac components}\end{array}$$

$$\text{phase} = \frac{\text{angle}(fft(u)/N)}{\mathrm{e}^{-j2\pi f_o nt/N}}180/\pi \quad \text{in degrees}$$

The DFT block has five outputs. Each output is the phasor (magnitude and phase in degrees) of the selected harmonic.

Note that in the requested harmonics box an array must be entered with the numbers of the desired harmonics (0 is the dc component).

4.3.7.6 Parameters

Fundamental frequency: Value of the fundamental frequency of the signal.

Window size: The length of the data window of samples of the input signal.

Requested harmonics: Array contains the harmonics components (magnitude and phase) that will be displayed in the output of the block. The harmonics are expressed as a multiple of the fundamental frequency.

4.3.7.7 Recommended Solver

Ode 45 (Dormant-Prince) can be used if the DAB is used. No time step is needed for this case. If only discrete signals are used in the model, the discrete solver can be used.

4.3.7.8 Example

In the following example, a signal composed by a dc component, the fundamental frequency, and the third harmonic is applied to the DFT block. The analytical expression for the input signal is the following:

$$F(\mathrm{tx}) = 5 + 10 \times \cos(2 \times \mathrm{pi} \times 60 \times \mathrm{tx}) + 30 \times \cos(3 \times 2 \times \mathrm{pi} \times 60 \times \mathrm{tx\text{-}pi}/180 \times 150)$$

The multiples of the fundamental frequency requested are: 0, 1, 2, 3, 4. It can be noted from the figure that the harmonics that are not zero are the DC, fundamental, and third components (Fig. 4.29).

4.3.8 *Orthogonal Components*

4.3.8.1 Purpose

Captures the data window of the sampled signal and computes the orthogonal components of the signal (magnitude and phase).

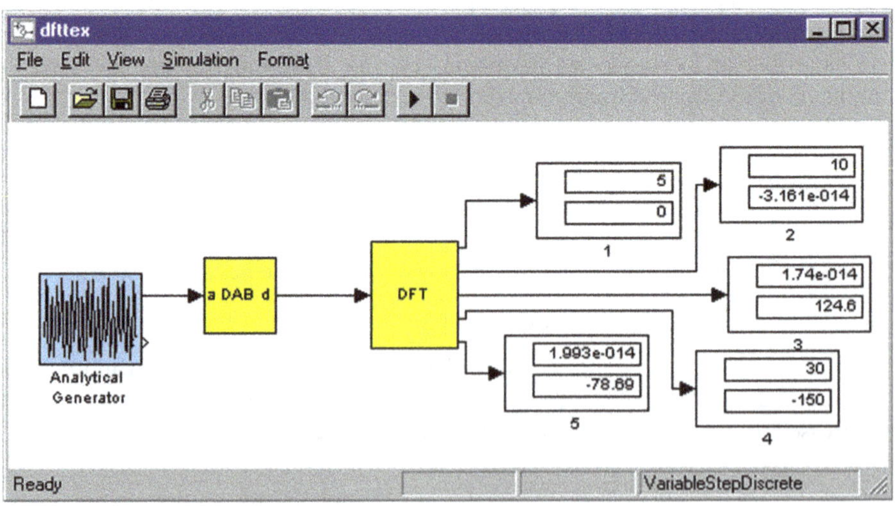

Fig. 4.29 Example for DFT block

4.3.8.2 Inputs

Data window of a sampled signal.

4.3.8.3 Outputs

The orthogonal components of the signal. Amplitude and angle (in degrees).

4.3.8.4 Dialog Box

See Fig. 4.30.

4.3.8.5 Description

This block captures the data window of a sampled signal and computes the orthogonal components of the signal in polar form (magnitude and phase in degrees). The predefined filters include Fourier and Walsh functions. The free expression filters may also be used. For the Fourier and Walsh filters, the full-cycle and half-cycle expressions were implemented. The user must input the number of samples per cycle.

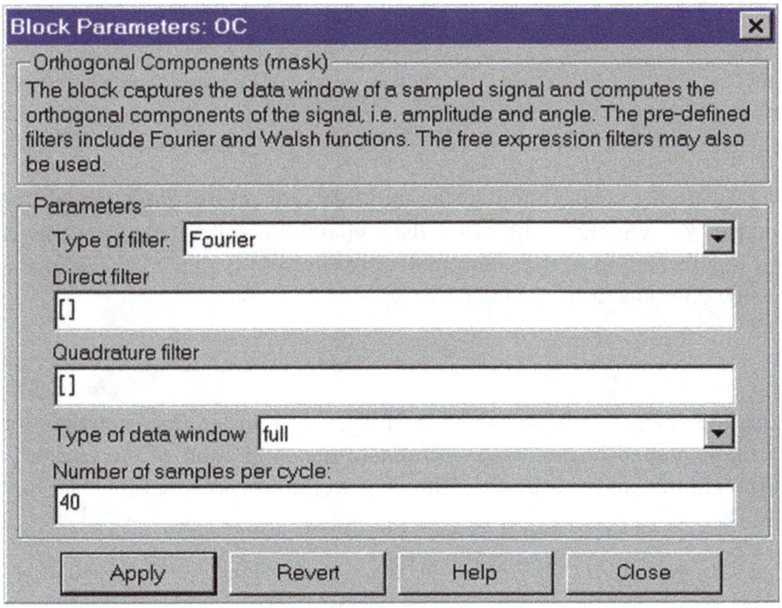

Fig. 4.30 Dialog box of orthogonal components (OC)

To obtain the orthogonal components from the sampled signal, three types of filter were implemented in this block: Fourier, Walsh, and Free expression.

Fourier-Based Orthogonal Components Filter

Suppose a signal seen by an overcurrent block is a sinusoidal signal of frequency f given by

$$x(t) = \sqrt{2}X \sin(2\pi f t + \phi)$$

This signal is conventionally represented by a phasor

$$\overline{X} = X\mathrm{e}^{j\phi} = X\cos\phi + jX\sin\phi \tag{4.1}$$

Assume that the sample rate of the signal is N times per cycle, then the sampled set,

$$x_k = \sqrt{2}X \sin\left(\frac{2\pi}{N}k + \phi\right) \tag{4.2}$$

Applying the discrete Fourier transform, the fundamental component of the sampled signal is

$$X_{\mathrm{C}} = \frac{2}{N}\sum_{k=0}^{N-1} x_k \cos\left(\frac{2\pi}{N}k\right) \tag{4.3}$$

$$X_{\mathrm{S}} = \frac{2}{N}\sum_{k=0}^{N-1} x_k \sin\left(\frac{2\pi}{N}k\right) \tag{4.4}$$

X_{C} and X_{S} are the cosine and the sine multiple sums in the fundamental frequency component $\overline{X}_1$. N is the width of the data window. It follows that the conventional phasor representation of the sinusoidal signal in (4.1) is related to the fundamental frequency component of the DFT by

$$\overline{X} = X_{\mathrm{C}} + jX_{\mathrm{S}} \tag{4.5}$$

Equations (4.3), (4.4), (4.5) are the full-cycle Fourier phasor estimation algorithm implemented in this block. To obtain the estimation of the phasor in polar form (magnitude and phase), the following well-known expressions are used:

$$\begin{aligned} X &= \sqrt{X_{\mathrm{C}}^2 + X_{\mathrm{S}}^2} \\ \Phi &= \tan^{-1}\left(\frac{X_{\mathrm{S}}}{X_{\mathrm{C}}}\right)\cdot\frac{180}{\pi}\ \text{(degrees)} \end{aligned} \tag{4.6}$$

The Fourier algorithm needs the data that to come from a whole cycle data window. Hence, the algorithm gives the first correct value after a cycle, i.e., the full-cycle Fourier phasor calculation algorithm introduces the delay of a full-cycle. In order to make a faster estimation; the half-cycle algorithm was implemented. In this algorithm, the orthogonal components are computed using the following expressions:

$$X_C = \frac{2}{N/2} \sum_{k=0}^{N/2-1} x_k \cos\left(\frac{2\pi}{N} k\right)$$

$$X_S = \frac{2}{N/2} \sum_{k=0}^{N/2-1} x_k \sin\left(\frac{2\pi}{N} k\right)$$

To obtain the polar form of the phasor estimation, (4.6) are used.

Walsh-Based Orthogonal Components Filter

Another way to obtain the orthogonal components is by using the Walsh-based orthogonal components filter. The first four Walsh functions are shown in Fig. 4.31.

The mathematical expressions of those Walsh functions are

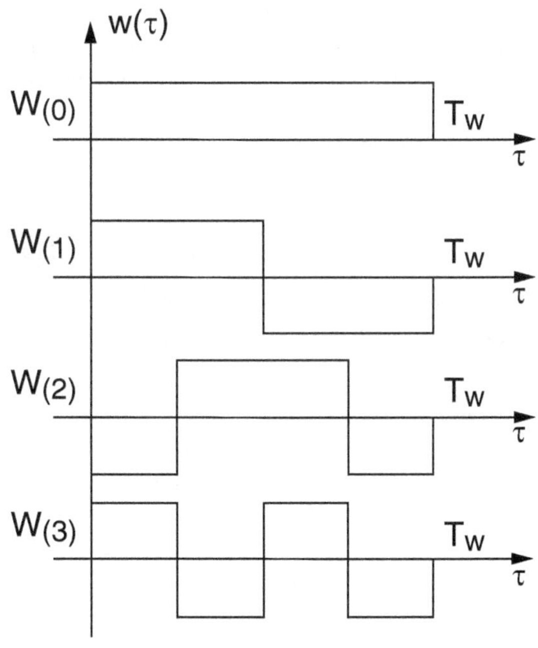

Fig. 4.31 First four order Walsh functions

$$
\begin{aligned}
W_0(t) &= 1 \qquad 0 < t < T_W \\
W_1(t) &= \begin{cases} 1 & 0 < t < T/2 \\ -1 & T/2 < t < T \end{cases} \\
W_2(t) &= \begin{cases} 1 & T_W/2 < t < T_W/2 \quad \text{or} \quad T_W/2 < \mathrm{t} < 3T_W/4 \\ -1 & 0 < t < T_W/4 \quad \text{or} \quad 3T_W/4 < \mathrm{t} < T_W \end{cases} \\
W_3(t) &= \begin{cases} 1 & 0 < t < T_W/4 \quad \text{or} \quad T_W/2 < \mathrm{t} < 3T_W/4 \\ -1 & T_W/2 < t < T_W/2 \quad \text{or} \quad 3T_W/4 < \mathrm{t} < T_W \end{cases}
\end{aligned}
$$

From Fig. 4.42, we know that the waveform of $-W_2(t)$ is similar to that of $\cos(\omega t)$, and the waveform of $W_1(t)$ is similar to that of $\sin(\omega t)$. Accordingly, the full-cycle Walsh-based orthogonal components are

$$
X_C = -\frac{\pi}{2N}\sum_{k=0}^{N-1} x_k W_2\left(\frac{2\pi}{N}k\right) \qquad X_S = \frac{\pi}{2N}\sum_{k=0}^{N-1} x_k W_1\left(\frac{2\pi}{N}k\right)
$$

Walsh filters have also a half-cycle version:

$$
X_C = -\frac{\pi}{N}\sum_{k=0}^{N/2-1} x_k W_2\left(\frac{2\pi}{N}k\right) \qquad X_S = \frac{\pi}{N}\sum_{k=0}^{N/2-1} x_k W_1\left(\frac{2\pi}{N}k\right)
$$

As in the Fourier filter, (4.6) are used to obtain the polar form of the phasor.

General orthogonal components filter

If we use filters other than the Fourier or Walsh, to calculate the orthogonal components, different filters will be needed to calculate the direct axis orthogonal component X_c and the quadrature axis orthogonal component X_s. These discrete filters can be written in the form of discrete transfer functions:

$$
X_c = f_c(z)x_k = \frac{a_o z^{p-1} + a_1 z^{p-2} + \cdots + a_{n-1}}{z^{p-1}} x_k
$$

$$
X_s = f_s(z)x_k = \frac{b_o z^{p-1} + b_1 z^{p-2} + \cdots + b_{n-1}}{z^{p-1}} x_k
$$

where p is the length of the filters.

The coefficients of the filter must satisfy:

Orthogonal condition:

$$\sum_{k=0}^{p} a_k b_k = 0$$

Unitary condition:

$$\max_t \left[\sum_{k=0}^{p} a_k \cos\left(\frac{2\pi}{N} + t\right) \right] = 1$$

$$\max_t \left[\sum_{k=0}^{p} b_k \cos\left(\frac{2\pi}{N} + t\right) \right] = 1$$

4.3.8.6 Parameters

Type of filter: There are three types of filters: Fourier, Walsh, and free expression.

Direct filter: Array with the coefficients of the direct orthogonal filter. This array only is needed when the free expression option is selected.

Quadrature filter: Array with the coefficients of the quadrature orthogonal filter. This array only is needed when the free expression option is selected.

Type of data window: The size of the data window. It can be full or half cycle.

Number of samples per cycle: The sampling frequency.

4.3.8.7 Recommended Solver

Ode45 (Dormand-Prince).l

4.3.8.8 Example

In the following model, the OC block is used to obtain the orthogonal components of a sinusoidal signal of amplitude one and a phase of zero degrees. The fundamental frequency is 60 Hz and the sampling frequency is 60 samples/cycle (Figs. 4.32, 4.33, and 4.34).

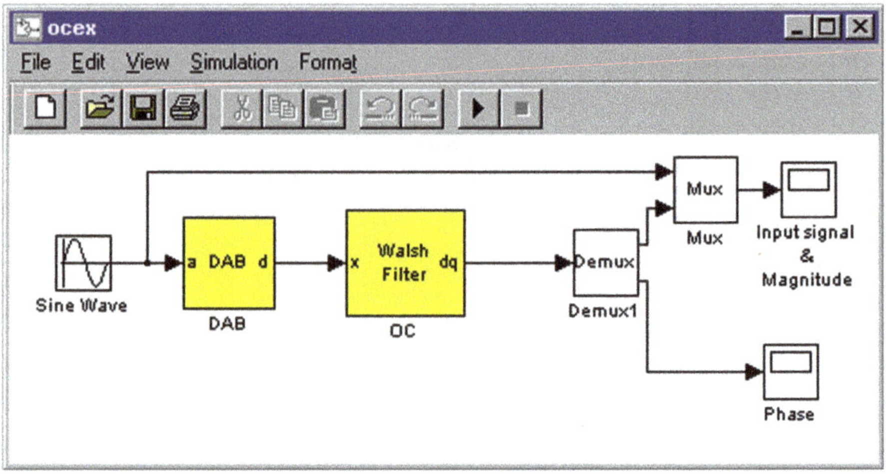

Fig. 4.32 Example for OC block

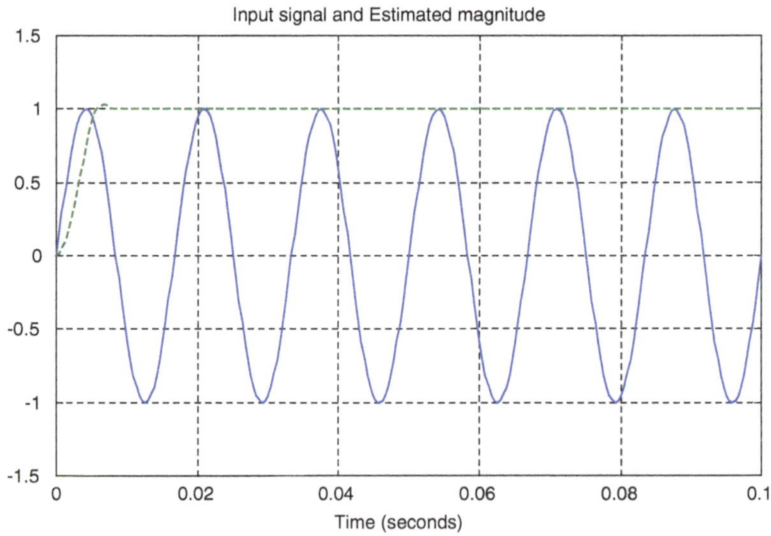

Fig. 4.33 Simulation of OC (magnitude of the signal)

4.3.9 Symmetrical Components

4.3.9.1 Purpose

Computes the symmetrical components from three-phase signals.

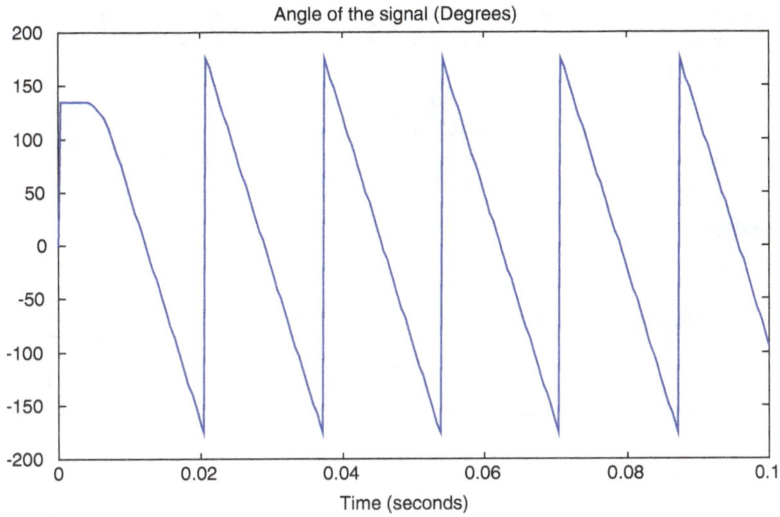

Fig. 4.34 Simulation of OC (angle of the signal)

4.3.9.2 Inputs

The inputs are the three-phase signals. They can be given in instantaneous form (three inputs) or in orthogonal form (six inputs) given in polar representation (phase in degrees).

4.3.9.3 Outputs

The symmetrical components of the three-phase signals are given in instantaneous or orthogonal form depending on the form selected for the inputs. There will be three outputs if the instantaneous form or six outputs if the orthogonal form is selected.

4.3.9.4 Dialog Box

See Fig. 4.35.

4.3.9.5 Description

The inputs of the block are the three-phase signals and the outputs are the symmetrical components. The inputs can be instantaneous (three inputs) or orthogonal components, i.e., magnitudes and angles of the signals (six inputs). The outputs reflect the signal type specified for inputs.

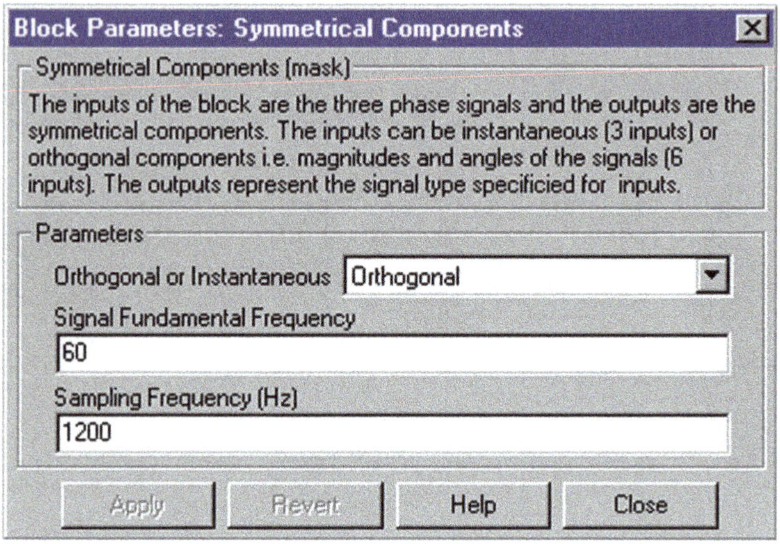

Fig. 4.35 Dialog box of symmetrical components (SC)

If the inputs are orthogonal components, the following equations are used to get the symmetrical components:

$$\begin{bmatrix} X_0 \\ X_1 \\ X_2 \end{bmatrix} = \frac{1}{3}\begin{bmatrix} 1 & 1 & 1 \\ 1 & a & a^2 \\ 1 & a^2 & a \end{bmatrix}\begin{bmatrix} X_a \\ X_b \\ X_c \end{bmatrix}$$

where,

X_0, X_1 and X_2 are zero-, positive-, and negative-sequence components, respectively, X_a, X_b, and X_c are phase a, b, and c phasors, respectively, and $a = \mathrm{e}^{j120^{\circ}}$.

If the inputs are instantaneous values, the following formulae are used to get the instantaneous symmetrical components.

$$\begin{aligned} x_0(n) &= \frac{1}{3}[x_a(n) + x_b(n) + x_c(n)] \\ x_1(n) &= \frac{1}{3}\left[x_a(n) - x_b\left(n - \frac{m}{6}\right) + x_c\left(n - \frac{m}{3}\right)\right] \\ x_2(n) &= \frac{1}{3}\left[x_a(n) + x_b\left(n - \frac{m}{3}\right) - x_c\left(n - \frac{m}{6}\right)\right] \end{aligned}$$

where:

x_0, x_1 and x_2 are zero-, positive-, and negative-sequence instantaneous values, respectively,

x_a, x_b, and x_c are phase a, b, and c instantaneous values, respectively,

n represents the present moment, and
m is the number of the samples in one fundamental cycle of the input variable.

Provided $m/3$ and $m/6$ are integers, the calculation of both sets of equations is simple. Where this condition is not true, the following method can be used.

Assuming $\frac{m}{6} = k + r$, where k is an integer and r a fraction, the following linear interpolation applies

$$x\left(n - \frac{m}{6}\right) = x(n - k)(1 - r) + x(n - k - 1)r$$

Terms including $m/3$ and $2m/3$ are dealt with similarly.

The fundamental frequency of the signal as well as sampling frequency must be provided.

4.3.9.6 Parameters

Orthogonal or instantaneous: The form of the input signal can be orthogonal components (magnitude and phase in degrees) or instantaneous. In the latter case, the inputs are three data windows with samples of a three-phase signal.

Signal fundamental frequency: Fundamental frequency for the signal in Hertz, typically 60 or 50 Hz.

Sampling frequency: Rate at which the sampling of the signal is made.

4.3.9.7 Recommended Solver

The recommended solver is ode-45 (Dormand-Prince) solver, as this block usually requires the DAB. For this solver, a time step does not need to be specified.

4.3.9.8 Example

The following model obtains the symmetrical components from a three-phase balanced signal of amplitude one (Figs. 4.36 and 4.37).

4.3.10 Triggering Element

4.3.10.1 Purpose

Implements a fault detector.

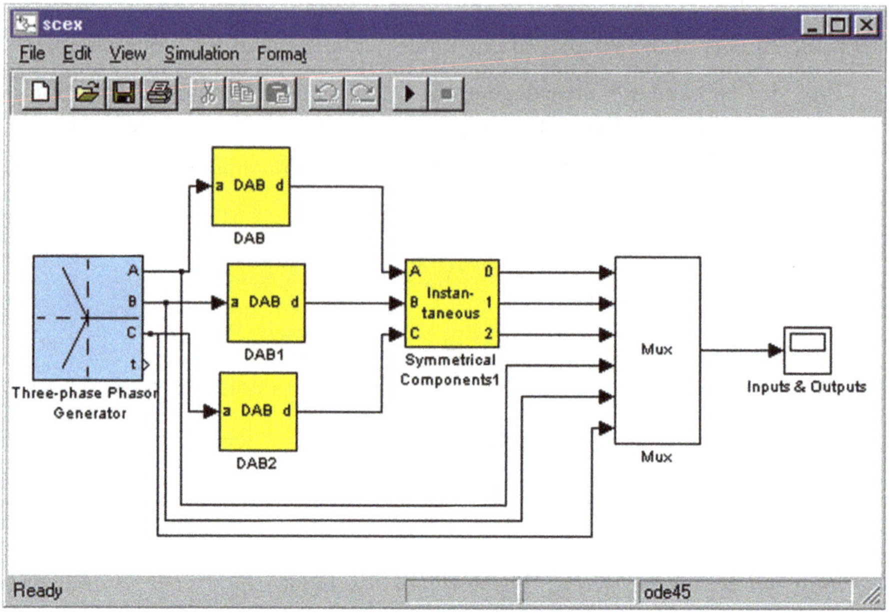

Fig. 4.36 Example for SC block

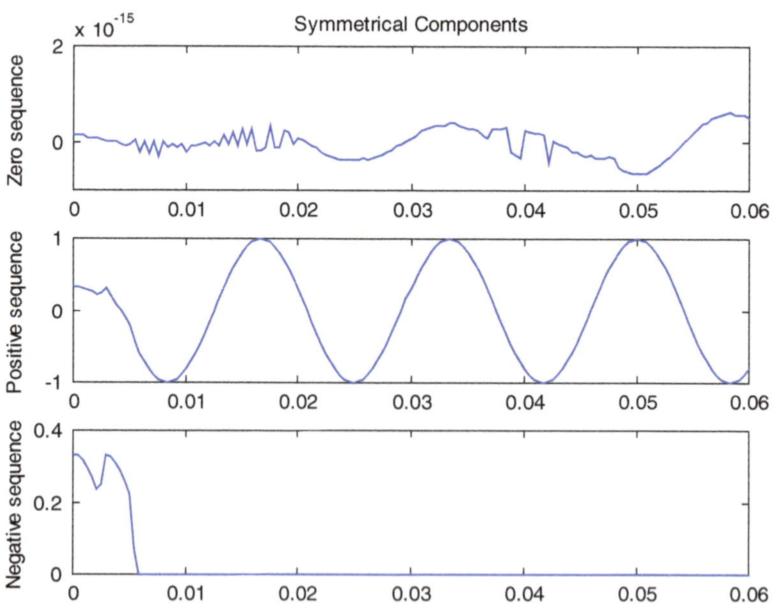

Fig. 4.37 Simulation results for SC

4.3.10.2 Input

An array with the most recent sample and $n-1$ previous samples of a signal, where n is the size of the data window.

4.3.10.3 Output

A signal that goes from low to high after there is a sudden change in the magnitude of the input signal and the activating counter reaches its preset value. If the output goes again to low after the magnitude of the input signal is stable, the deactivating counter reaches its preset value.

4.3.10.4 Dialog Box

See Fig. 4.38.

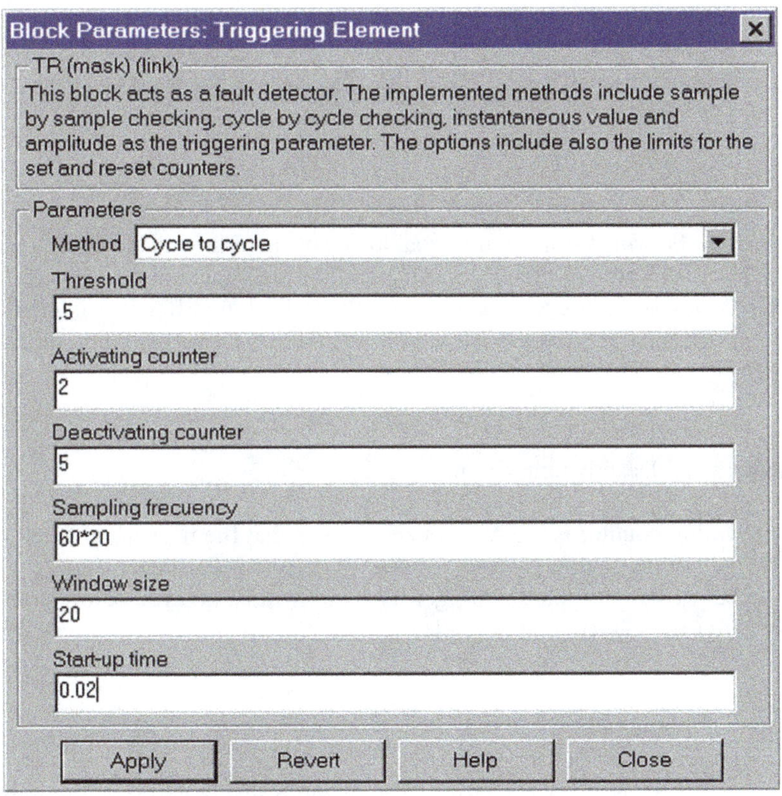

Fig. 4.38 Dialog box of triggering element (TE)

4.3.10.5 Description

This block is fed by a data window of a given signal and sets its Boolean output high if a fault is detected.

There are three methods available to detect a fault: value, sample to sample, and cycle to cycle. These methods are described below.

Value

This method compares the absolute value of the present sample of the signal to the threshold. If the value is greater than the threshold, the counter is activated. If the activated counter reaches its preset value, and the signal is still greater than the threshold, the output is set high.

Sample to Sample

A comparison between the two most recent samples of the signal is performed. If the absolute value of this difference is greater than the threshold, the activating counter is started and if this difference is still greater than the threshold when the activating counter reaches its preset value, then the output is set to one.

Cycle to Cycle

In this case, the window size must equal to one cycle of the fundamental frequency. In this case, the value of the recent sample of the signal and the value of the signal measured a cycle before are compared. If their absolute difference is greater than the threshold and this situation remains during the activating counting, the output is set high.

Activating and Deactivating Counters

The activating counter is the number of samples that the triggering signal will be delayed before going to high, after the fault is detected. The deactivating counter is the number of samples that the triggering signal will be delayed before going back to low, after the input signal is stable.

4.3.10.6 Parameters

Method: Method used for triggering. The methods available are value, sample to sample, and cycle to cycle.

Threshold: A value to determine if there is a fault.

Activating counter: Number of samples that the triggering signal will be delayed before tripping, after a fault is detected.

Deactivating counter: Number of samples that the triggering signal will need to go back to zero after the input signal is stable.

Sampling frequency: The rate of sampling of the input signal which must be the same as in the DAB.

Window size: The length of the data window.

Start-up time: After this moment, the block is activated.

4.3.10.7 Recommended Solver

This block has discrete states only, so the discrete solver can be used if there is no need to use the DAB. In this case, the time step should be selected so that all the harmonics of the signal can be detected. If the model uses the DAB, as in the example shown below, the ode-45 (Dormand-Prince) solver is recommended. For this solver, a time step does not need to be specified.

4.3.10.8 Example

In the following example, a fault occurs at $t = 0.03$ s and the magnitude of the input signal goes from 1 to 2 at that time. There are high frequency components added due to the fault. The DAB filters these components and feeds the TR block with a data window of size 20. The triggering element uses a cycle to cycle method. The threshold is 0.5, the activating counter is 2, and the deactivating counter is 5. It can be noted in Fig. 4.40 that the triggering signal goes to high short after the fault occurs (the delay is given by the activating counter) and goes to low again at around $t = 0.11$ s. Note that after that time the difference of the sample at the present time and the sample of one cycle before is less than the threshold specified in the parameters, i.e., the fault is stable (Figs. 4.39 and 4.40).

4.3.11 Universal Comparator

4.3.11.1 Purpose

Performs signal comparisons of three types: signal-to-signal, signal-to-threshold, and signal-to-time.

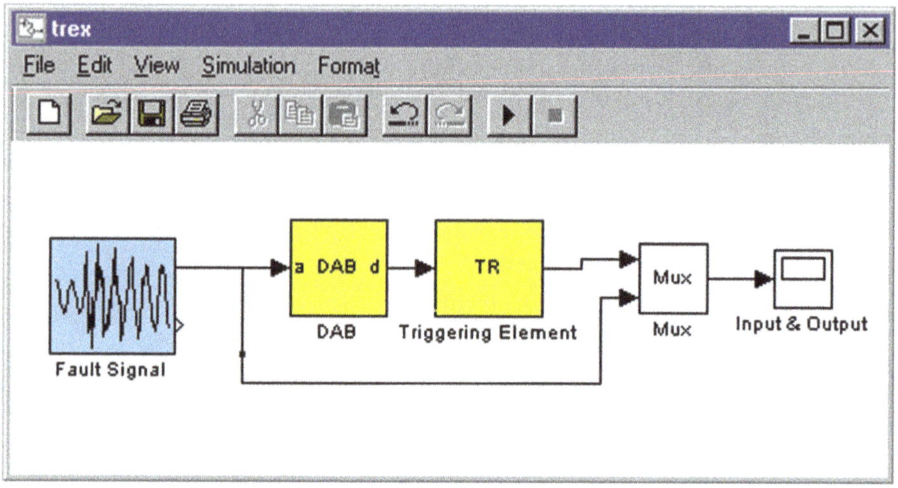

Fig. 4.39 Example for TE block

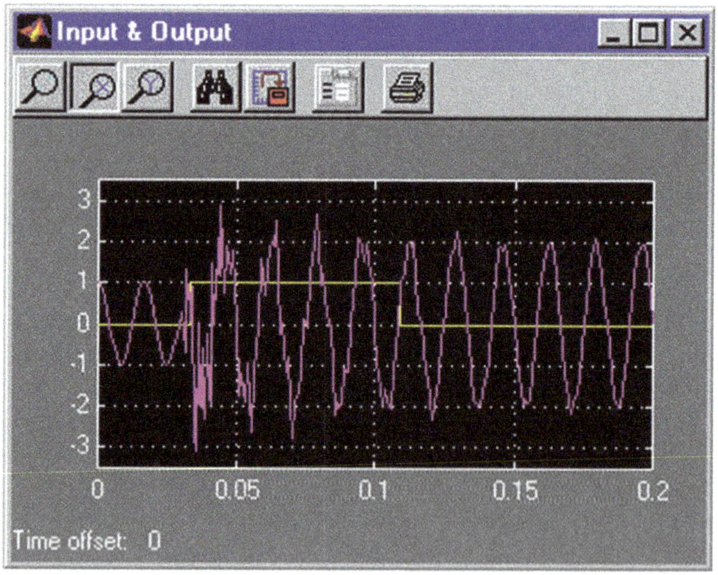

Fig. 4.40 Simulation result for TE

4.3.11.2 Inputs

If the signal–signal comparison option is selected, two inputs feed this block. Otherwise, the block needs only one input. In all cases, the signals are given in instantaneous form.

4.3.11.3 Output

If the signal-to-signal or signal-threshold comparison is selected, the output is one if the comparison specified in the dialog box is true and zero otherwise. If the signal-time comparison is selected, the output is one or zero depending on the location of the input signal in the signal–time characteristic selected and of the *I–t* emulation method.

4.3.11.4 Dialog Box

See Fig. 4.41.

4.3.11.5 Description

The Universal Comparator has either one or two inputs. If the signal-time comparison or the signal-threshold options are selected, only one input is needed. Otherwise, two inputs must feed the block. The UC block has one output: The Boolean result of

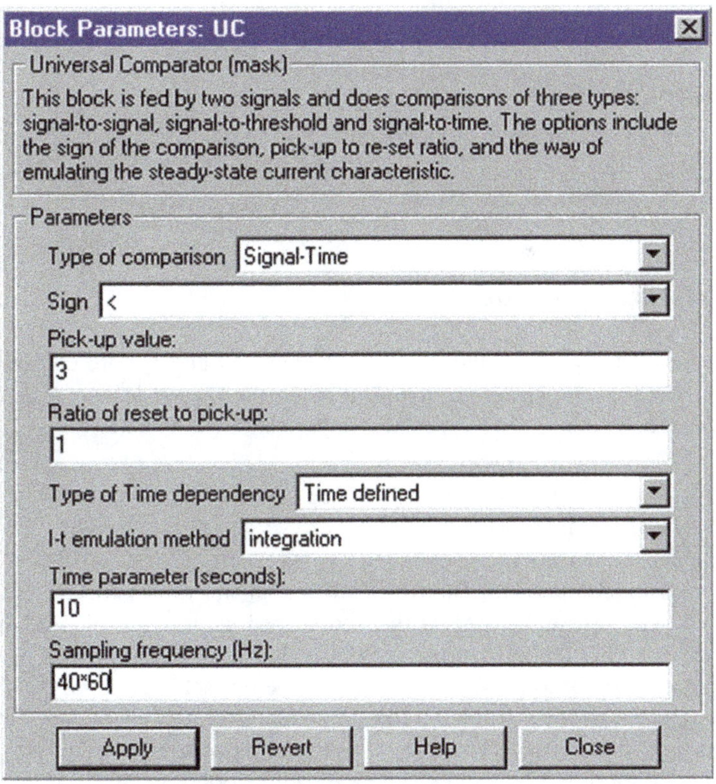

Fig. 4.41 Dialog box of universal comparator (UC)

the comparison. The options include the sign of the comparison, pick-up to reset ratio, and the way of emulating the steady state current characteristic.

Signal-to-Threshold Comparison

In this type of comparison, S_1 is compared against a constant. The previous state of the output of the block is also considered. If this previous state is one, then the threshold is multiplied by the ratio of reset to pick up and then the comparison is made.

We have:

If $C(t-\Delta t)=0$ then
$C(t)=S_1>\text{threshold}$
Else
$C(t)=S_1>(\text{threshold}\times k_1)$
End
Or
If $c(t-\Delta t)=0$ then
$C(t)=S_1<\text{threshold}$
Else
$C(t)=S_1<(\text{threshold}\times k1)$
End
Typically:
$0<k_1<1$ for $S_1>$ threshold comparisons
and
$k_1>1$ for $S_1<$ threshold comparisons

Signal-to-Signal Comparison

In this type of comparison, the two input signals (S_1 and S_2) are compared. If the Boolean result of the comparison is true, then the output is one. Otherwise, the output is zero. There are two types of comparison: $S_1>S_2$ and $S_1<S_2$.

Signal-to-Time Comparison

In this type of comparison, for the input signal S_1, the operating (t_{op}) time is calculated. The operating time is then compared against the elapsed time since the signal became greater than the pick-up value. If t is greater than t_{op}, then the output of the block becomes one. The elapse time is reset every time the signal goes below the pick-up value. The operating time is calculated every sampling period. The comparison is also performed every sampling period.

The operating time is calculated using the corresponding signal–time characteristic functions. There are six types of signal–time characteristic functions: Time defined, standard inverse, very inverse, extremely inverse, RI inverse, and long time inverse. These functions are shown below and the curves that represent them, for different values of the time parameter (k), are shown in Figs. 4.44, 4.55, and 4.56, respectively. Note that the x-axis represents the times of the pick-up value. The y-axis represents the time.

Time defined

$$t_{op} = k$$

Standard inverse

$$t_{op} = \frac{0.14 \cdot k}{I_n^{0.02} - 1}$$

Very inverse

$$t_{op} = \frac{13.5 \cdot k}{I_n - 1}$$

Extremely inverse

$$t_{op} = \frac{80.0 \cdot k}{I_n^2 - 1}$$

RI inverse

$$t_{op} = 3.1 \cdot k\left(1 + \frac{2.2}{I_n^{2.2}}\right)$$

Long time inverse

$$t_{op} = \frac{120 \cdot k}{I_n - 1}$$

where:

t_{op} = operation time
I_n = normalized input signal. $I_n = S_1$/(pick-up value)
k = time parameter (Figs. 4.42, 4.43, and 4.44)

I−t *Emulation Method*

For the signal–time comparisons, there are two options for the input signal: consider the present value of the signal or the average value (integration) of the signal since the instant of time that S_1 became greater than the pick-up value.

These two methods are shown below.

Present value: In this case, the present value of the input signal is substituted into the corresponding equation for the $I-t$ characteristic functions.

Integration: In this case, the average of the input signal is calculated by using the following expression:

$$I_{prom} = \frac{1}{(t_1 - t_0)} \int_{t_0}^{t_1} I_n dt$$

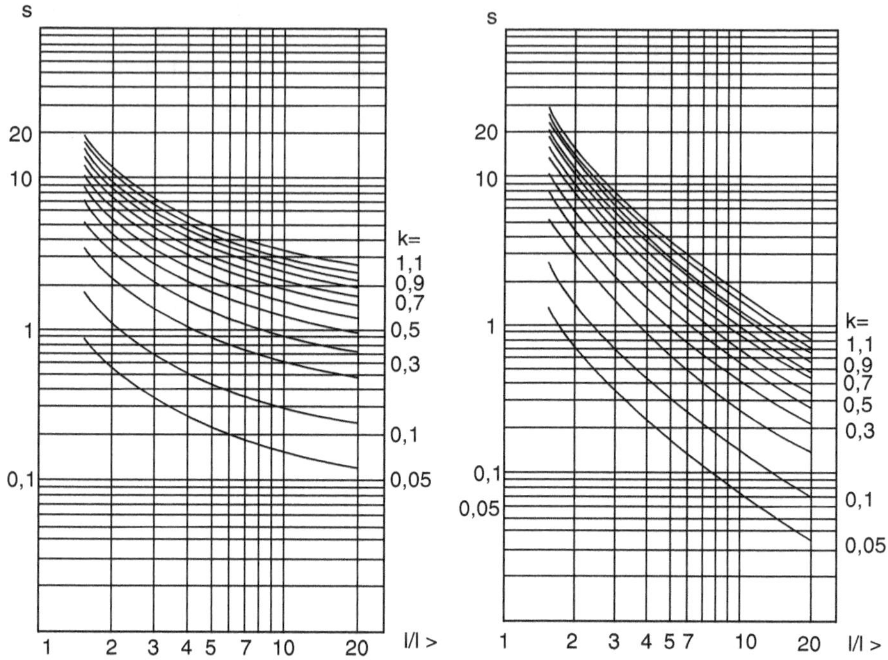

Fig. 4.42 Inverse and very inverse functions

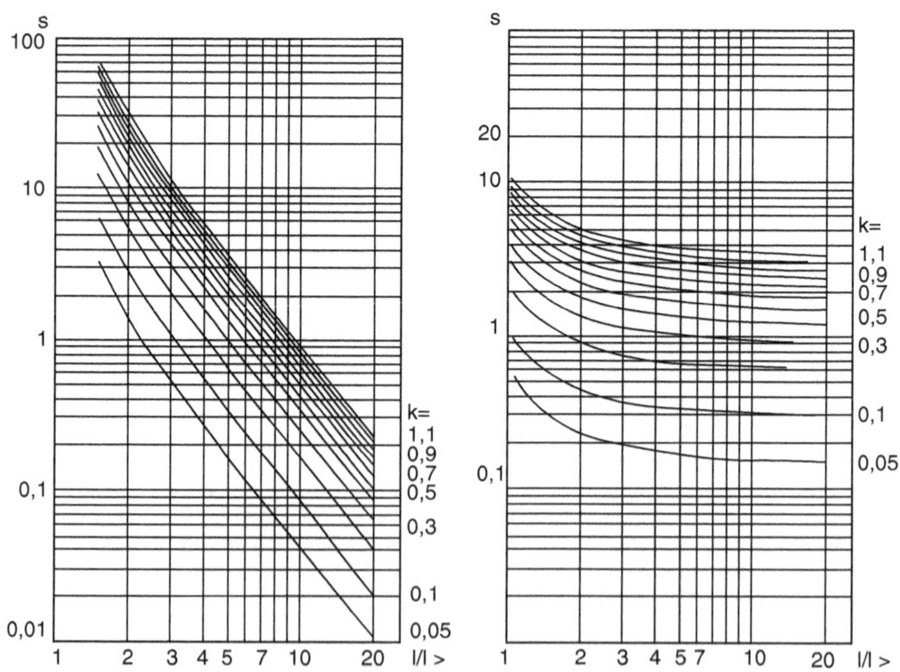

Fig. 4.43 Extremely inverse and RI inverse curves

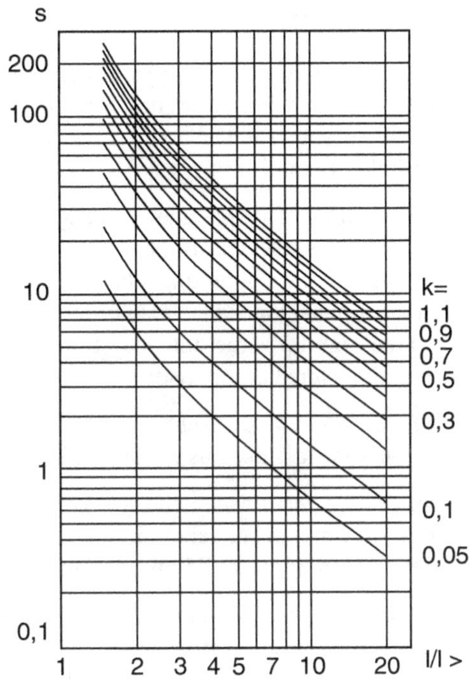

Fig. 4.44 Long time inverse curves

where

$t_1 =$ current time
$t_0 =$ time when the signal becomes greater than the pick-up value

As the signal is discrete, to calculate the previous integral, the trapezoidal rule is used. And we have the following expression:

$$I_{avg}(t) = \frac{1}{(t_1 - t_0)} \begin{pmatrix} I_n(t_0) + \frac{\Delta t}{2} I_n(t_0 + \Delta t) + I_n(t_0 + \Delta t) + \frac{\Delta t}{2} I_n(t_0 + 2\Delta t) + \cdots \\ +I_n(t_1 - \Delta t) + \frac{\Delta t}{2} I_n(t_1) \end{pmatrix}$$

where:

$\Delta t = \dfrac{1}{f_s}$

$f_s =$ sampling frequency

This average of the input signal is substituted into the corresponding equation for the time of operation.

4.3.11.6 Parameters

Type of comparison: The type of comparison to be performed by the block. There are three types: Signal–Signal, Signal–Threshold, and Signal–Time.
Sign: The sign for the comparison (> or <) for the signal–signal and signal–threshold comparisons.

Pick-up value: This parameter is the value of the threshold if the signal–threshold comparison is selected. If the signal–time comparison is selected, this value is used to normalize the signal–time characteristic curves. This value is the minimum value that the input signal can take in order to calculate the operating time. For values less than the pick-up value, the operating time is infinite. This parameter is not used in the signal–signal comparison.

Ratio of reset to pick-up: This parameter is used only in the signal–threshold comparison. It is used as explained in the description section.

Type of time dependency: the type of signal–time characteristic curve. There are six types of time dependency implemented: time defined, standard inverse, very inverse, extra inverse, RI inverse, and long time inverse.

I–t emulation method: there are two types of emulation available: present value and integration. These methods are explained in the description section.

Time parameter: Constant, given in seconds, used to calculate the operating time in the signal–time comparison.

Sampling frequency: The rate of sampling of the signal.

4.3.11.7 Recommended Solver

Ode45 (Dormand-Prince) can be used.

4.3.11.8 Examples

In the following example, the UC is used to compare a signal against a threshold. The sign of the comparison is >, the threshold is 0.8, and the ratio of reset to pick-up is 0.5 (Fig. 4.45).

The corresponding input–output plot is shown in Fig. 4.46.

Another example using the UC block is shown below. The options are: signal–time comparison, type of time dependency: standard inverse, pick-up value: 1, ratio of reset to pick-up: 1, *I–t* emulation method: integration, and time parameter of 0.1 s. Note that the input signal goes to zero at $t = 0.7$ s and back to one at $t = 0.72$ s. This resets the UC block and delays the operating time (Fig. 4.47 and 4.48).

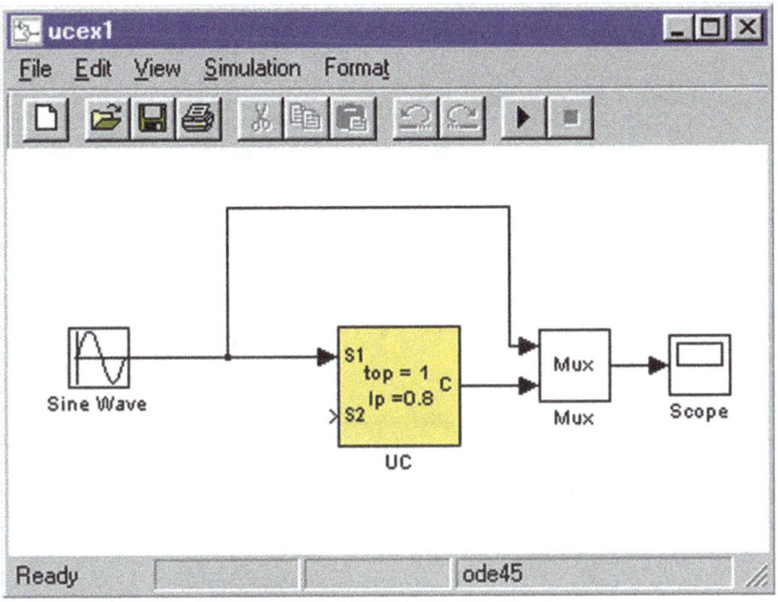

Fig. 4.45 Example 1 for UC block

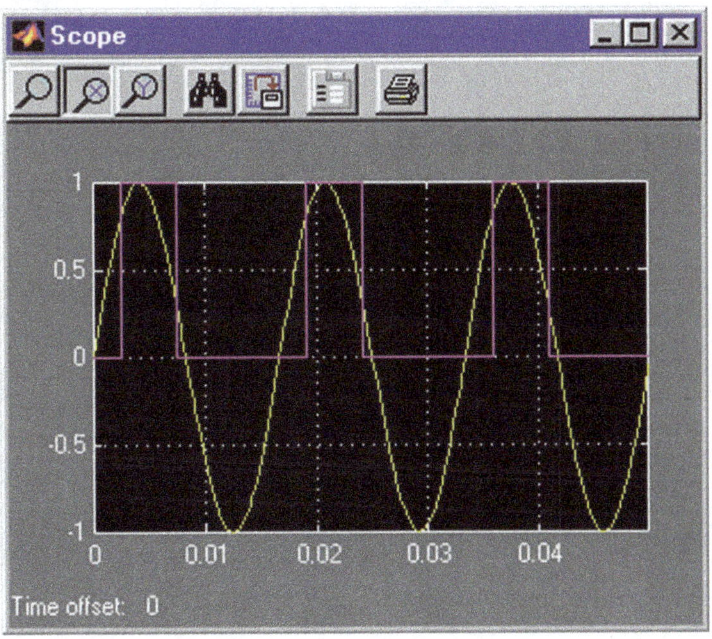

Fig. 4.46 Input–output plot for simulation of UC Example 1

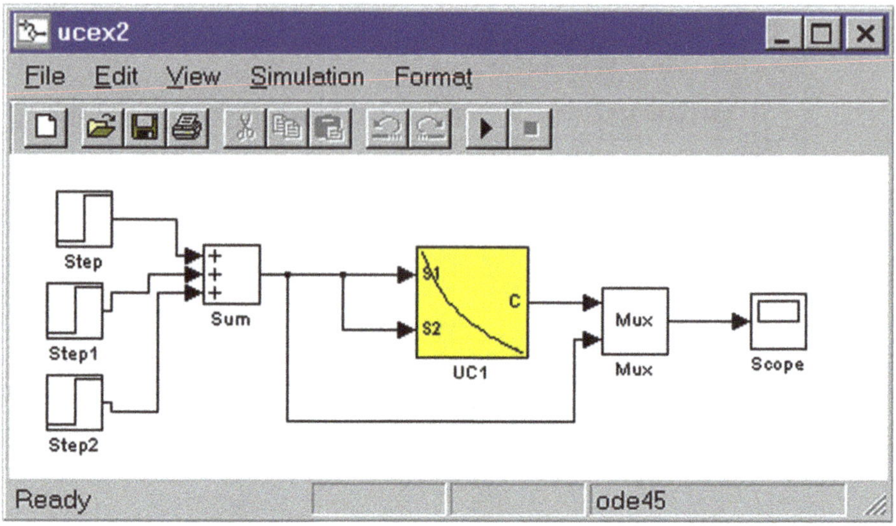

Fig. 4.47 Example 2 for UC block

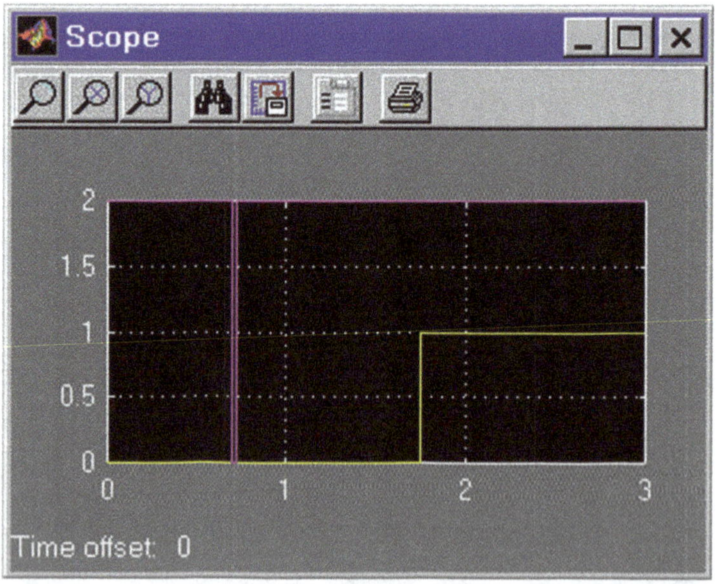

Fig. 4.48 Input–output plot for UC Example 2

4.3.12 *Phase Selection*

4.3.12.1 Purpose

Implements a fault classification algorithm

4.3.12.2 Inputs

The inputs of this block are the three-phase voltage and currents in phasor form, i.e., magnitude and angle (in degrees).

4.3.12.3 Outputs

There is only one output. It is a number that identifies the type of fault detected. If there is no fault, the output is zero.

4.3.12.4 Dialog Box

See Fig. 4.49.

4.3.12.5 Description

This block takes the orthogonal components of the three-phase current and voltages. It recognizes if a fault exists based on the inputs. The threshold for the current, voltage, and impedance must be provided in the same units as the input currents and voltages.

The output is a number that represents the type of the detected fault. The numbers and their corresponding fault types are shown in Table 4.2.

The methods used to classify a fault are the following:

Current-only, voltage-only, impedance-only, current and voltage, current or voltage, current and impedance, current or impedance, current and voltage and impedance, current or voltage or impedance.

In the voltage/current/impedance-only methods, the information of one type of signal (voltage, current, or impedance) is used to classify the fault. In the methods that use the AND operator, the two or three types of signals are used to detect and classify the fault. Only if the fault is detected for the two or three types of signals, the block is capable to classify the fault. In the methods that use the OR operator, the fault can be detected in either one of the two or three types of signal, so the block is able to classify the fault.

In general, the methods that use the AND operator are selected and give a more dependable fault classification than the methods that use the OR operator. But the

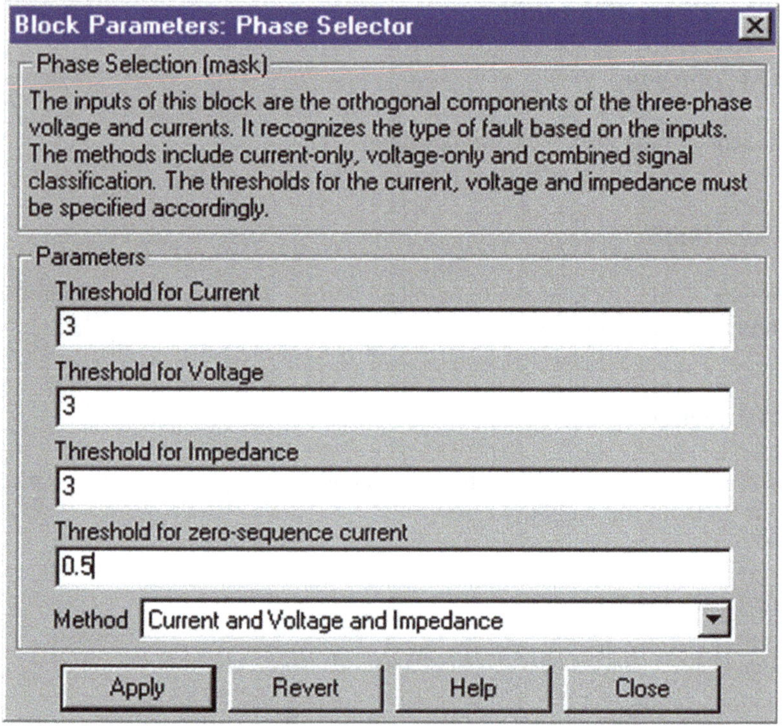

Fig. 4.49 Dialog box of phase selector (PS)

Table 4.2 Fault types of phase selection element

Number	Type of fault
1	A-G
2	B-G
3	C-G
4	A-B-G
5	A-C-G
6	B-C-G
7	A-B
8	B-C
9	A-C
10	A-B-C-G
11	A-B-C

latter methods give a more secure response, i.e., the probabilities that a fault is not detected are less than those in the former methods.

In some cases, if the transmission line is very long, the voltage method may not be suitable, so current method is used instead. In some other cases, where the transmission line is quite short, the voltage or impedance method may provide better classification.

4.3.12.6 Parameters

Threshold for current/voltage/impedance/zero sequence current: Limit value for detection of the fault (maximum for the currents, and minimum for the voltage and impedance). If the value of the voltage, current, or impedance is outside the corresponding threshold, a fault is detected.

Method: The method that the block uses to classify the fault. The methods available are: current, voltage, impedance, current and voltage, current or voltage, current and impedance, current or impedance, voltage and impedance, voltage or impedance, current and voltage and impedance, and current or voltage or impedance.

4.3.12.7 Recommended Solver

For this block, the discrete solver can be used if there is no need to use the DAB to obtain the orthogonal components. In this case, the time step should be selected so that all the harmonics of the signal can be detected. If the model uses the DAB, as in the example shown below, the ode-45 (Dormand-Prince) solver is recommended. For this solver, a time step does not need to be specified.

4.3.12.8 Example

In the following example, an A-B-G fault is simulated . The PS block detects a fault of type 4, which corresponds to an A-B-G fault. The current-only method is used in this case (Figs. 4.50, 4.51, and 4.52).

4.3.13 Vector Group Compensator for 2-Winding Transformers

4.3.13.1 Purpose

Computes the differential and restraining currents from the instantaneous values of six-phase currents of a 2-winding transformer.

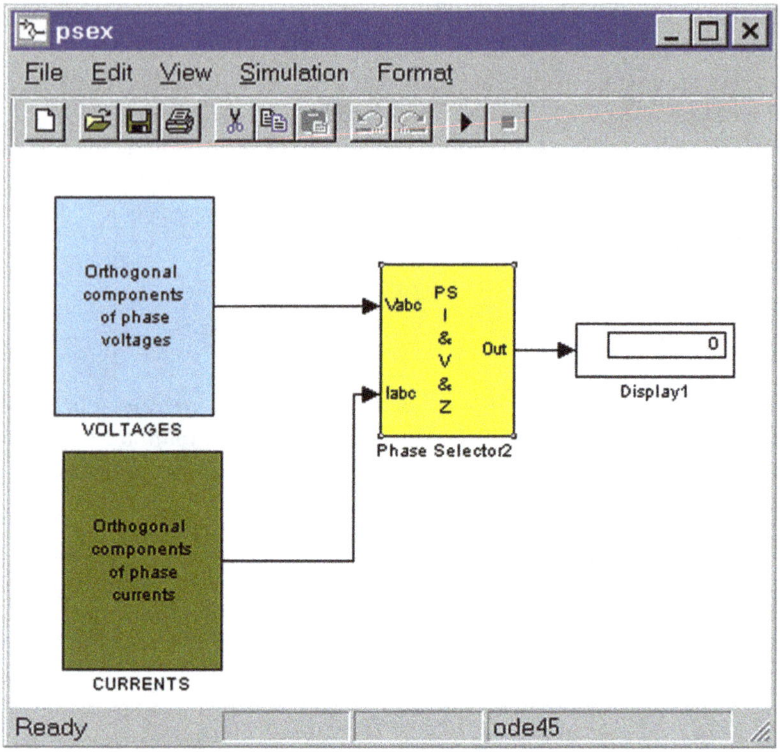

Fig. 4.50 Example for PS block

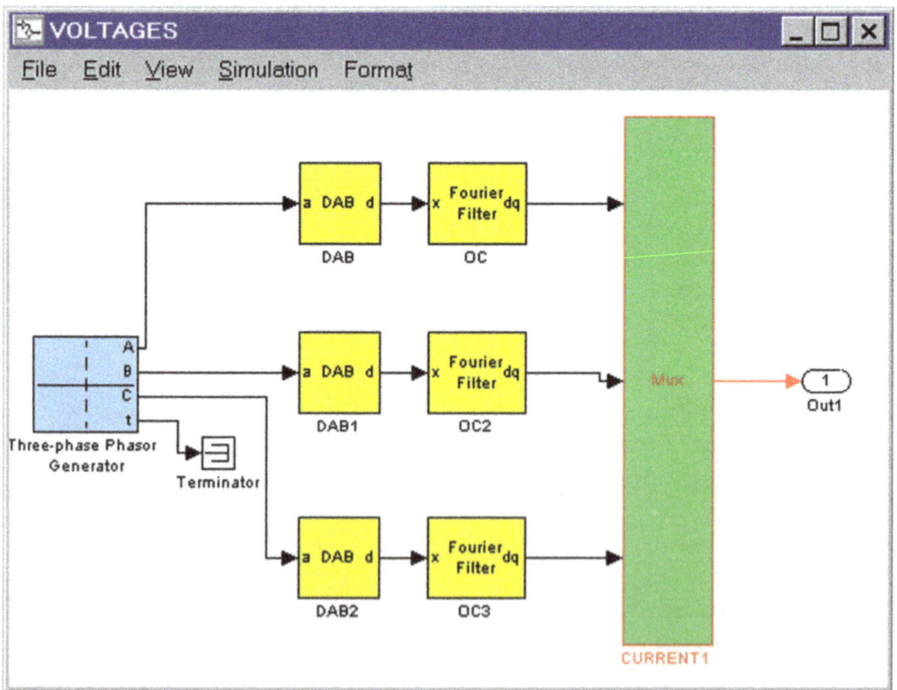

Fig. 4.51 Calculation of the orthogonal components of phase voltages

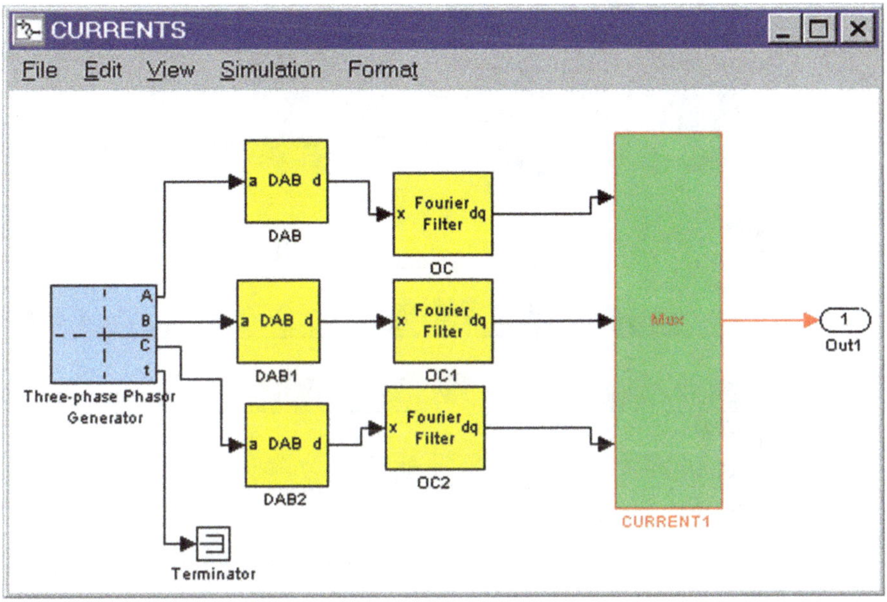

Fig. 4.52 Calculation of the orthogonal components of phase currents

4.3.13.2 Inputs

This block has three inputs. They are the instantaneous values of the three-phase currents.

4.3.13.3 Outputs

The outputs are the instantaneous values of the differential and restraining currents for each phase.

4.3.13.4 Dialog Box

See Fig. 4.53.

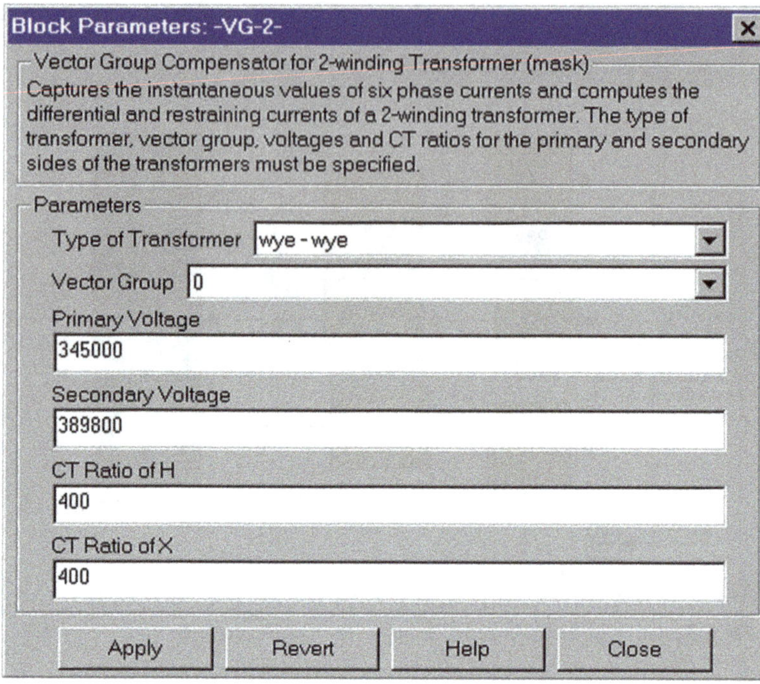

Fig. 4.53 Dialog box of Vector Group Compensator for 2-winding Transformer (VG-2)

4.3.13.5 Description

The purpose of this block is to obtain the differential and restraining currents from the phase currents measured at both sides of the 2-winding transformer. From the outputs of the block, the location of a fault can be determined.

The VG-2 block has six inputs: three-phase currents from the high voltage side and three from the low voltage side of the transformer. All the currents are in instantaneous values. The outputs are the instantaneous values of the restraining and differential currents (six outputs).

The user must input the type, vector group, primary and secondary voltages, and CT ratios that correspond to the two-winding transformer that is connected to the VG-2 block.

As an example of how the differential and restraining currents are calculated, the corresponding equations for a Δ-Y transformer (vector group 1) are shown below:

$$i_{Da} = \frac{1}{\sqrt{3}}(i_{Ha} - i_{Hc}) + CHX \cdot i_{Xa}$$

$$i_{Db} = \frac{1}{\sqrt{3}}(i_{Hb} - i_{Ha}) + CHX \cdot i_{Xb}$$

$$i_{Dc} = \frac{1}{\sqrt{3}}(i_{Hc} - i_{Hb}) + CHX \cdot i_{Xc}$$

$$i_{Ra} = \frac{1}{2}\left[\frac{1}{\sqrt{3}}(i_{Ha} - i_{Hc}) - CHX \cdot i_{Xa}\right]$$

$$i_{Rb} = \frac{1}{2}\left[\frac{1}{\sqrt{3}}(i_{Hb} - i_{Ha}) - CHX \cdot i_{Xb}\right]$$

$$i_{Rc} = \frac{1}{2}\left[\frac{1}{\sqrt{3}}(i_{Hc} - i_{Hb}) - CHX \cdot i_{Xc}\right]$$

4.3.13.6 Parameters

Type of transformer: The type of connections for the transformer. Four options are available: wye-delta, delta-wye, wye-wye, and delta-delta.

Vector group: Parameter that affects the phase shifts between input and output signals. The vector groups available are 0, 1, 3, 5, 6, 7, 9, 11.

Primary voltage: Voltage level of the primary windings of the transformer.

Secondary voltage: Voltage level of the secondary windings of the transformer.

CT ratio of H: Ratio of the Current Transformers connected to the primary windings of the transformer.

CT ratio of X: Ratio of the Current Transformers connected to the secondary windings of the transformer.

4.3.13.7 Recommended Solver

Ode15s (stiff-NDF)

4.3.13.8 Example

In the following example, a fault occurs in phase A of a delta-wye transformer at $t = 0.03$ s (Fig. 4.54). The differential and restraining currents obtained by the VG-2 block are shown in Figs. 4.55 and 4.56, respectively.

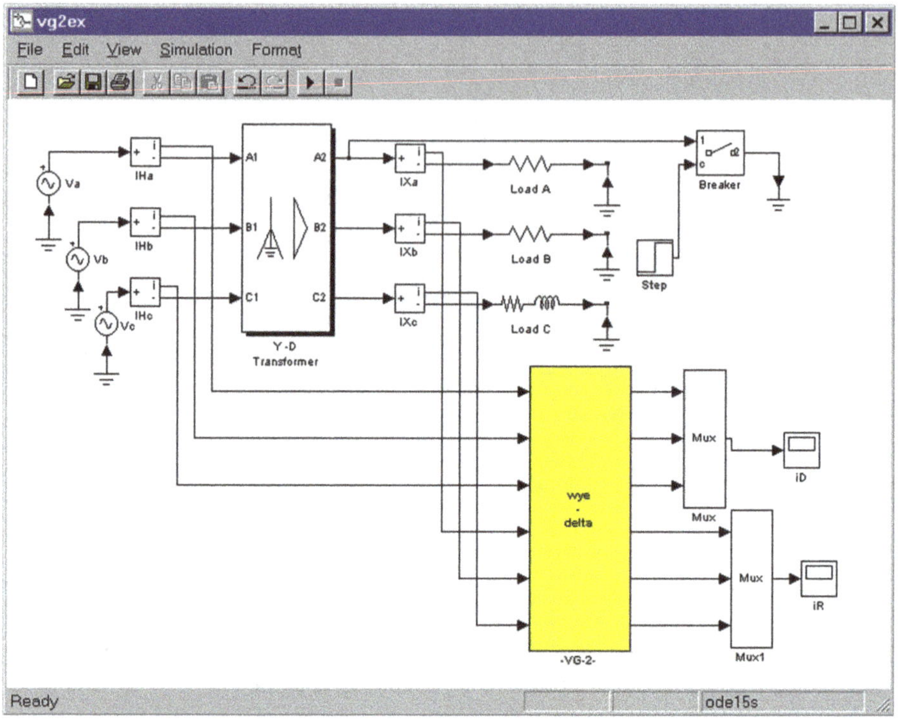

Fig. 4.54 Example for the VG-2 block

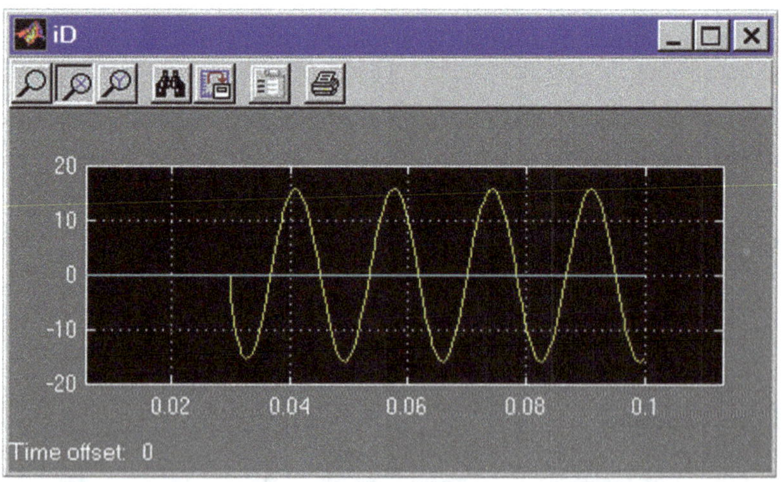

Fig. 4.55 Differential currents obtained by VG-2 block

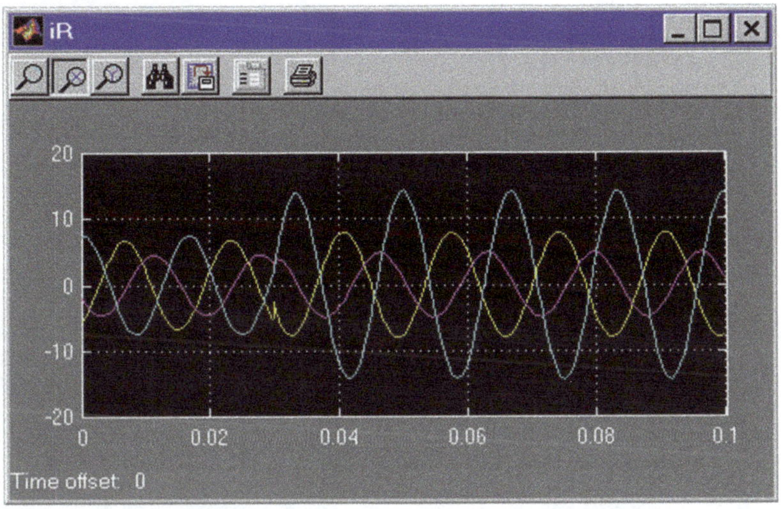

Fig. 4.56 Restraining currents obtained by VG-2 block

4.3.14 Zone Comparator

4.3.14.1 Purpose

Performs the zone comparison function.

4.3.14.2 Inputs

The inputs for this block are the resistance and reactance signals.

4.3.14.3 Outputs

This block has five outputs. Each output corresponds to a zone defined in the dialog box. An output is set to high if the impedance signal is within its limits.

4.3.14.4 Dialog Box

See Fig. 4.57.

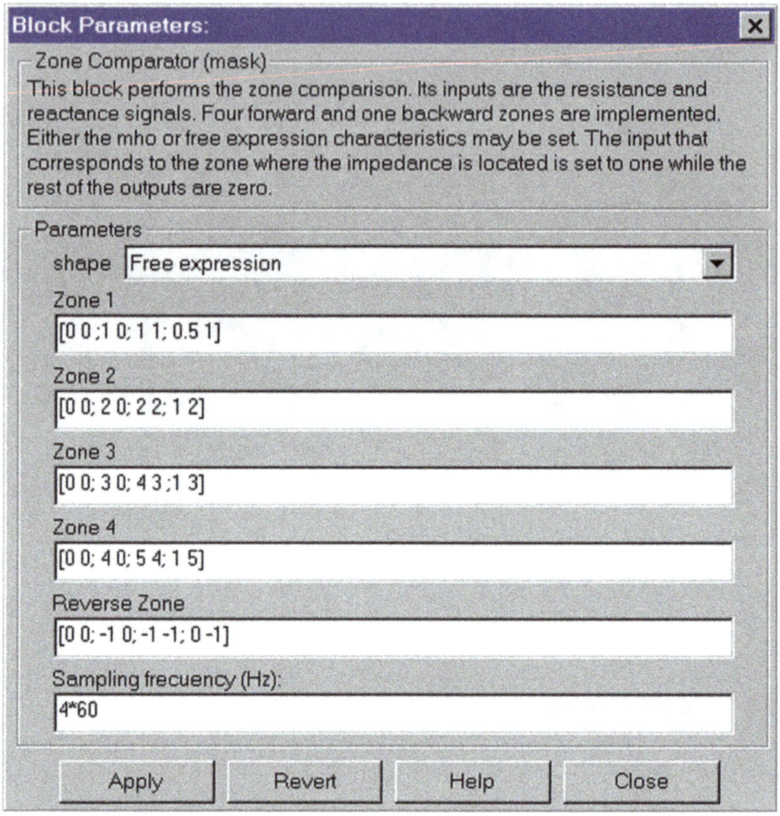

Fig. 4.57 Dialog box of Zone Comparator

4.3.14.5 Description

The inputs of this block are resistance and reactance signals. Four forward zones and one backward zone are implemented. Either the mho or free expression characteristics may be set. The block has five outputs (one for each zone). At each simulation step, the output that corresponds to the zone where the impedance is located is set to one, while the rest of the outputs are zero.

For the mho characteristic, the location of the center of each zone must be provided. Note that in this case, the circles for each zone are tangent to the origin and the location of the center of each circle is enough to define each zone.

For the free expression characteristic, for each zone a two-column matrix with the points of the polygons of each zone must be provided. Each row represents one point. The first column is the x-axis coordinate and the second the y-axis coordinate. The points of the polygon in the array must be adjacent.

The sampling frequency must be provided.

4.3.14.6 Parameters

Shape: There are two options: Mho and Free expression. In the former, the zones are circles tangent to the origin, while in the latter, the zones can be any polygon.

Zone 1: If the Mho option is selected, it is a two-element array that contains the coordinates for the center of the circle for zone 1. If the Free expression option is selected, it is an array two-column matrix, which contains the points of the polygon for zone 1. Note that the points must be given for adjacent vertices.

Zone 2: If the Mho option is selected, it is a two-element array that contains the coordinates for the center of the circle for zone 2. If the Free expression option is selected, it is an array two-column matrix, which contains the points of the polygon for zone 2. Note that the points must be given for adjacent vertices.

Zone 3: If the Mho option is selected, it is a two-element array that contains the coordinates for the center of the circle for zone 3. If the Free expression option is selected, it is an array two-column matrix, which contains the points of the polygon for zone 3. Note that the points must be given for adjacent vertices.

Zone 4: If the Mho option is selected, it is a two-element array that contains the coordinates for the center of the circle for zone 4. If the free expression option is selected, it is an array two-column matrix, which contains the points of the polygon for zone 4. Note that the points must be given for adjacent vertices.

Reverse zone: If the Mho option is selected, it is a two-element array that contains the coordinates for the center of the circle for the reverse zone. If the free expression option is selected, it is an array two-column matrix, which contains the points of the polygon for the reverse zone. Note that the points must be given for adjacent vertices.

Sampling frequency: The rate of sampling of the signal in Hertz.

4.3.14.7 Recommended Solver

Ode45 (Dormand-Prince).

4.3.14.8 Example

In the following example, the way the ZC operates is illustrated. The zones are defined as follows:

Zone 1: [0 0; 1 0; 1 1;0 1]
Zone 2: [0 0; 2 0; 2 2;0 2]
Zone 3: [0 0; 3 0; 3 3;0 3]
Zone 4: [0 0; 4 0; 4 4; 0 4]
Reverse zone: [0 0; −1 0; −1 −1; 0 −1] (Fig. 4.58)

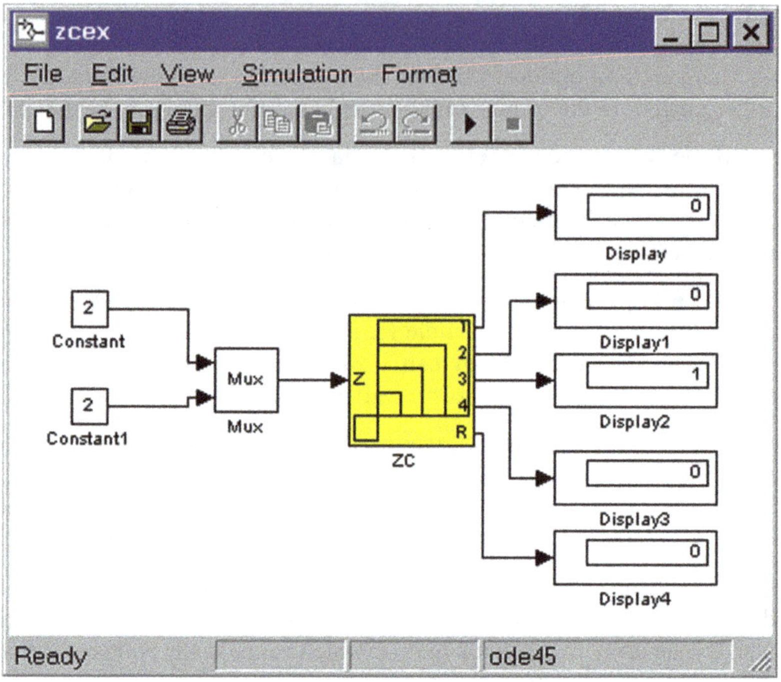

Fig. 4.58 Example for the Zone Comparator block

4.4 Interfacing Power System and Relay Models

4.4.1 Analytical Generator

4.4.1.1 Purpose

Generates a signal by providing the equation describing the output value.

4.4.1.2 Output

The first output is the instantaneous value of the signal generated. The second output is the time signal, which goes from low to high at the start-up time.

4.4.1.3 Dialog Box

See Fig. 4.59.

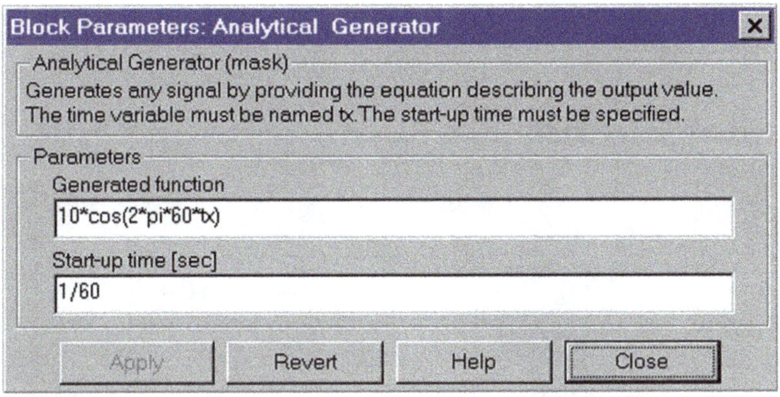

Fig. 4.59 Dialog box of Analytical Generator

4.4.1.4 Description

The block generates a signal using its analytical description. Also, the start-up time of the generator is specified. The actual time variable used to generate the output waveform is shifted in time by the start-up time. The output remains zero before the start-up time. The start-up time must be given in seconds.

4.4.1.5 Parameters

Generated function: Analytical expression for the function to be generated. The default expression is shown above.

Start-up time: Time when the signal is activated. The default is 1/60 s.

4.4.1.6 Recommended Solver

Method: ode45 (Dormand-Prince).

Time step: variable step.

4.4.1.7 Example

The following analytical function will be generated:

$$10 \times \cos(2 \times \mathrm{pi} \times 60 \times \mathrm{tx}) \times \exp(-\mathrm{tx}/0.03) + 2 \times \sin(4 \times \mathrm{pi} \times 60 \times \mathrm{tx}) + 3 \times \sin(6 \times \mathrm{pi} \times 60 \times \mathrm{tx})$$

The start-up time is 1/60 s (Figs. 4.60 and 4.61).

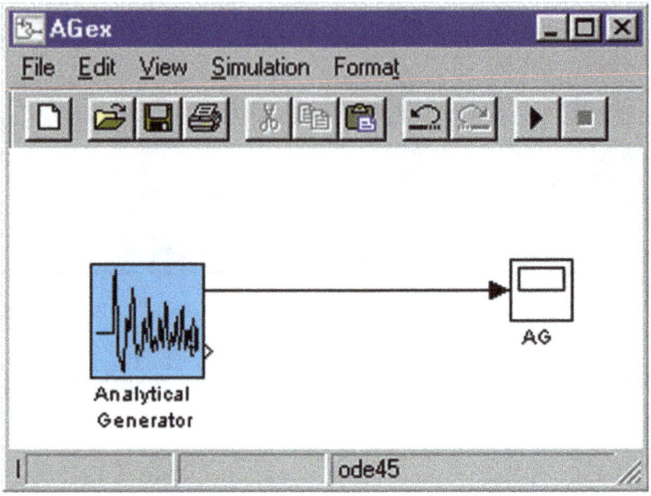

Fig. 4.60 Example for the Analytical Generator

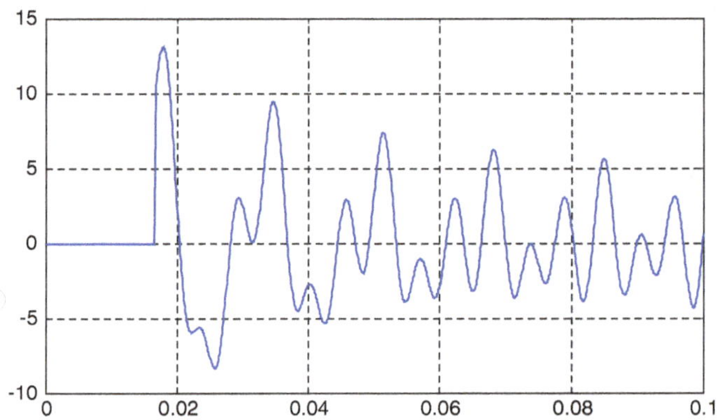

Fig. 4.61 Simulation results for Analytical Generator

4.4.2 Fault Signal Generator

4.4.2.1 Purpose

Generates a short circuit signal using assumed analytical model.

4.4.2.2 Outputs

There are two outputs for this block. The first one is the instantaneous value of the fault signal itself as specified in the parameters. The second one is the time signal, which goes from low to high at the start-up time. Both outputs are zero before the start-up time.

4.4.2.3 Dialog Box

See Fig. 4.62.

Description

This block generates a fault signal using parameters of its analytical approximation, i.e., amplitudes, phases, and frequency of the pre-fault and fault signal, fault time, DC time constant, as well as amplitudes, frequencies, and time constants of the oscillatory components.

The output signal is generated according to the following equations:

$$A_{dc} = A_{pre} \cos\left(2\pi f_0 t_0 + \varphi_{pre}\right) - A_{post} \cos\left(2\pi f_0 t_0 + \varphi_{post}\right)$$

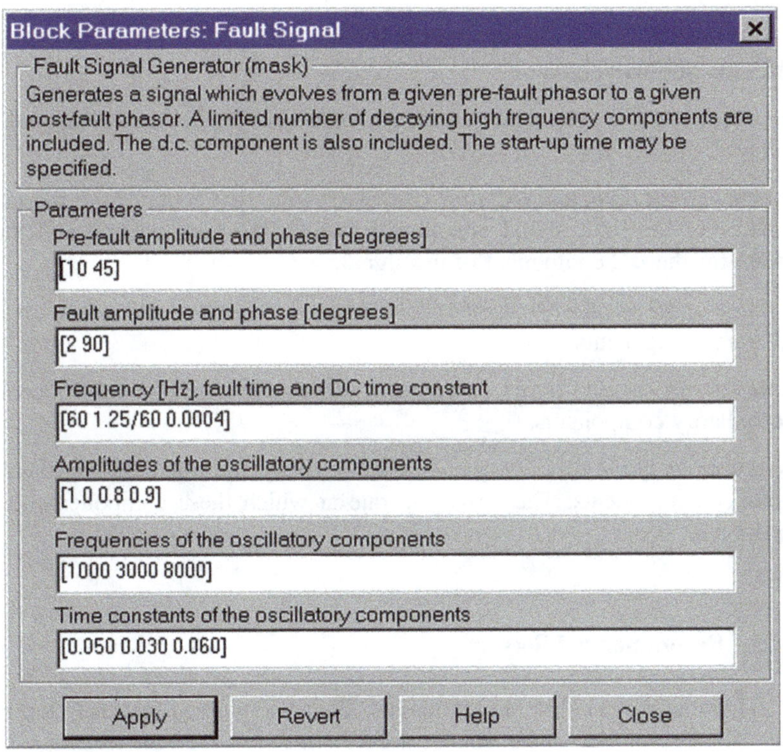

Fig. 4.62 Dialog box of Fault Signal Generator

$$x(t) = \left\{ \begin{array}{cc} A_{\text{pre}} \cos\left(2\pi f_0 t + \varphi_{\text{pre}}\right) & t < t_0 \\ A_{\text{post}} \cos\left(2\pi f_0 t + \varphi_{\text{post}}\right) + A_{\text{dc}} e^{-(t-t_0)/T_{\text{dc}}} + \sum_{k=1}^{p} A_k e^{(-t-t_0)/T_k} \sin\left(2\pi f_k (t + t_0)\right) & t > t_0 \end{array} \right\}$$

where:

A_{pre} and φ_{pre} are the pre-fault amplitude and phase, respectively.
A_{post} and φ_{post} are the fault amplitude and phase, respectively.
$x(t)$ is the output signal.
f_o is the frequency of the pre-fault and fault signal $x(t)$.
t_o is the fault incidence time.
A_k is the amplitude of the kth oscillatory component.
f_k is the frequency of the kth oscillatory component.
T_k is the time constant of the kth oscillatory component.
A_{dc} is the amplitude of the dc component.
T_{dc} is the time constant of the dc component.

4.4.2.4 Parameters

Pre-fault amplitude and phase: Amplitude and phase (in degrees) of the phasor that defines the pre-fault signal.

Fault amplitude and phase: Amplitude and phase of the phasor of the signal during the fault.

Frequency, fault time, and DC time constant: Array that contains the fundamental frequency of the fault signal, the time at which the fault occurs, and the time constant for the DC component of the signal.

Amplitude of the oscillatory components: Array that contains the amplitudes of the oscillatory components.

Frequencies of the oscillatory components: Array that contains the frequencies of the oscillatory components.

Time constants of the oscillatory components: Array with the time constants of the oscillatory components. They give the rate at which these components decrease after the fault occurs.

4.4.2.5 Recommended Solver

Ode45(Dormand-Prince) is recommended. There is no need for time step for this solver. The discrete solver can also be used. In this case, the time step must be such that all the harmonics are generated properly.

4.4.3 Example

Generate a fault signal with the following characteristics:

Pre-fault amplitude: 10
Pre-fault phase(degrees): 45
Fault amplitude: 4
Fault phase(degrees): 90
Fault time: 1.25/60 s
DC time constant: 0.0004
Oscillatory components: Table 4.3, Figs. 4.63 and 4.64.

Table 4.3 Oscillatory components

Frequency	Amplitude	Time constant
800	1.0	0.05
1000	0.8	0.03
3000	0.9	0.06

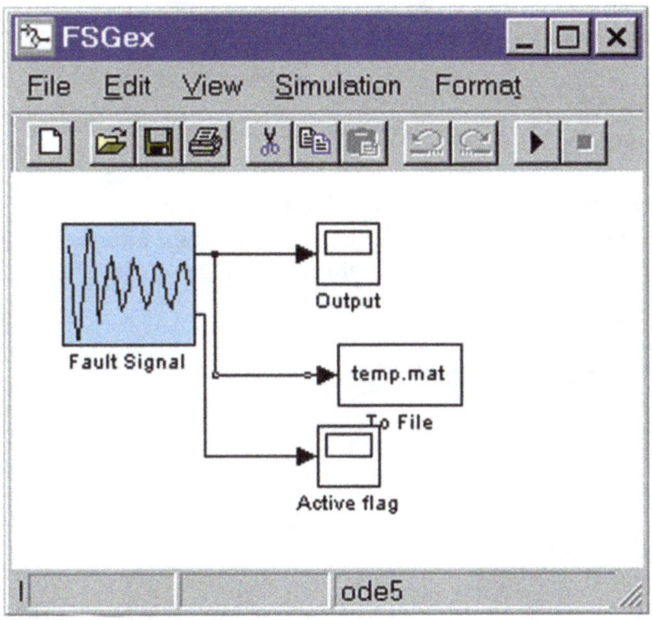

Fig. 4.63 Model for the FSG example

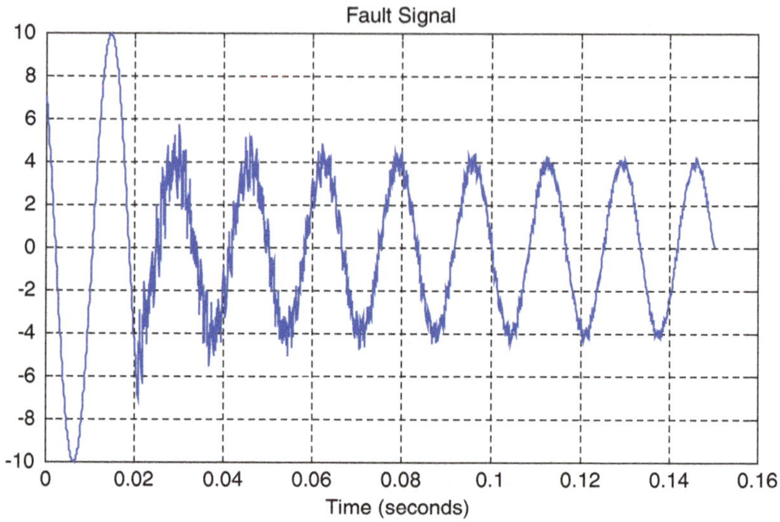

Fig. 4.64 Simulation results for FSG

4.4.4 Phasor Generator

4.4.4.1 Purpose

Generates a sinusoid waveform according to the parameters specified.

4.4.4.2 Outputs

This block has two outputs. The first one is the instantaneous value of the sinusoidal signal specified by the parameters. The second one is the time signal, which is set to high at the start-up time. Before that time, this output is zero.

4.4.4.3 Dialog Box

See Fig. 4.65.

4.4.4.4 Description

The block generates a waveform given by its phasor description. The output signal is defined by the input parameters: amplitude, phase (in degrees), frequency, and start-up time.

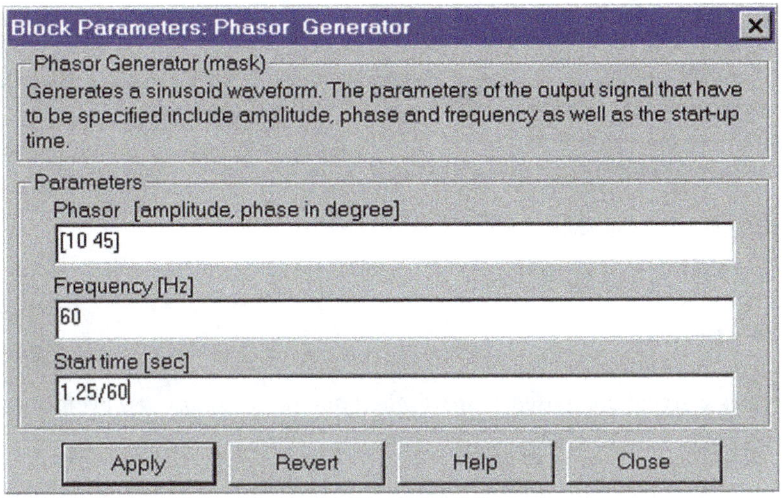

Fig. 4.65 Dialog box of Phasor Generator

The output of the block is generated according to the following equation:

$$x(t) = A \cdot \cos(2\pi f t + \phi)$$

where:

A is the amplitude of the signal
f is the frequency of the signal
ϕ is the phase of the signal

4.4.4.5 Parameters

Phasor: Amplitude and phase (in degrees) of the phasor.

Frequency [*Hz*]: Frequency of the sinusoid waveform.

Start-up time [*s*]: The signal is generated after this time. The output is zero before this time.

4.4.4.6 Recommended Solver

The ode45 (Dormand-Prince). In this case, there is no time step needed.

4.4.4.7 Example

The Phase Generator is used to generate a sinusoid signal with a magnitude of 10 and phase of 45°. The start-up time is 1.25/60 s (Fig. 4.66).

The results of the simulation are shown in Fig. 4.67.

4.4.5 Spectrum Generator

4.4.5.1 Purpose

Generates a signal composed of the limited number of sinusoid waveforms.

4.4.5.2 Outputs

The first output (x) is the instantaneous value of a signal composed of the sinusoid waveforms specified in the dialog box. The second output (t) is a signal that is set high when the Spectrum Generator is activated. Before this time, this output is zero.

4.4.5.3 Dialog Box

See Fig. 4.68.

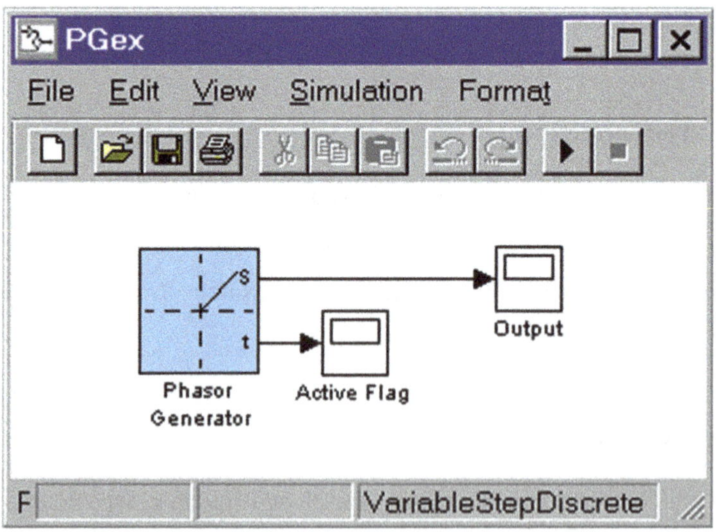

Fig. 4.66 Example for the Phase Generator

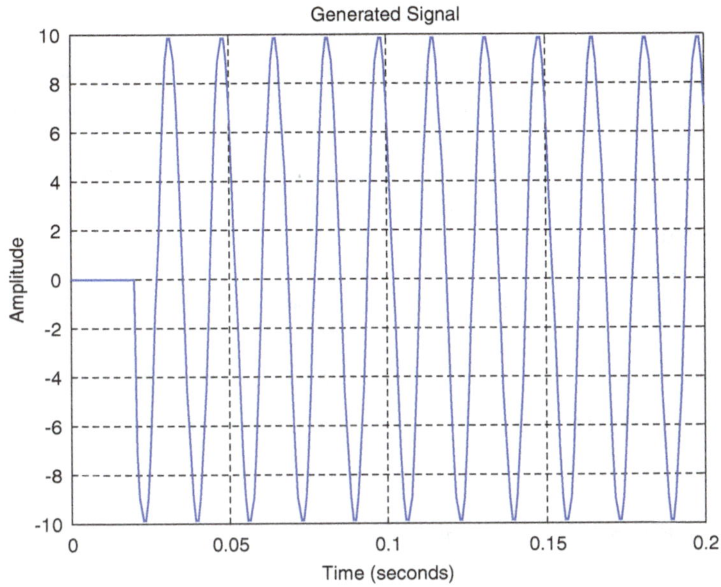

Fig. 4.67 Simulation results for Phasor Generator

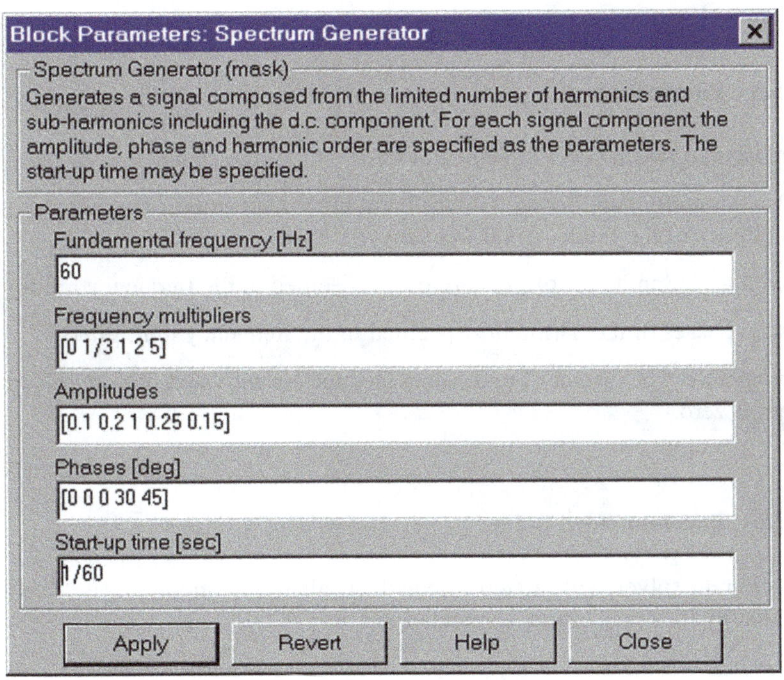

Fig. 4.68 Dialog box of Spectrum Generator

4.4.5.4 Description

The block generates its output signal as a sum of cosine functions of given amplitudes, frequencies, and phase angles. Also, the start-up time of the generator is specified.

The frequency multipliers are represented as an array which contains the frequencies of the signal components as multiples of the fundamental frequency. Note, that the arrays for the magnitudes and phases must match the frequency multiplier array.

The output waveform is generated using the following equation:

$$y(t) = \sum_k A_k \cos(2\pi mft + \phi_k) \quad \text{for } t > t_o$$
$$y(t) = 0 \quad \text{for } t < t_o$$

where:

f is the fundamental frequency in Hz
m is the frequency multiplier
A is the amplitude of the kth component of the output waveform
ϕ is the phase of the kth component of the output waveform (radians)
t_o is the start-up time (s)

4.4.5.5 Parameters

Fundamental frequency: Fundamental frequency of the signal.

Frequency multipliers: Array with the harmonics that compose the signal expressed as multiples of the fundamental frequency.

Amplitudes: Amplitudes of the harmonics specified in the frequency multipliers.

Phases: Phases of the harmonics specified in the frequency multipliers.

Start-up time: The instant of time when the block is activated. Before this time, the output is zero.

4.4.5.6 Recommended Solver

The discrete solver with a step selected, so all the harmonics generated can be detected by the other blocks connected.

4.4.5.7 Example

Generate a signal composed of the following harmonics: Table 4.4.

The fundamental frequency is 60 Hz and the start-up time is 1/60 s. It can be seen in Fig. 4.69, that the icon for the SG block shows the amplitudes of the components of the signal in the frequency domain. In Fig. 4.70, the signal generated is shown. Note that the signal is zero before the start-up time (1/60 s).

4.4.6 Three-Phase Phasor Generator

4.4.6.1 Purpose

Generates three sinusoid waveforms according to parameters specified.

Table 4.4 Signal harmonics

Frequency multipliers	Amplitude	Phase (degrees)
0	0.2	0
1/3	0.2	45
1	1	0
3	0.1	90
5	0.15	60
7	0.1	10

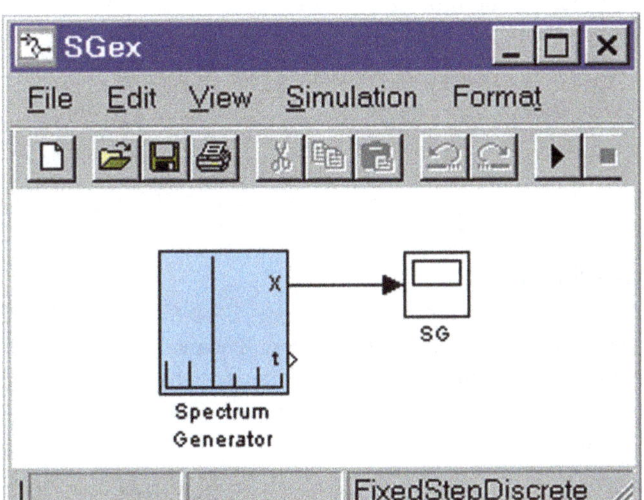

Fig. 4.69 Example for the Spectrum Generator

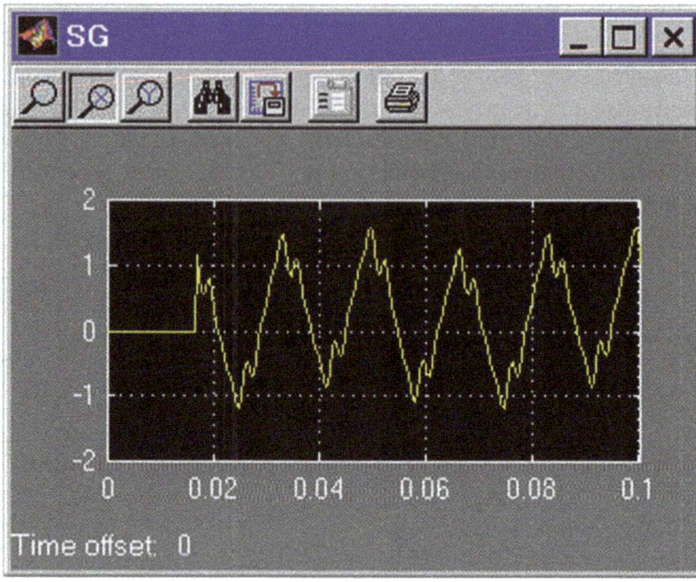

Fig. 4.70 Simulation results for Spectrum Generator

4.4.6.2 Outputs

The first three outputs of this block are the instantaneous values of the symmetrical or unsymmetrical phase quantities (phase A, B, and C) or the symmetrical components (0, +, and – sequences). The fourth output is the time signal, which goes from low to high at the start-up time.

4.4.6.3 Dialog Box

See Fig. 4.71.

4.4.6.4 Description

The block generates three sinusoid waveforms. The parameters of the output signals to be specified include amplitude and phase for each phasor as well as the frequency and the start-up time. The phasors may be set either as phase or symmetrical component values.

Let:

A_k be the amplitude of the kth phasor; $k = A, B$, or C
ϕ_k be the phase of the kth phasor; $k = A, B$, or C

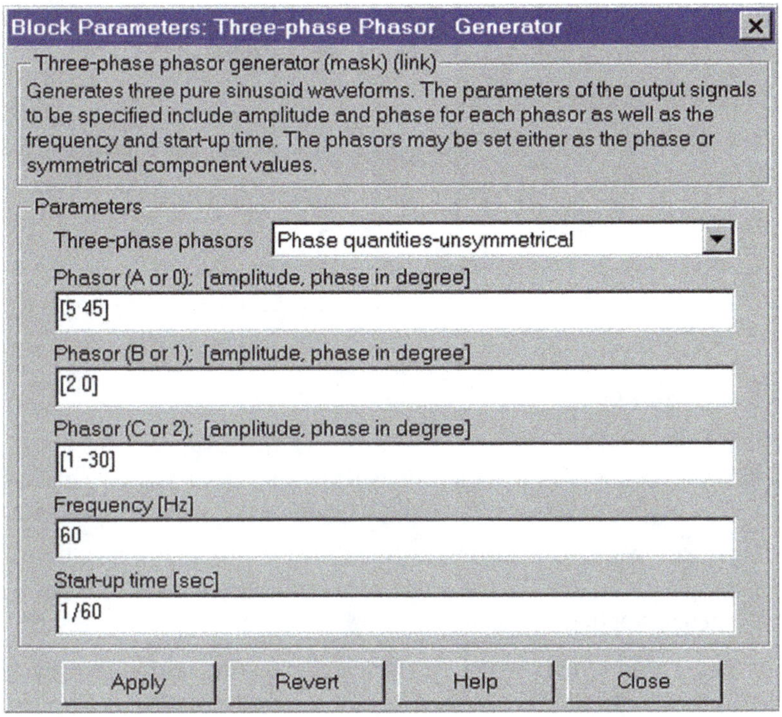

Fig. 4.71 Dialog box of Three-phase Phasor Generator

There are three options for the output signals.

1. Phase quantities—symmetrical

$$\begin{aligned} &\text{amplitude\_}A = A_A, \quad &&\text{phase\_}A = \phi_A \\ &\text{amplitude\_}B = A_A, \quad &&\text{phase\_}B = \phi_A - 120 \\ &\text{amplitude\_}C = A_A, \quad &&\text{phase\_}C = \phi_A + 120 \end{aligned}$$

2. Phase quantities—unsymmetrical

$$\begin{aligned} &\text{amplitude\_}A = A_A, \quad &&\text{phase\_}A = \phi_A \\ &\text{amplitude\_}B = A_B, \quad &&\text{phase\_}B = \phi_B \\ &\text{amplitude\_}C = A_C, \quad &&\text{phase\_}C = \phi_C \end{aligned}$$

3. Symmetrical components
   Let:

$$a = \mathrm{e}^{j2\pi/3}$$

$$\begin{bmatrix} ph\_A \\ ph\_B \\ ph\_C \end{bmatrix} = \begin{bmatrix} 1 & 1 & 1 \\ 1 & a & a^2 \\ 1 & a^2 & a \end{bmatrix} \begin{bmatrix} Ae^{j\phi_A} \\ Be^{j\phi_B} \\ Ce^{j\phi_C} \end{bmatrix}$$

For the three cases above, the waveforms for A, B, and C are generated using the following equations:

$$\begin{aligned} x_A(t) &= \text{amplitude\_}A\cos(2\pi ft + \text{phase\_}A) \\ x_B(t) &= \text{amplitude\_}B\cos(2\pi ft + \text{phase\_}B) \\ x_C(t) &= \text{amplitude\_}C\cos(2\pi ft + \text{phase\_}C) \end{aligned}$$

4.4.7 Parameters

Three-phase phasors: The options for the three sinusoid signals generated are: symmetrical or unsymmetrical three-phase quantities or symmetrical components.

Phasor (a or 0): Amplitude and phase (in degrees) of the phasor that defines phase A if phase quantities are selected or the zero sequence component if the symmetrical components option is selected.

Phasor (b or 1): Amplitude and phase (in degrees) of the phasor that define phase B if the phase quantities-unsymmetrical option is selected or the positive sequence component if the symmetrical components option is selected. If the phase quantities-symmetrical option is selected, this information is not needed.

Phasor (c or 2): Amplitude and phase (in degrees) of the phasor that defines the phase C, if the phase quantities-unsymmetrical option is selected or the negative sequence component, if the symmetrical components option is selected. If the phase quantities-symmetrical option is selected, this information is not needed.

Frequency: Frequency of the signals to be generated.

Start-up time: Instant of time when the three-phase signal generator is activated.

4.4.7.1 Recommended Solver

Ode45 (Dormand-Prince). There is no time step needed for this solver. The discrete solver can also be used. In this case, the time step has to be selected according to the frequency of the signal generated.

4.4.7.2 Example

The three-phase generator is used to generate unsymmetrical three-phase signal. The parameters of the phase signals are the following: Table 4.5.

The start-up time is 1/60 s (Figs. 4.72 and 4.73).

Table 4.5 Parameters of phase signals

Phase	Magnitude	Phase (degrees)
A	5	45
B	2	0
C	1	−30

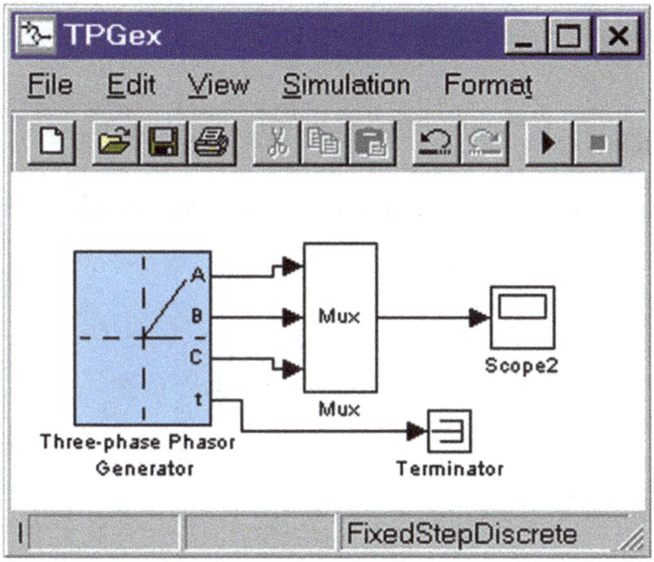

Fig. 4.72 Example for the Three-phase Generator

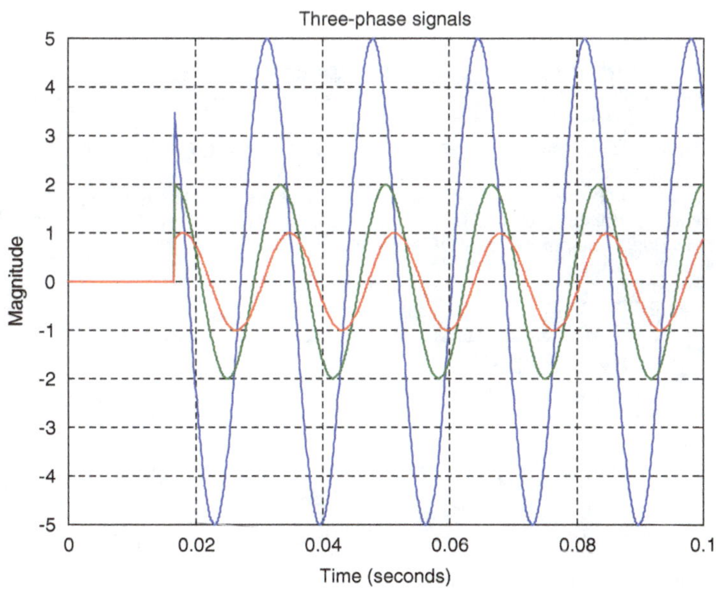

Fig. 4.73 Simulation results for Three-phase Generator

4.5 GUI and Analysis Tools

4.5.1 Phasor Display

4.5.1.1 Purpose

Display phasors dynamically on the complex plane.

4.5.1.2 Inputs

Up to six phasors (12 inputs) given in polar form (phase in degrees).

4.5.1.3 Output

Graph with the input phasors in the complex plane.

4.5.1.4 Dialog Box

See Fig. 4.74.

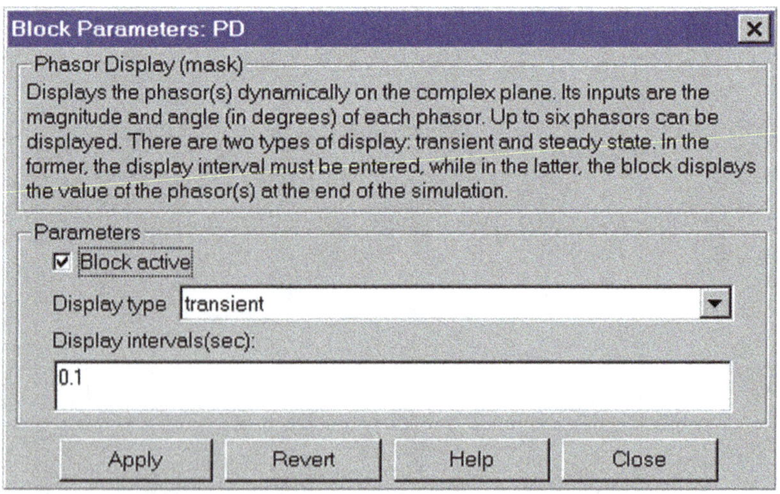

Fig. 4.74 Dialog box of Phasor Display

4.5.1.5 Description

The PD block shown in Fig. 4.75 is able to display up to six phasors in polar form (angles in degrees). The input signals must be multiplexed, so they are all combined in one input signal (Fig. 4.76).

The phasors can be displayed in two ways: steady state and transient. In the former, the phasors are displayed at the end of the simulation, i.e., the steady state of the phasor is shown. In the latter, the phasors are displayed at the specified display intervals. The display intervals must be entered in seconds. This block has an enable/disable option.

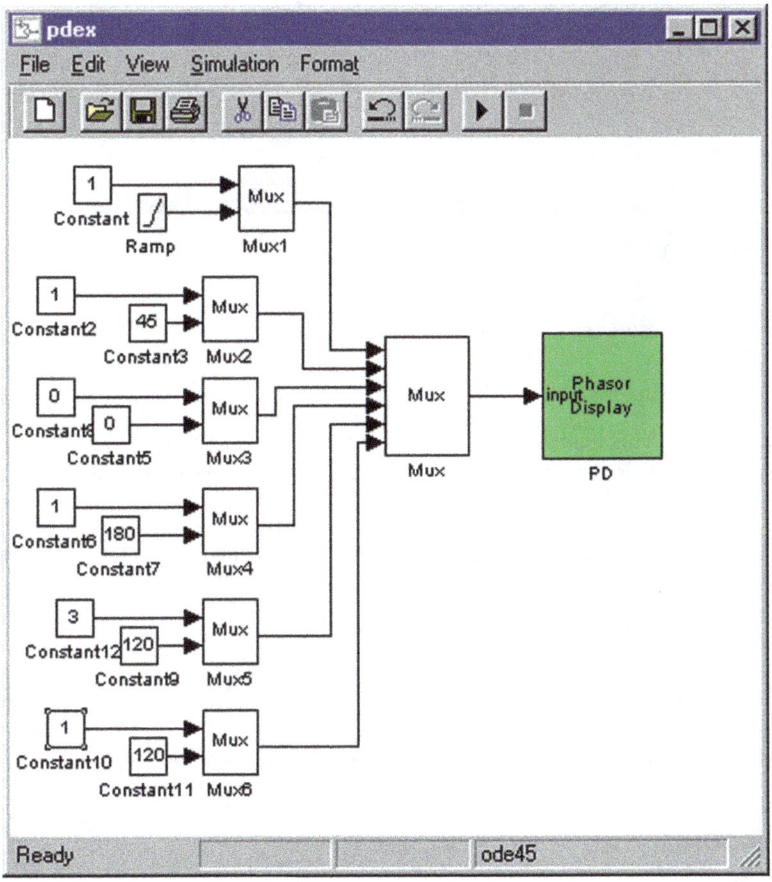

Fig. 4.75 Example for de PD block

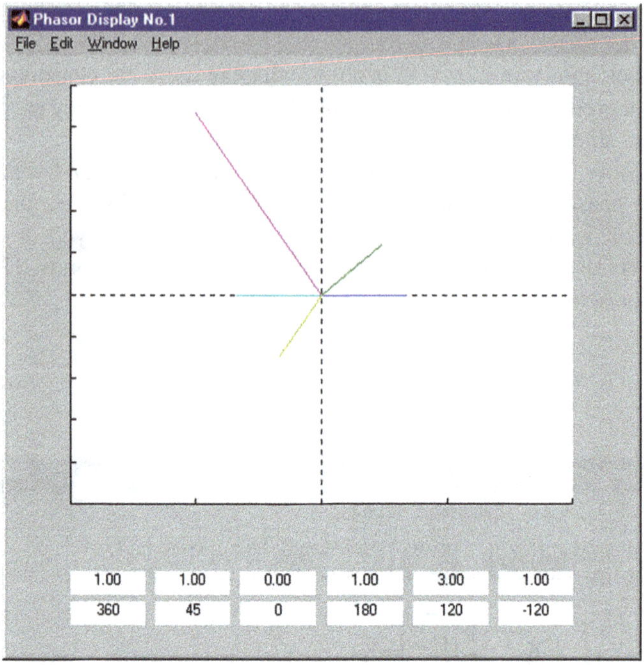

Fig. 4.76 Example for Phasor Display

4.5.1.6 Parameters

Block active: Enables the block if checked

Display type: This parameter has two options: transient or steady state. In the former, the phasors are updated at certain intervals, while in the latter the phasors are displayed at the end of the simulation time.

Display intervals: The intervals in seconds for displaying the phasors if the transient option is selected.

4.5.1.7 Recommended Solver

Ode45 (Dormand-Prince). No time step needs to be specified for this solver.

4.5.1.8 Example

The following example shows how this block works. The phasors at the end of the simulation time are shown in Fig. 4.75.

4.6 Summary

This chapter introduces the elements and modules used to build relay models, interfacing to signals that resemble power system measurements and display simulation results.

Chapter 5
Design and Implementation of Relay Communication Schemes and Trip Logic

5.1 Introduction

This chapter presents different communication schemes and trip logic used in distance relaying. The exercise examples illustrate how different communication schemes are used in a protective relay work. After completion of this exercise, the reader will have an understanding of how these schemes work.

5.2 Communication Schemes

5.2.1 Introduction

The models were developed in Simulink and simulate the following trip logic:

- PUTT (Permissive Underreaching Transfer Trip)
- PUTT + OZ (Permissive Underreaching Transfer Trip with additional overreaching measurement)
- BLOV + TB (blocking overreaching with current reversal logic)
- BLOV + UZ + TB (blocking overreaching with independent measuring zone and current reversal logic)
- BLUN (block underreaching)
- POTT + WEI + TB (permissive overreaching transfer tripping with weak infeed logic and current reversal logic)

5.2.2 Working with Software

Each communication scheme has the following components:

© Springer International Publishing Switzerland 2016

M. Kezunovic et al., *Design, Modeling and Evaluation of Protective Relays for Power Systems*, DOI 10.1007/978-3-319-20919-7_5

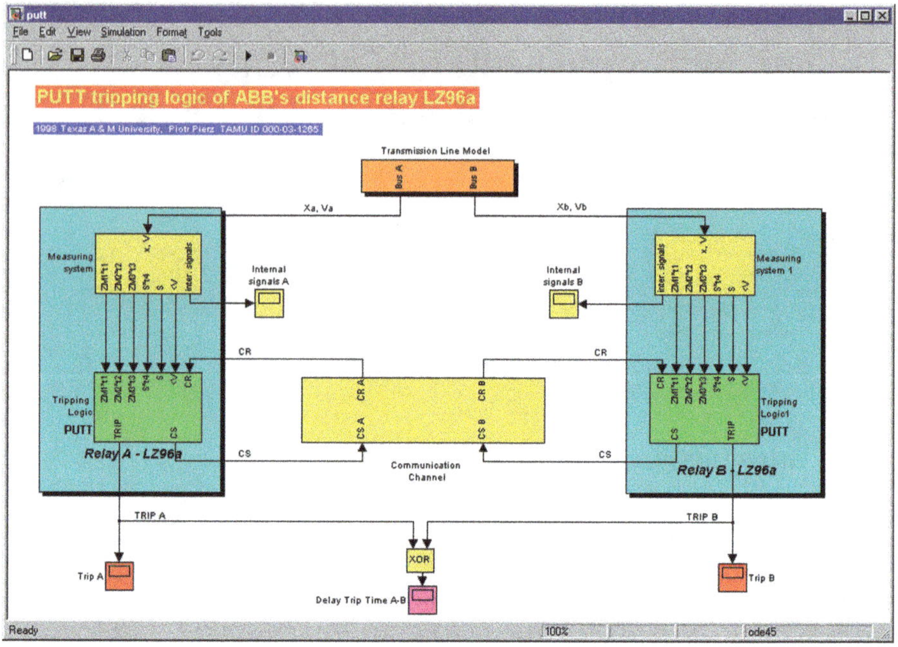

Fig. 5.1 Block diagram for transmission line protected by two relays (PUTT tripping logic)

- Transmission line model
- Two relays (each relay has a measuring system and trip logic)
- Communication channel
- Displays

The transmission line model, communication channel, and displays are practically the same for every communication scheme model. The only component that changes is the selected trip logic of the distance relay. There are other slight changes and they will be pointed out. The main structure of the models is shown in Fig. 5.1. The components that vary in the specific components of the model are described in the following sections.

5.2.2.1 Transmission Line Model

This model enables the user to specify a per unit fault location from the substation A, the time of fault occurrence (in seconds), and the per unit voltage after the fault where Va in the per unit voltage after the fault in substation A and Vb is the per unit voltage after the fault in substation B. The last parameter is implemented only for the PUTT and PUTT + OZ schemes. Knowing the fault location from substation A (which is equal to X_a varying from 0 to 1), the block calculates the fault location for the substation B ($X_b = 1 - X_a$). The fault locations and post-fault voltages with respect to substations A and B are the block. Prior to

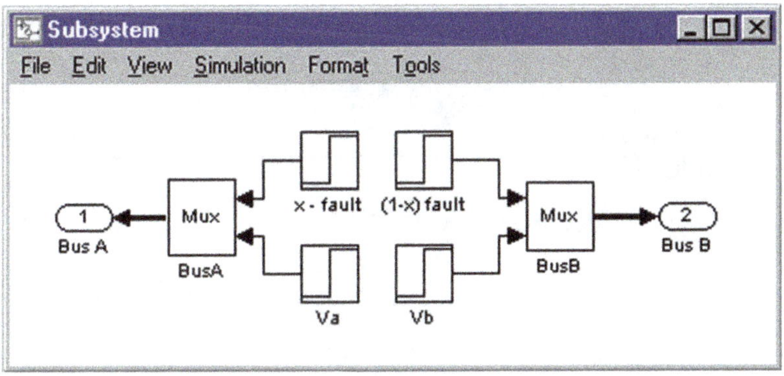

Fig. 5.2 Block diagram for model of a transmission line

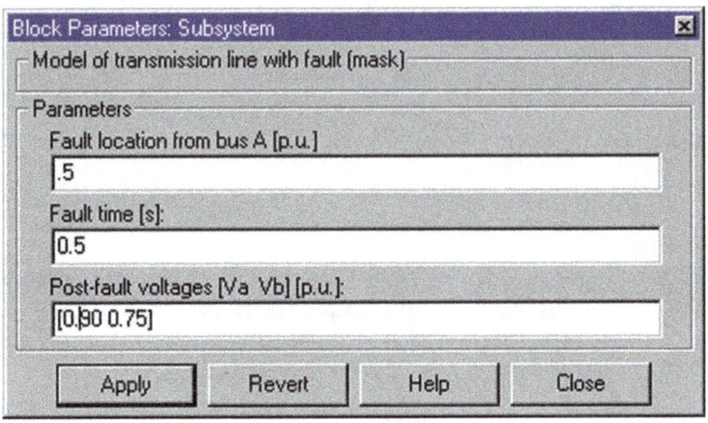

Fig. 5.3 Mask of transmission line model (user interface)

the fault, the fault location signals have a value of infinity (actually, the value is 1000 p.u. that is practically infinite for our purposes). The voltage signals are assumed to be 1.05 p.u. prior to the fault (Figs. 5.2 and 5.3).

The user must input the values of fault location, post-fault voltages at bus A and bus B, and time of fault occurrence. Note that the post-fault voltages are only needed for the PUTT and PUTT + OZ schemes.

5.2.2.2 Measuring System

This block senses the fault location signal and calculates the basic signals used by the relay. They include ZM1, ZM2, ZM3, S, <U, which are zone reach, times, and undervoltage threshold respectively.

There are slight differences between the measuring systems for the different tripping logic models. These measuring systems are shown below.

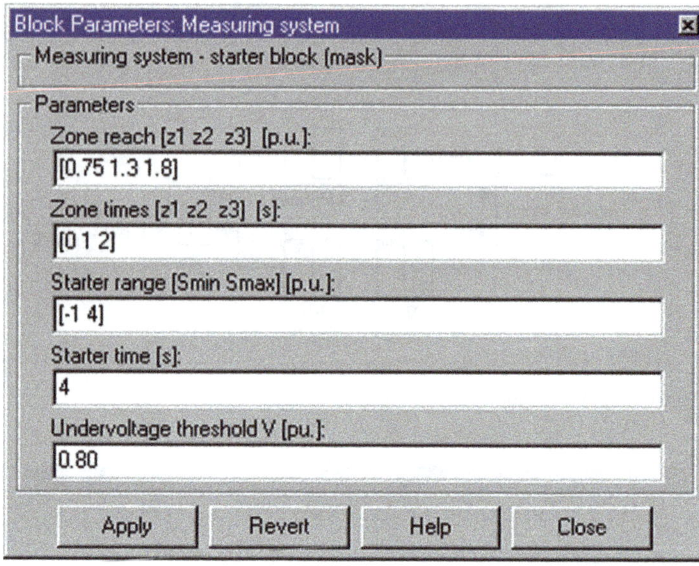

Fig. 5.4 User menu for the PUTT logic measuring system

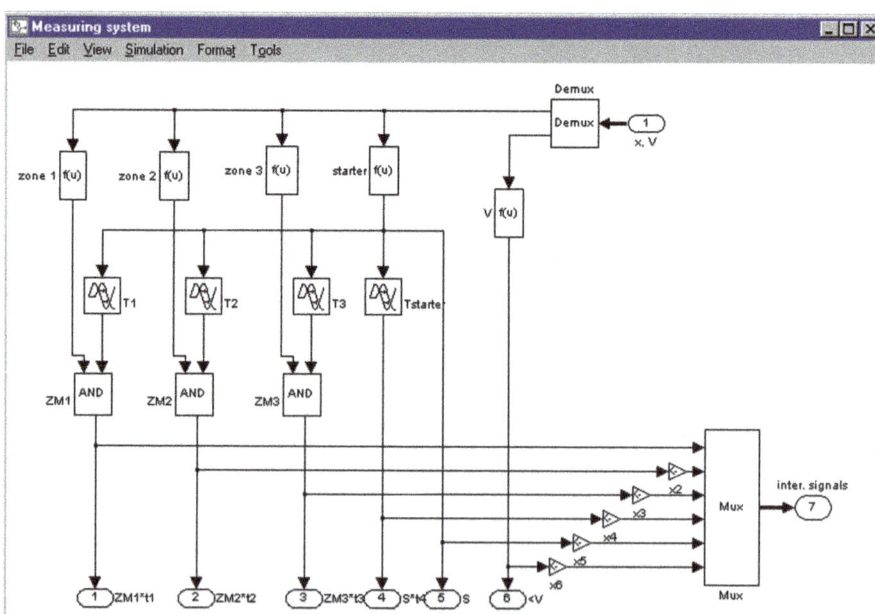

Fig. 5.5 Block diagram of the PUTT logic measuring system

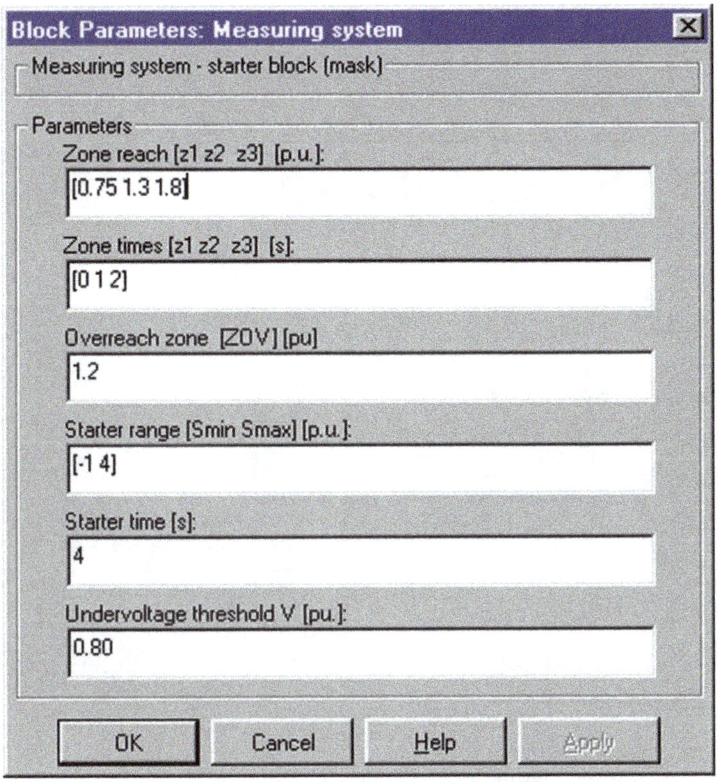

Fig. 5.6 User menu for the PUTT+OZ logic measuring system

Measuring System for PUTT Logic (Figs. 5.4 and 5.5)

Values provided by user:

- Per unit zone impedances (zone 1, zone 2, zone 3)
- Delay times (in seconds) of relay trip for each zone (for zone 1 time delay is usually zero—operates immediately)
- Boundaries of the starter (per unit values)
- Delay time of starter
- Threshold for undervoltage element

Measuring System for PUTT + OZ Logic (Figs. 5.6 and 5.7)

Values provided by user:

- Per unit zone impedances (zone 1, zone 2, zone 3)
- Delay times (in seconds) of relay trip for each zone (for zone 1 time delay is usually zero—operates immediately)
- Per unit Overreach Zone (ZOV)

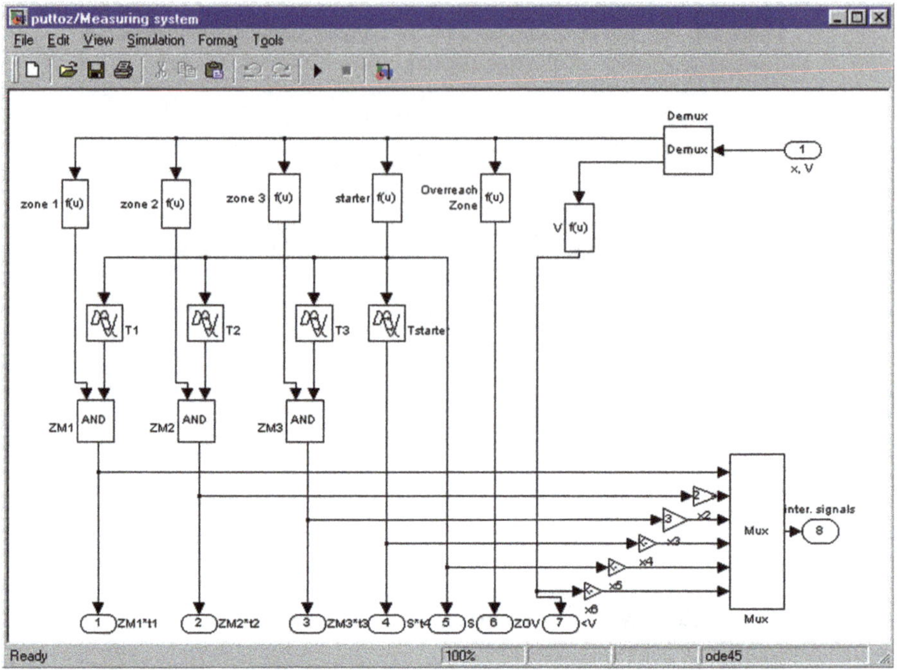

Fig. 5.7 Block diagram of the PUTT+OZ logic measuring system

- Boundaries of the starter (per unit values)
- Delay time of starter
- Threshold for undervoltage element.

Measuring System for BLOV + TB Logic (Figs. 5.8 and 5.9)

Values provided by user:

- Per unit zone impedances (zone 1, zone 2, zone 3)
- Delay times (in seconds) of relay tripping for each zone (for zone 1 time delay is usually zero—operates immediately)
- Boundaries of the starter (per unit values)
- Delay time of starter
- Per unit Overreach Zone (ZOV)
- ZMB in per unit
- Auxiliary delay time (t_{HF})

Measuring System for BLOV + UZ + TB Logic (Figs. 5.10 and 5.11)

Values provided by user:

- Boundaries of the starter (per unit values)

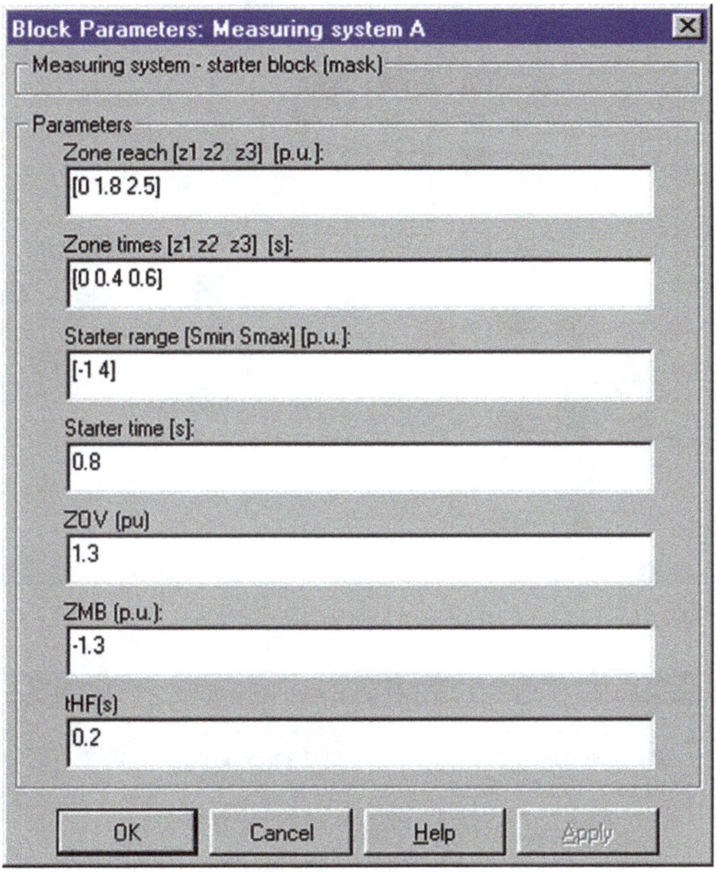

Fig. 5.8 User menu for the BLOV+TB logic measuring system

- Per unit zone impedances (zone 1, zone 2, zone 3)
- Delay times (in seconds) of relay trip for each zone (for zone 1 time delay is usually zero—operates immediately)
- Delay time of starter
- Per unit Overreach Zone (ZOV)
- Auxiliary delay time (t_{HF})
- ZMB in per unit

Measuring System for BLUN Logic (Figs. 5.12 and 5.13)

Values provided by user:

- Per unit zone impedances (zone 1, zone 2, zone 3)
- Delay times (in seconds) of relay tripping for each zone (for zone 1 time delay is usually zero—operates immediately)

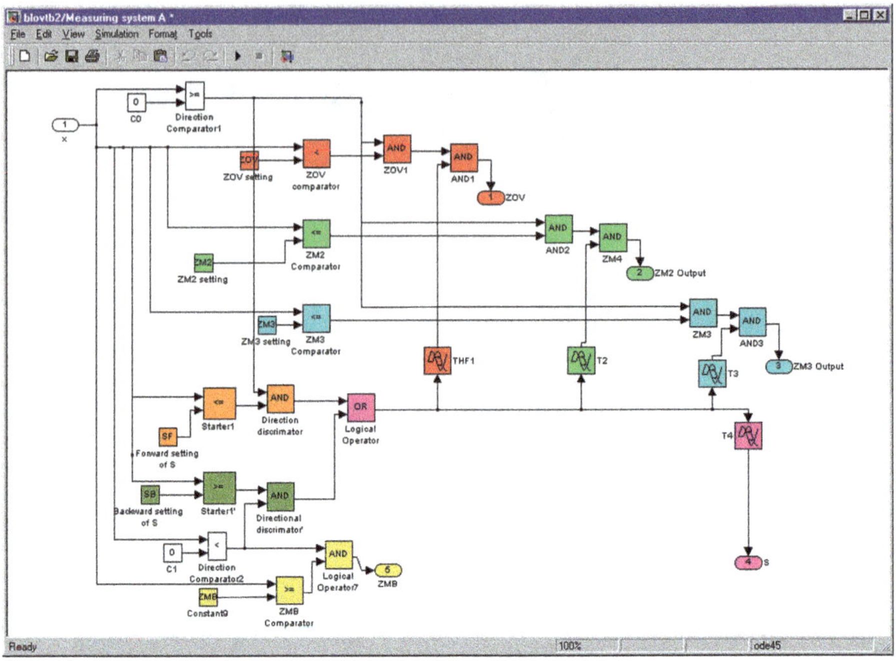

Fig. 5.9 Block diagram of the BLOV+TB logic measuring system

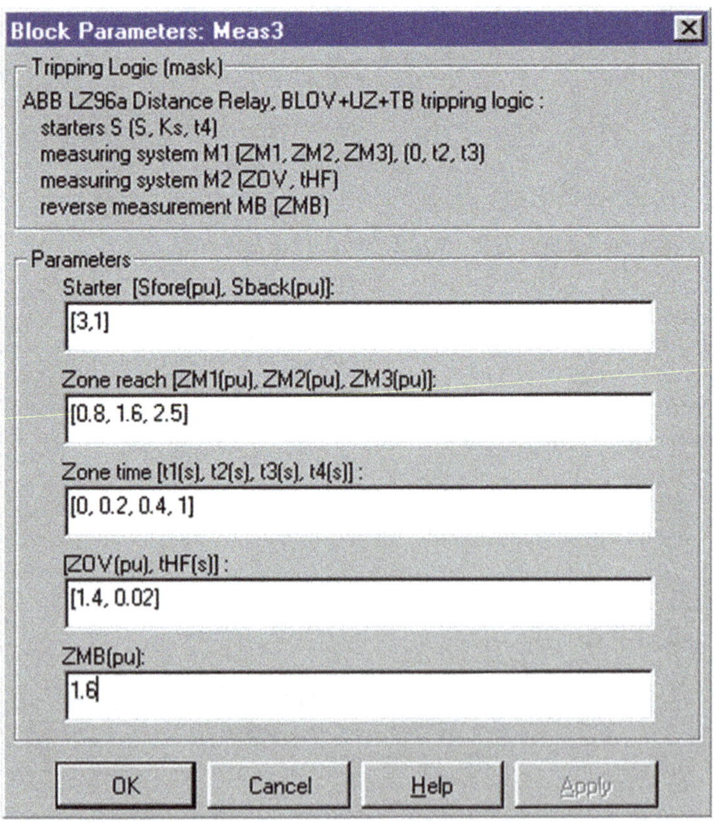

Fig. 5.10 User menu for the BLOV+UZ+TB logic measuring system

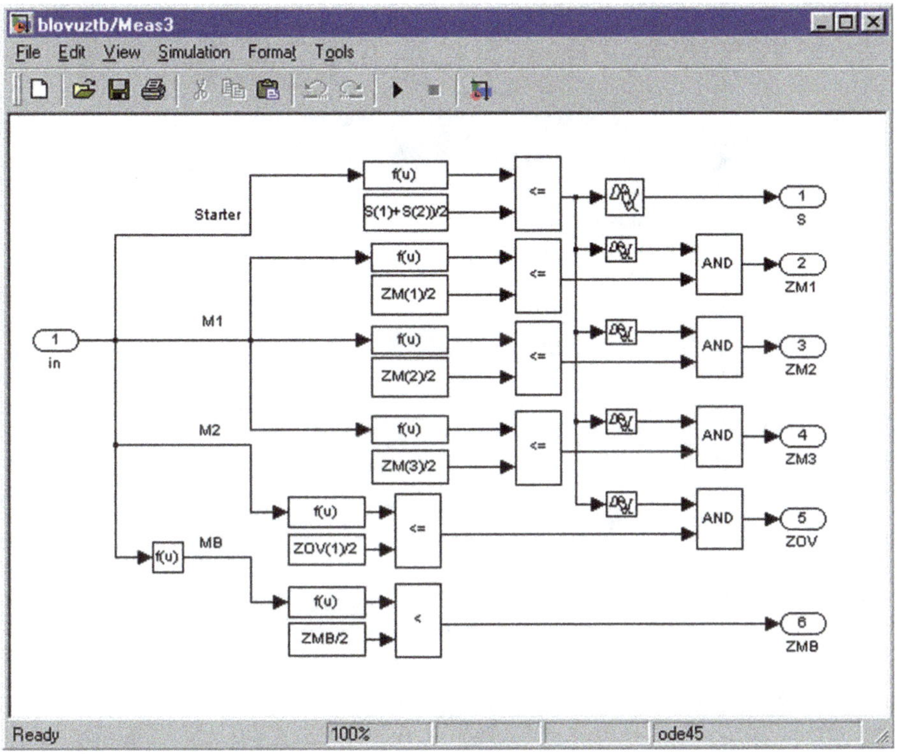

Fig. 5.11 Block diagram of the BLOV+UZ+TB logic measuring system

- Overreach zone, Boundaries of the starter (per unit values)
- Delay time of starter
- Per unit Overreach Zone (ZOV)
- ZMB in per unit
- Auxiliary delay time (t_{HF})

Measuring System for POTT + WEI + TB Logic (Figs. 5.14 and 5.15)

Values provided by user:

- Per unit zone impedances (zone 1, zone 2, zone 3)
- Delay times (in seconds) of relay tripping for each zone (for zone 1 time delay is usually zero—operates immediately)
- Boundaries of the starter (per unit values)
- Delay time of starter
- ZMB in per unit
- Undervoltage relay range in per unit

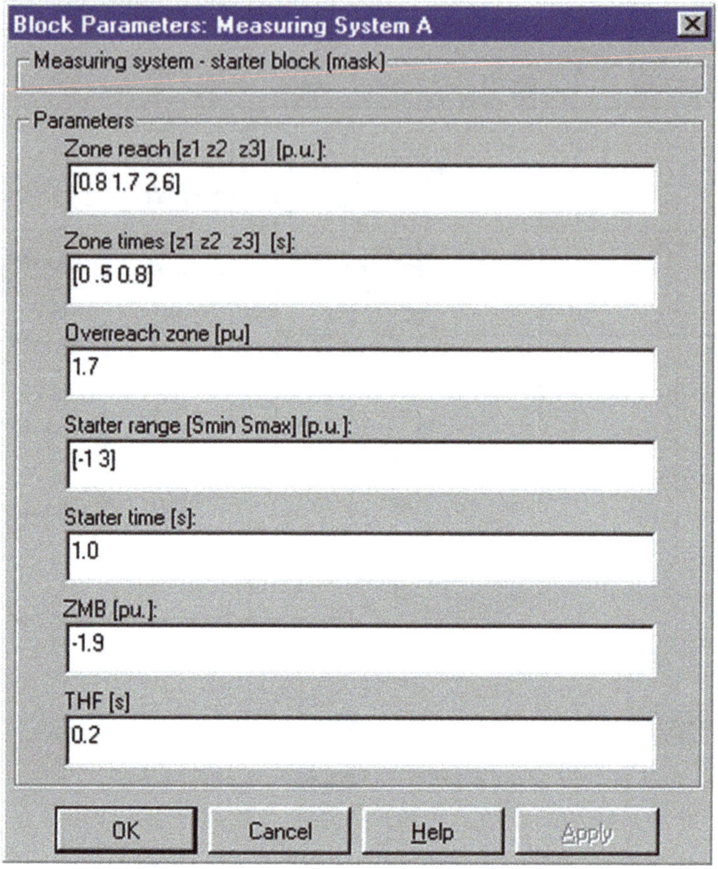

Fig. 5.12 User menu for the BLUN logic measuring system

5.2.2.3 Trip Logic

PUTT Logic

The PUTT logic first determines its output signal CS (Carrier Send). Second, it determines the TRIP signal based on the logic flags from the *Measuring System* and the CR (Carrier Receive) signal.

According to PUTT mode the trip signal is equal to:

$$\mathrm{TRIP} = \mathrm{ZM}_1 + (\mathrm{S} + \mathrm{U} <) \cdot \mathrm{CR} \; + \; \mathrm{ZM}_2 \cdot t_2 + \mathrm{ZM}_3 \cdot t_3 \; + \; \mathrm{S} \cdot t_4$$

The Simulink subsystem that simulates this logic is shown in Fig. 5.16.

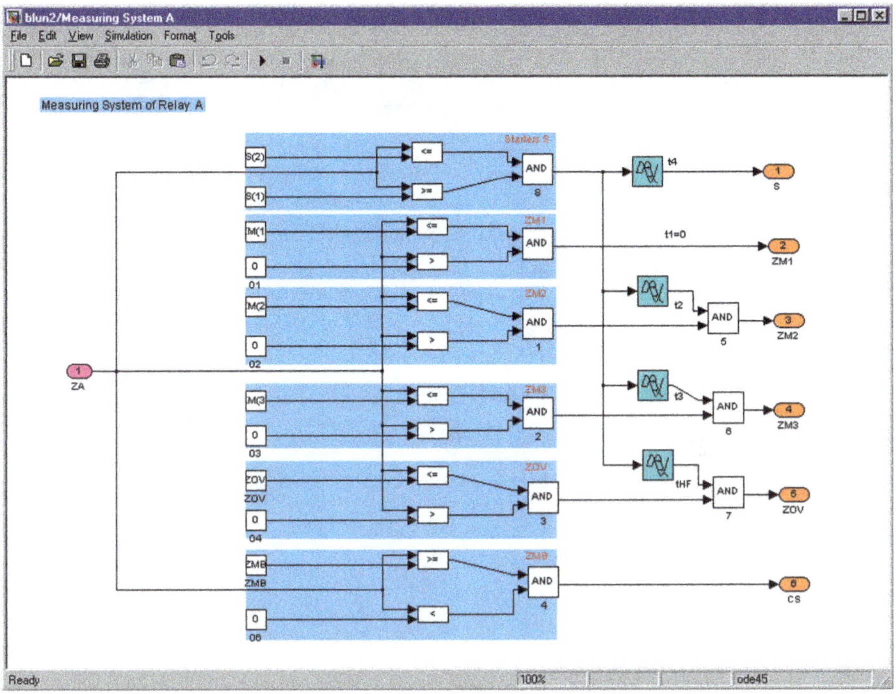

Fig. 5.13 Block diagram of the BLUN logic measuring system

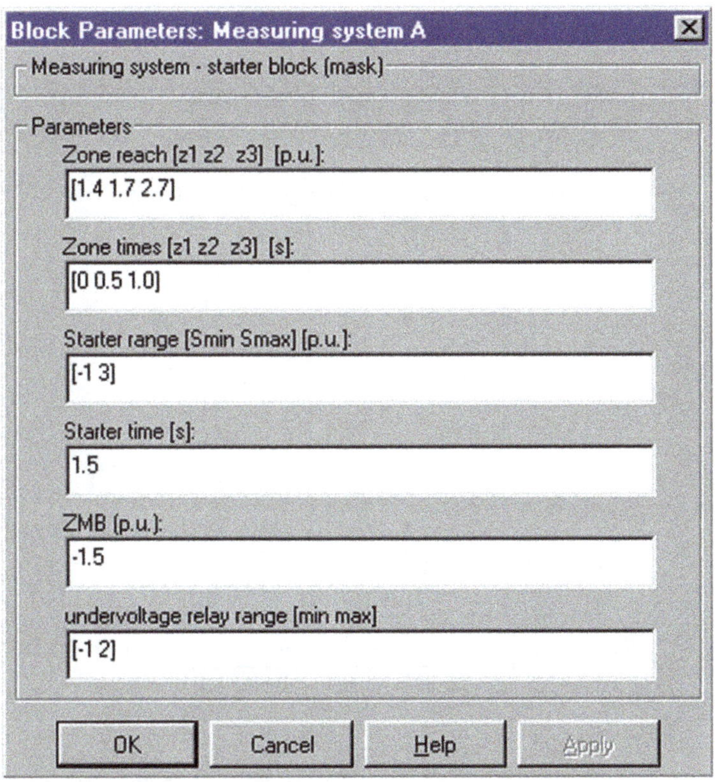

Fig. 5.14 User menu for the POTT+WEI+TB logic measuring system

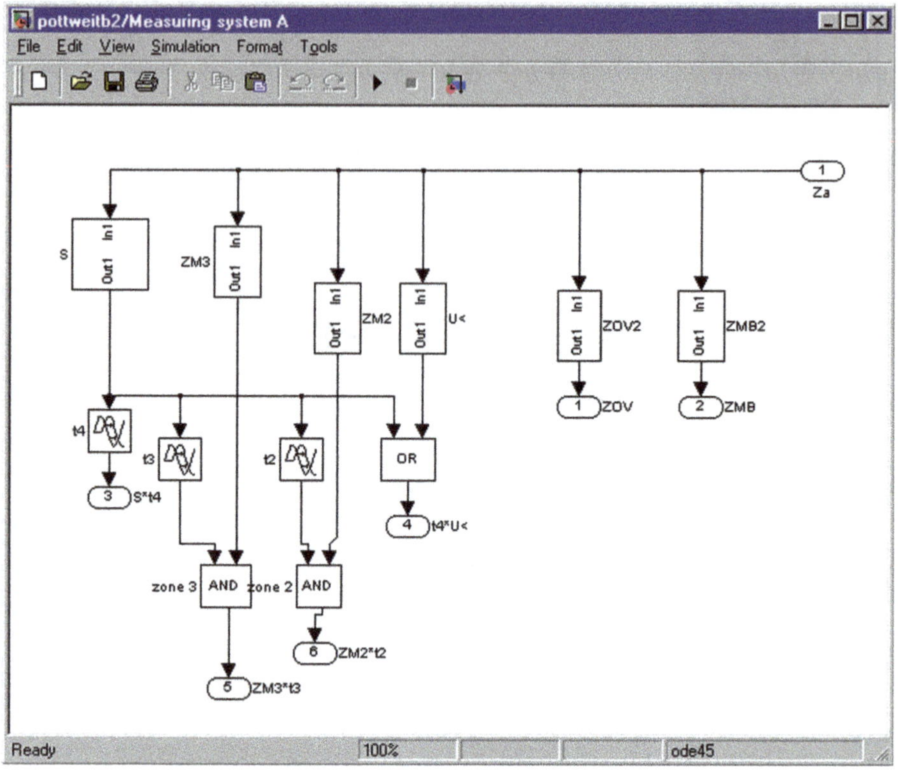

Fig. 5.15 Block diagram of the POTT+WEI+TB logic measuring system

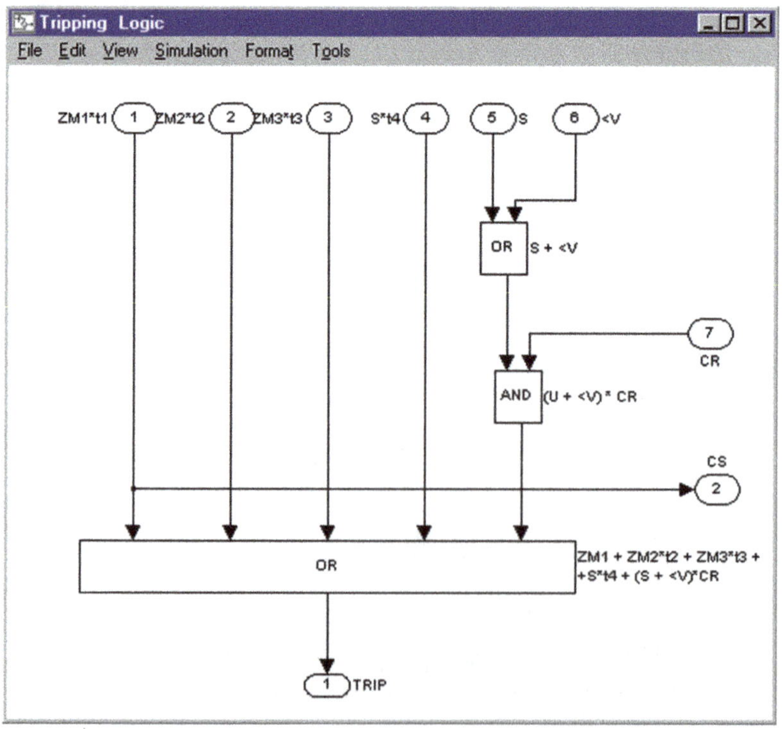

Fig. 5.16 The PUTT mode tripping logic block

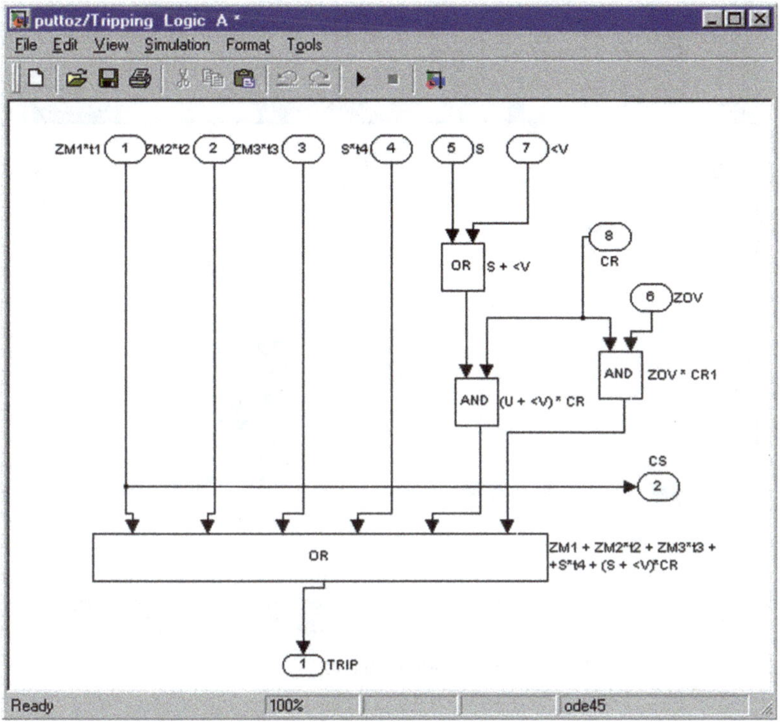

Fig. 5.17 The PUTT + OZ mode tripping logic block

PUTT + OZ Logic

The expression for the Trip signal for this PUTT + OZ logic is:

$$\mathrm{TRIP} = \mathrm{ZM}_1 + \mathrm{ZOV} \cdot \mathrm{CR} + \mathrm{ZM}_2 \cdot t_2 + \mathrm{ZM}_3 \cdot t_3 + \mathrm{S} \cdot t_4$$

The Simulink subsystem that simulates this logic is shown in Fig. 5.17.

BLOV + TB Logic

The expression for the Trip signal for this logic is:

$$\mathrm{TRIP} = \mathrm{ZOV} \cdot \overline{\mathrm{CR}} \cdot t_{\mathrm{HF}} + \mathrm{ZM}_2 \cdot t_2 + \mathrm{ZM}_3 \cdot t_3 + \mathrm{S} \cdot t_4$$

The Simulink subsystem that simulates this logic is shown in Fig. 5.18.

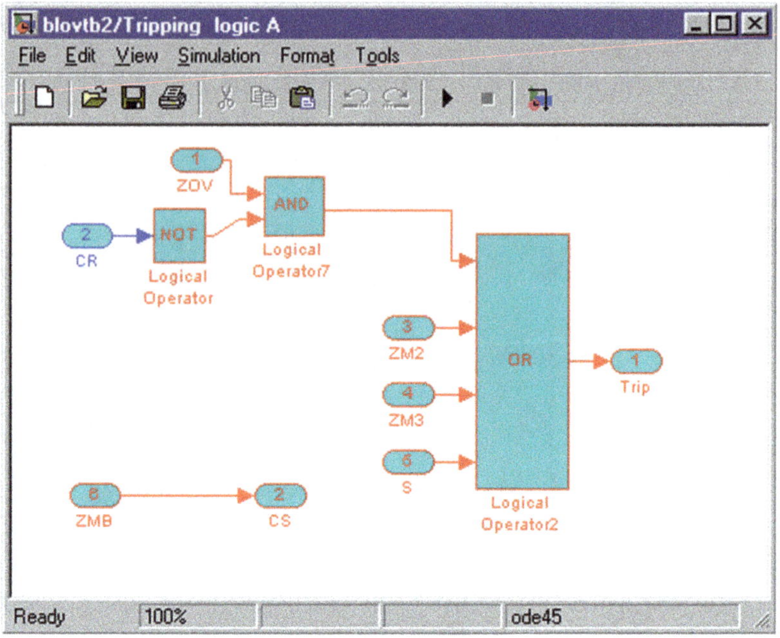

Fig. 5.18 The BLOV + TB mode tripping logic block

BLOV + UZ + TB Logic

The expression for the Trip signal for this logic is:

$$\mathrm{TRIP} = \mathrm{ZM}_1 + \mathrm{ZOV} \cdot \overline{\mathrm{CR}} \cdot t_{\mathrm{HF}} + \mathrm{ZM}_2 \cdot t_2 + \mathrm{ZM}_3 \cdot t_3 + \mathrm{S} \cdot t_4$$

The Simulink subsystem that simulates this logic is shown in Fig. 5.19.

BLUN Logic

The expression for the Trip signal for this logic is:

$$\mathrm{TRIP} = \mathrm{ZM}_1 + \mathrm{ZOV} \cdot \overline{\mathrm{CR}} \cdot t_{\mathrm{HF}} + \mathrm{ZM}_2 \cdot t_2 + \mathrm{ZM}_3 \cdot t_3 + \mathrm{S} \cdot t_4$$

The Simulink subsystem that simulates this logic is shown in Fig. 5.20.

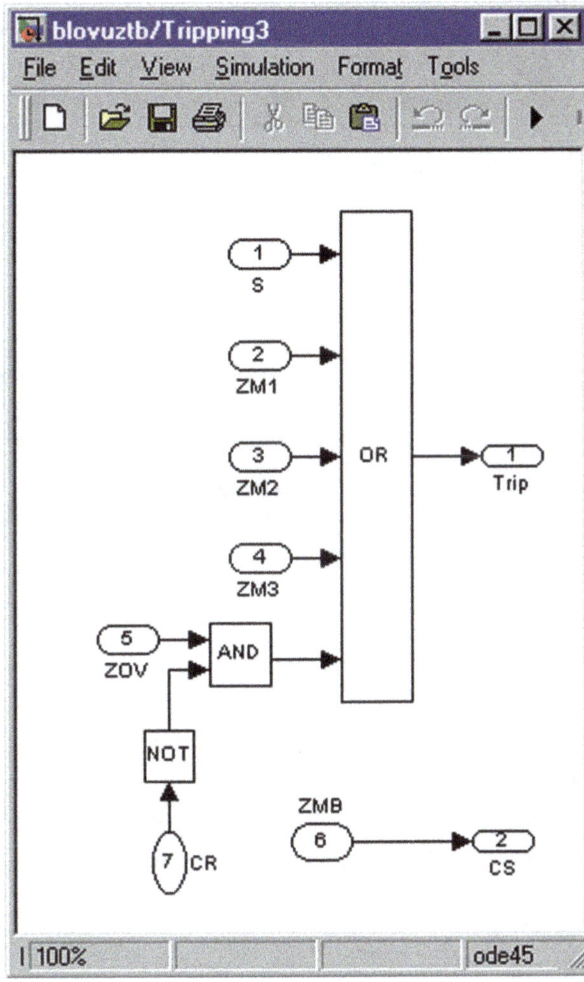

Fig. 5.19 The BLOV + UZ + TB mode tripping logic block

POTT + WEI + TB Logic

The expression for the Trip signal for this logic is:

$$\mathrm{TRIP} = \mathrm{ZOV} \cdot \mathrm{CR} + (\mathrm{S} + \mathrm{U} <) \cdot \overline{\mathrm{ZOV}} \cdot \overline{\mathrm{ZMB}} \cdot \mathrm{CR} + \mathrm{ZM}_2 \cdot t_2 + \mathrm{ZM}_3 \cdot t_3 + \mathrm{S} \cdot t_4$$

The Simulink subsystem that simulates this logic is shown in Fig. 5.21.

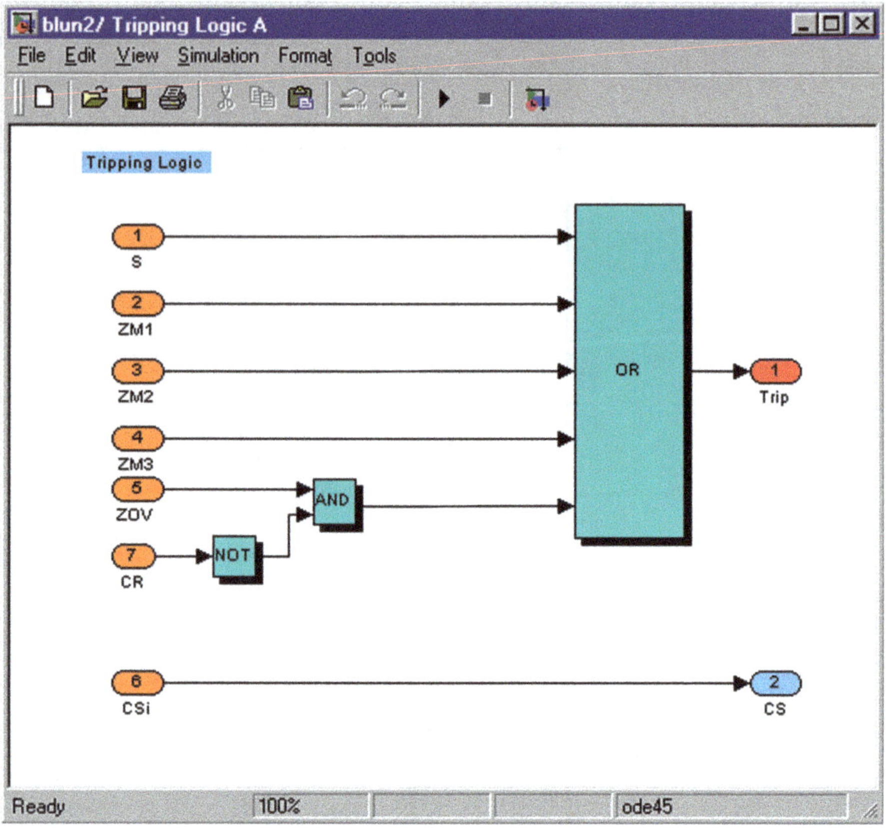

Fig. 5.20 The BLUN mode tripping logic block

5.2.2.4 Communication Channel

This block provides communication between relays at both ends of the line. The user can enter following parameters of communication channel:

- Communication channel delay time [ms]
- A to B communication channel failure
- B to A communication channel failure
- Logic value of erroneous signal (0 or 1) in failure mode operation case (Figs. 5.22 and 5.23)

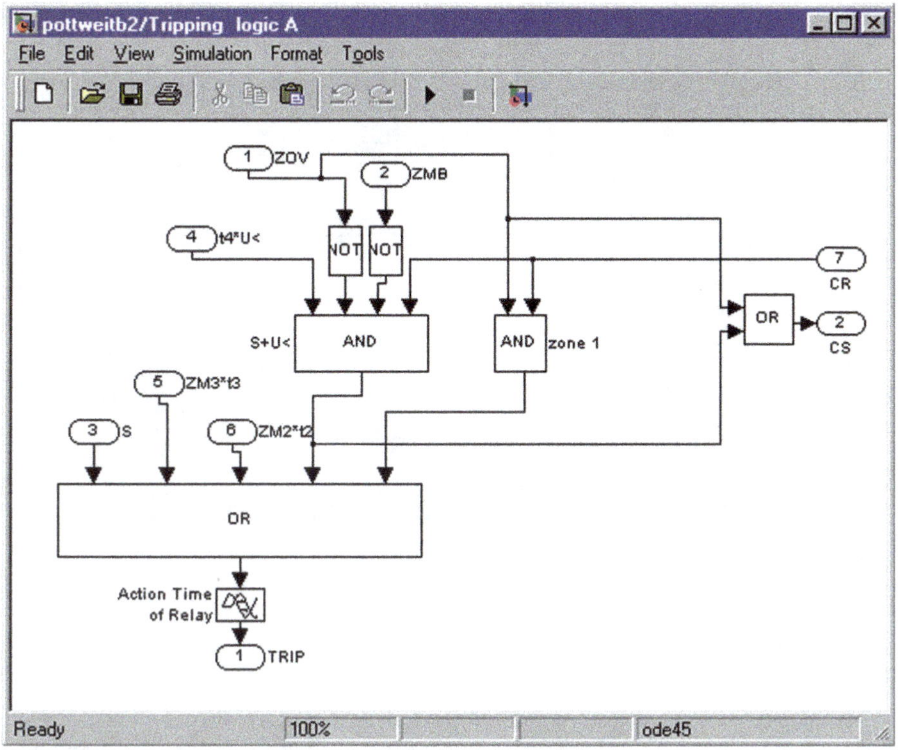

Fig. 5.21 The POTT + WEI + TB mode tripping logic block

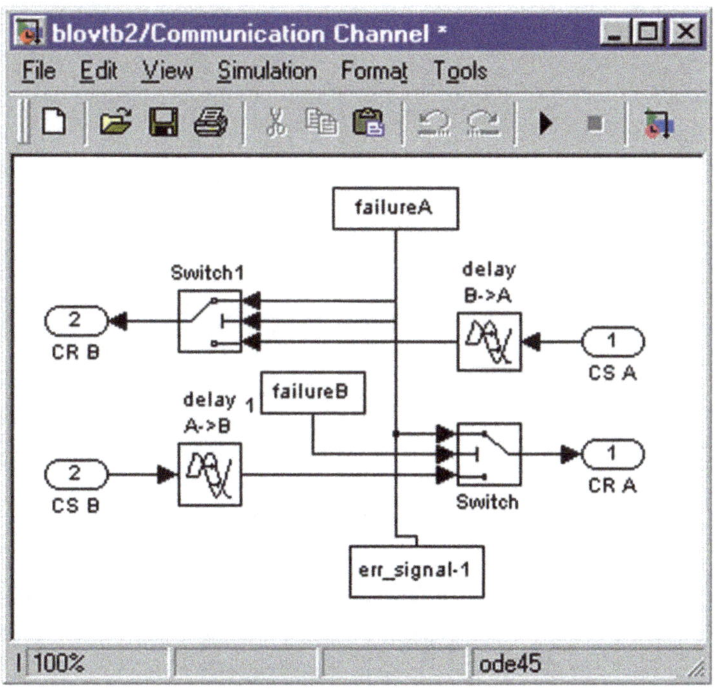

Fig. 5.22 Block diagram of communication channel

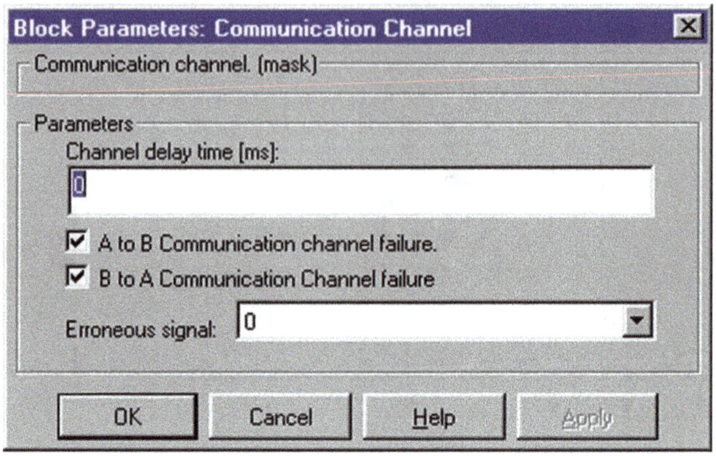

Fig. 5.23 Mask of communication channel

5.3 Summary

This chapter presents different communication schemes and tripping logics used in distance relaying. After this chapter, the reader should have a deep understanding of the communication schemes used in power protection.

Chapter 6
Design and Implementation of Overcurrent, Pilot, and Distance Protection

6.1 Introduction

This chapter discusses the different protection systems for transmission lines including overcurrent, differential, pilot, and distance protection.

6.2 Line Protection System: Overcurrent Relaying

6.2.1 Introduction

This section describes the design of the line protection system for two-terminal lines with at least the following protection functions:

- Three-phase directional instantaneous overcurrent relays for the primary protection.
- Three-phase time overcurrent relays for the backup protection.
- Residual time overcurrent relay.

It is assumed that the following signals are available: I_A, I_B, I_C—currents, V_A, V_B, V_C—voltages. Detail description of the algorithm, i.e., the way of converting the aforementioned input signals into the output signal, TRIP, is provided.

The design was implemented as a MATLAB model using the *Relay Elements* library for modeling protective relays. The algorithm was organized into subsystems for performing the needed functions.

© Springer International Publishing Switzerland 2016

M. Kezunovic et al., *Design, Modeling and Evaluation of Protective Relays for Power Systems*, DOI 10.1007/978-3-319-20919-7_6

6.2.2 *Theoretical Background*

6.2.2.1 Importance of Transmission Lines

Transmission lines provide connections between the various parts of the power system and the associated equipment. Being the points of connection means that power from the generation sources must flow across the lines, either under normal conditions or abnormal conditions. In addition, because of their length and open air construction, transmission lines usually are the most exposed part of the power system to events that can cause faults.

Most faults in an electric utility system with a network of overhead lines are one-phase-to-ground faults, resulting primarily from lightning-induced transient high voltage or from falling trees. In the overhead distribution systems, momentary tree contact caused by wind is another major cause of faults. Ice, freezing snow, and wind during severe storms can cause many faults and much damage. Even though single-phase-to-ground are the most common type of faults (80 %), there are some other types of faults such as phase-to-phase-to-ground (10 %), phase-to-phase (8 %), and three-phase (2 %).

Protection of transmission lines is very important and therefore very complex. The proper operation and protection of the whole power system depends on the decision to keep the line working or to remove the line from operation. At this point, backup protection emerges as an important issue to ensure that the decision about the status of the line is correct.

6.2.2.2 Backup Protection

Backup is defined as "protection that operates independently of specified components in the primary protective system." It may duplicate the primary protection or may be intended to operate only if the primary protection fails or is temporarily out of service.

Backup protection can be seen in two basic forms: redundant and remote. Redundant case is when the backup protection is provided as an additional protection in the primary protection zone and sometimes extending into the adjacent system. Remote backup is the case when the backup protection is overlapping the primary relays in one protection area acting from an adjacent protection area.

In this example, a directional-instantaneous overcurrent relay is the primary protection and a time overcurrent relay is the backup protection acting as a redundant system.

6.2.2.3 Directional Overcurrent Relays

System faults usually, but not always, provide significant changes in the system measurements, which can be used to distinguish between tolerable and intolerable system conditions. These changing quantities include overcurrent, over- or undervoltage, power, power factor or phase angle, power or current direction, impedance, frequency, temperature, physical movements, pressure, and contamination of the insulation quantities.

Directional overcurrent relays sense the *direction of current* (*or power*) *flow* at a specific location, which indicates the direction of the fault.

A directional-sensing unit requires a *reference quantity*, which should be reasonable constant against which the current in the protected circuit can be compared. For relays intended to provide operation for phase-type faults, one of the system voltages can be used as a reference. For all practical purposes, most system voltages do not change their phase positions significantly during a fault. In contrast, line currents can shift around 180° for faults on one side of the circuit and CTs relative to a fault on the other side of the CTs. The reference quantity is commonly called the "polarizing" quantity.

6.2.2.4 Principle of Operation

The interpretation of relay polarity is illustrated in Fig. 6.1. In Fig. 6.1a, the maximum operating torque or energy occurs when the current flowing to the polarity of the CT windings (I_{pq}) leads by 30° the voltage reference (V_{rs}). As

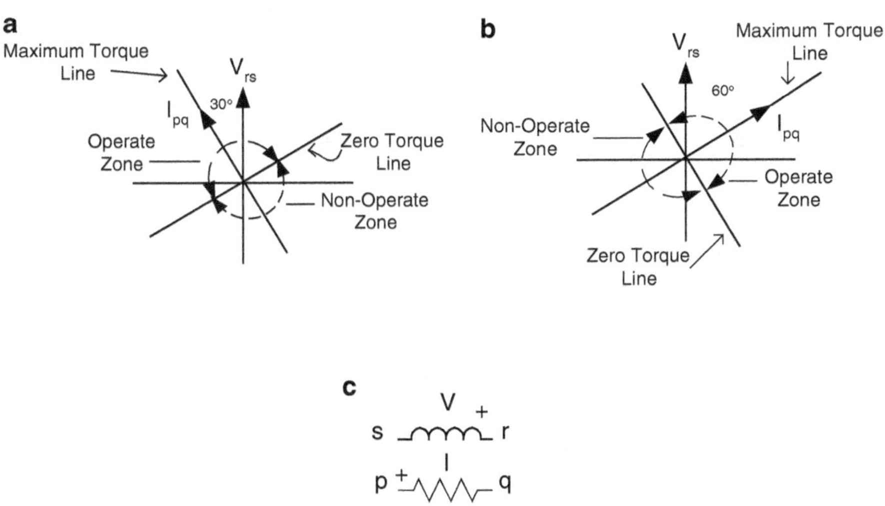

Fig. 6.1 Typical directional relay characteristics. (**a**) 30° type unit. (**b**) 60° type unit. (**c**) Voltage polarized directional relay

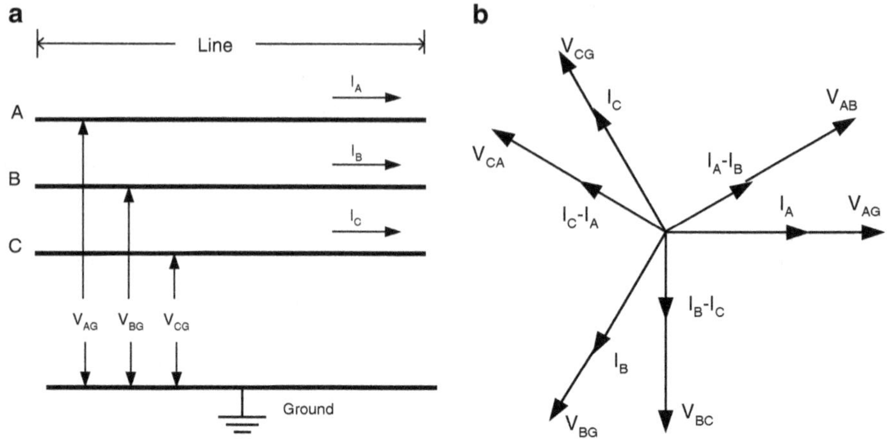

Fig. 6.2 (**a**) Circuit diagram showing the location and the assumed directions of current and voltage drops. (**b**) Phasor diagrams showing current and voltage magnitudes and phase relations

seen, the unit will operate for currents from almost 60° lagging the reference voltage V_{rs} to almost 120° leading. The operating (tripping) zone is represented by the half plane, bordered on one side by the zero-torque (non-operating) line and extending in the direction that contains both the reference (polarizing) and operating quantities. The operating torque at any angle is a function of the cosine of the angle between the current (I_{pq}) and the maximum-torque line, as well as the magnitudes of the operating quantities.

Scheme of Fig. 6.1 is used: (a) for phase-fault protection and for ground-fault protection (the 60° unit of Fig. 6.1), and (b) with a $3V_0$ reference.

Several phase voltages exist within the power system and are available for consideration as the reference quantity for directional relaying; those voltages are shown in Fig. 6.2. In that figure, the phasor diagram is for a typical three-phase circuit operating with balanced or symmetrical quantities.

6.2.2.5 Phase-Fault Directional Sensing

Five different connections for phase-fault directional sensing have been used over the years. These are outlined in Table 6.1.

For a number of years, connections 4 and 5 have been used almost exclusively. These two connections are fundamentally the same, and they are known as the "90° connection." The only difference between them is the angle that the system current lags the system voltage for maximum operating torque or energy. Either 60° or 45° is the typical angle of the fault current for the maximum energy or torque.

Table 6.1 Connection chart for phase-fault directional sensing

Connection	Phase A	Phase B	Phase C	Maximum torque occurs when
	I	I	I	
	V	V	V	
1	I_A	I_B	I_C	I lags 30°
	V_{AC}	V_{BA}	V_{CB}	
2	$I_A - I_B$	$I_B - I_C$	$I_C - I_A$	I lags 60°
	V_{AC}	V_{BA}	V_{CB}	
3	I_A	I_B	I_C	I lags 60°
	$-V_C$	$-V_A$	$-V_B$	
4	I_A	I_B	I_C	I lags 45°
	V_{BC}	V_{CA}	V_{AB}	
5	I_A	I_B	I_C	I lags 60°
	V_{BC}	V_{CA}	V_{AB}	

6.2.2.6 The 90°–60° Connection

The 90° connection applies a power system voltage that lags the power system *unity power factor current* by 90°. These voltages and currents are obtained from the power system through voltage and current transformers. Typical three-phase connections are shown in Fig. 6.3.

The phase A directional unit receives I_A, and from the system phasors of Fig. 6.2, we can see that the 90° lagging voltage is V_{BC}. The phase B directional unit receives I_B where the lagging voltage is V_{CA}, and the phase C directional unit receives I_C where the 90° lagging voltage is V_{AB}. These are also shown in Table 6.1 for connections 4 and 5.

In Table 6.1, the currents are connected so that when I_A, I_B, and I_C are flowing in the direction indicated by the "trip direction" arrow, the secondary currents flow through the directional units from polarity to nonpolarity. With the trip direction of the currents established in the directional unit current coils, the voltages V_{BC} on unit A, V_{CA} on unit B, and V_{AB} on unit C must be connected from polarity to nonpolarity on the directional unit voltage coils, as shown. The phasor diagram of Fig. 6.3 applies the directional unit characteristic of Fig. 6.1a to the power system phasors. The maximum-torque line leads the voltage by 30°, so with V_{BC} polarity to nonpolarity on the relay voltage winding, the maximum-torque line is drawn 30° leading, as illustrated in the phasor diagram of Fig. 6.3. This is 60° lagging the unity power factor position of the current phasor I_A. Therefore, whenever the phase A current in the power system lags by 60°, the directional unit will operate at maximum torque with the lowest pickup value and the highest sensitivity. Because most system faults provide relatively large currents, the range of possible operation is for power system currents from almost 30° leading to 150° lagging in the trip direction. Similar relations exist for the other two units using I_B and I_C phase currents.

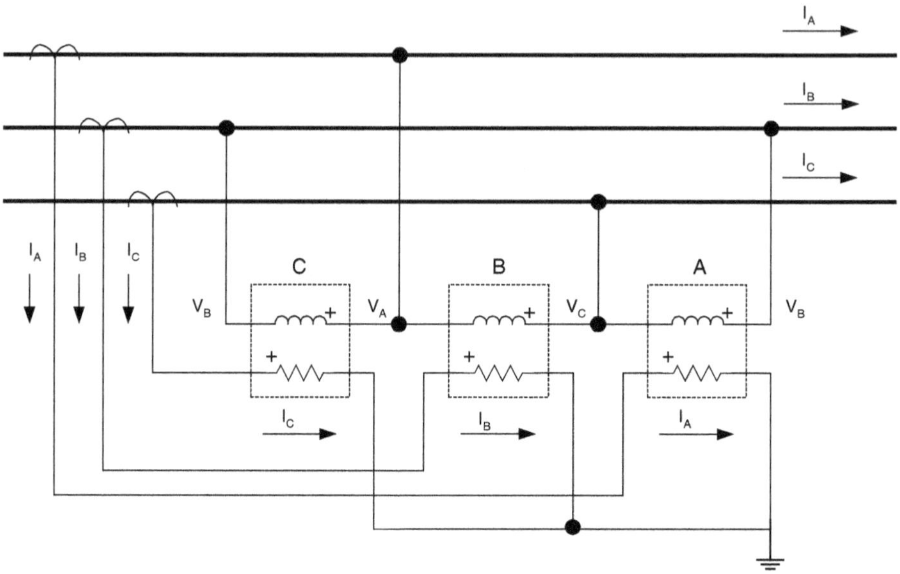

Fig. 6.3 Typical three-phase connection for phase-fault directional sensing using the 30° unit from Fig. 6.1

Solid-state relays provide the possibility of restricting the operating zone. For most power system faults, the current will lag the voltage from close to say 5°–15° (large arc resistances at low voltages) to 80°–85° at the high voltages, thus restricting the operating zone by adjusting the zero-torque lines is usual.

6.2.2.7 Time Overcurrent Relays

The time-overcurrent type relay is one of the first protective relays invented. Originally, it was a watt-hour meter with contacts and restricted disk travel. Today, as it has been for many years, the design is entirely different, except that it uses the fundamental induction disk principle. This principle produces a fast operation at high currents and slow operation at light current; hence, an inverse time characteristic.

Over the years, various shapes of time curves have evolved, which are shown in Fig. 6.4. Solid-state versions of these relays essentially duplicate these curves and general characteristics, with lower burdens, wider application ranges, and adjustable time characteristics.

The inverse time characteristic allows the relay to discriminate between permissible and non-permissible conditions. Relays that discriminate by time protect the equipment with which they are associated and also act as backup protection for other relays. The major disadvantage is that there is a delay in the removal of the

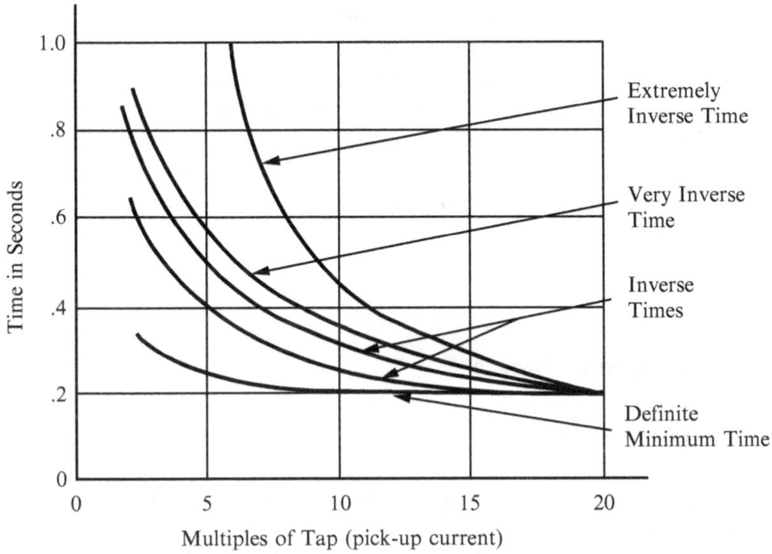

Fig. 6.4 Typical inverse-time overcurrent relay characteristics. For general comparison, the curves are fixed at 0.2 s at 20 times minimum pickup current

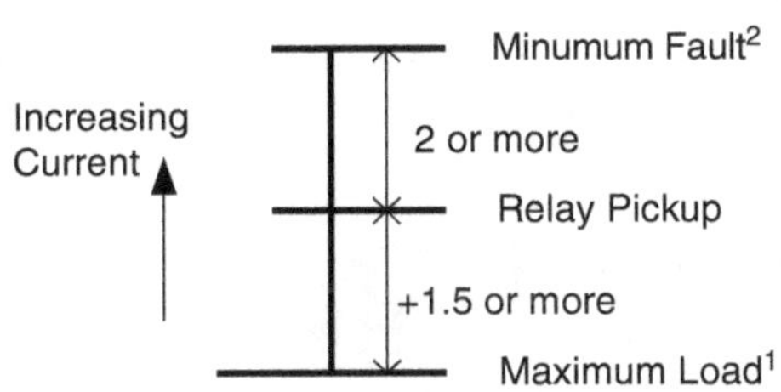

Fig. 6.5 Criteria for overcurrent relay pickup. [1]For phase relays. Maximum short time load (assuming asymmetrical offset, magnetizing inrush, cold load, and unusual operation). For ground relays. Maximum zero sequence unbalance, phase-to-ground (neutral) loads. [2]Nominally $I_{3\phi}$ for phase faults, $I_{\phi\phi} = 0.866 I_{3\phi}$. Or $3I_0$ for ground faults

fault that may increase damage to the faulty equipment and increase the possibility of damage of healthy equipment, which is carrying the fault current.

The minimum-operating criteria for overcurrent relays are shown in Fig. 6.5. These relays may operate instantaneously, with fixed or inverse time delays (as shown in Fig. 6.4).

The operation of overcurrent relays at the border of a protection zone is not as precise as in the case of the differential protection. Hence, they may underreach or overreach for faults near this border. Using a proper time delay can solve this problem.

6.2.2.8 Residual Time Overcurrent Relay

The setting and coordination procedure for residual relays, both inverse-time and instantaneous overcurrent, is the same as for phase relays. The pickup value in residual relays for the inverse-time units must be set above the tolerable zero-sequence unbalance on the line, and single-phase-to-ground fault data are used. This relay uses the zero sequence current instead of line current to determine if a trip should be issued.

6.2.3 Simulation Models

6.2.3.1 Description of the System

A block diagram of the system implemented in MATLAB is depicted in Fig. 6.6.

The system consists of a radial line that has at one end a three-phase voltage source and at the other end a three-phase load. The system has a breaker unit to connect or disconnect the source. All the power elements are marked in yellow. In addition, the system has CTs and VTs at both the sending end and receiving end; these elements are shown in blue. The signal processing block and the protective relays block are marked in green and dark green, respectively. The block for Faults is marked in red and all the scopes (measurement elements) are marked in white.

A more detailed description of each block is presented below.

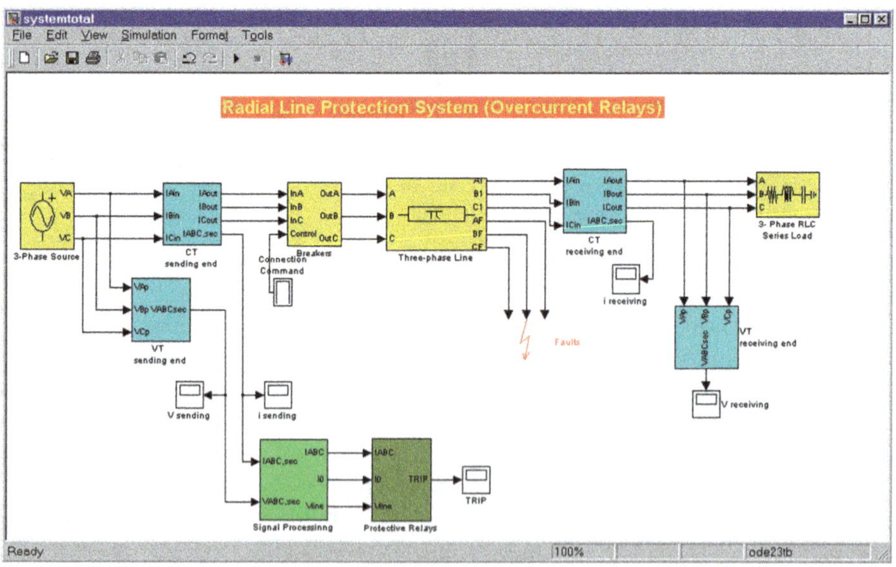

Fig. 6.6 Block diagram of a power system implemented for simulation purpose

6.2.3.2 Three-Phase Voltage Source

The purpose of this subsystem is to generate a set of three single-phase voltage sources 120° apart in an ABC sequence. The impedance connected in series with the source is included to represent internal impedance (Figs. 6.7 and 6.8).

The only parameter to define by the user is the rms voltage for each phase. In case the user wants to modify the internal impedance of the source or the phase angle between the voltages, the option "look under mask" should be used.

6.2.3.3 Current Transformer Block

The current sensing module performs the function of three current transformers (CTs) and is depicted in Fig. 6.9.

The inputs are the line currents I_A, I_B, and I_C, the outputs are the same currents scaled down, and isolated from the power system (Fig. 6.10).

The input value named *Maximum Normal Current* is to determine the "gain" of the CTs, this value can be seen as the base current value. The objective of this gain is to simulate the turn ratio of the CT. In this case the secondary current is scaled to 1 A.

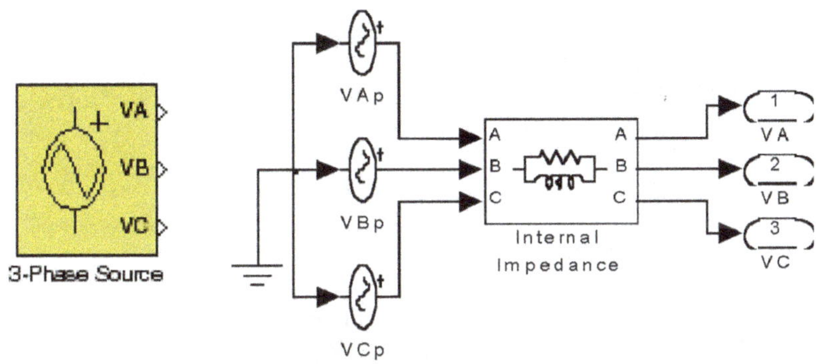

Fig. 6.7 Icon of the three-phase voltage source subsystem and models in mask view

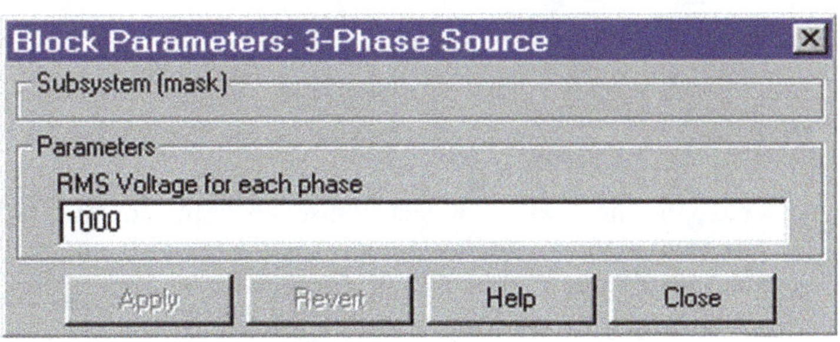

Fig. 6.8 Dialog box of the three-phase voltage source

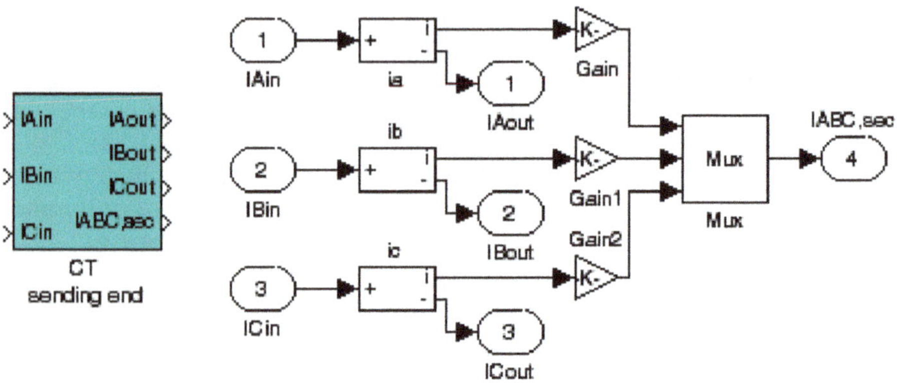

Fig. 6.9 Icon of the current transformers subsystem and models in mask view

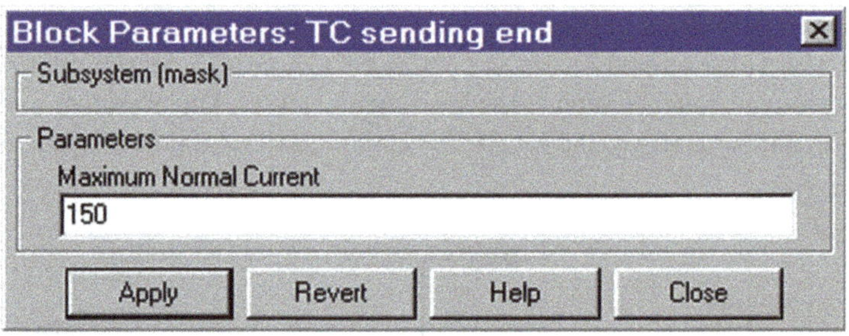

Fig. 6.10 Dialog box of the current transformer subsystem

6.2.3.4 Voltage Transformer Block

The voltage sensing module performs the function of three voltage transformers (VTs) in Y-Y connection, then the inputs are V_{AN}, V_{BN}, and V_{CN}. The outputs are the same voltages scaled down, isolated from the power system without phase shift (Figs. 6.11 and 6.12).

The input value named *Nominal Voltage* is to determine the "gain" of the VTs; this value can be seen as the Base Voltage value.

6.2.3.5 Breakers

The Breakers have the function to connect or disconnect the power source from/to the line. The connection command block controls all the three breakers (Fig. 6.13).

Since the breaker is modeled as a current source driven by the voltage appearing across its terminals, a large resistor should be connected in parallel to each breaker to avoid connections with open circuits or other current sources.

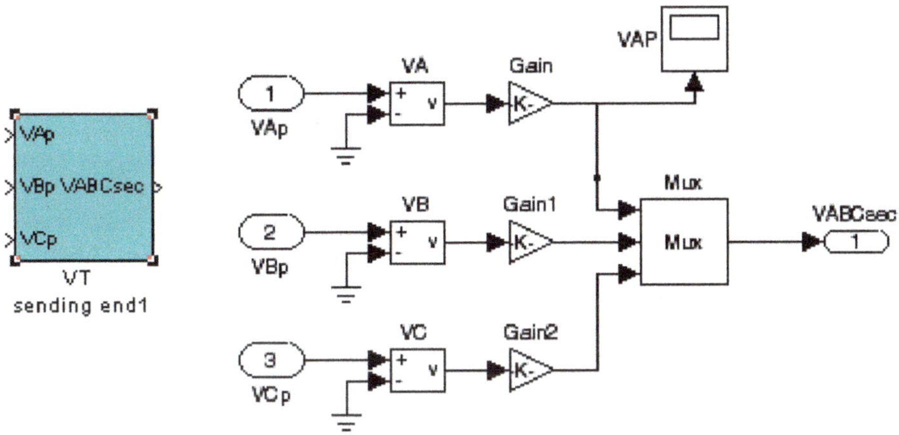

Fig. 6.11 Icon of the voltage transformers subsystem and models in mask view

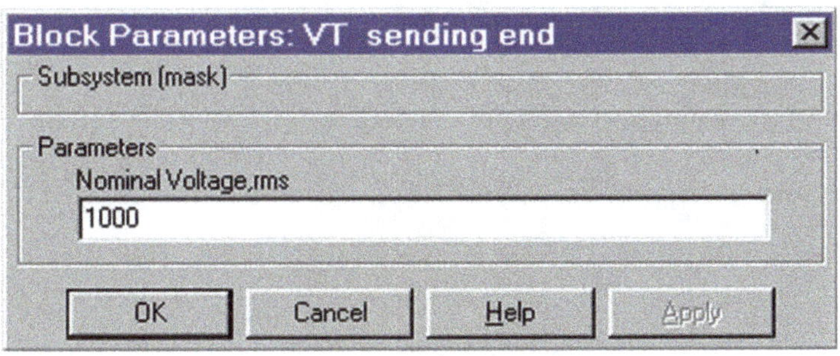

Fig. 6.12 Dialog box of the voltage transformer subsystem

6.2.3.6 Line Model

For the transmission line, a PI model is used; this helps to define the fault in a different place because this model uses the length of the line to calculate its parameters (Fig. 6.14).

Terminals A, B, and C represent the sending end of the transmission line, while terminals A1, B1, and C1 represent the receiving end of the transmission line. Terminals AF, BF, and CF are connections to the subsystem that represents the faults. Once the user defines the parameters for the transmission line, the only parameter to be adjusted is the fault location as can be seen in Fig. 6.15.

The load is defined using a block that represents a three-phase RLC series load. This block allows the setting of loads with different power factor. The information required for this block is shown in Fig. 6.16.

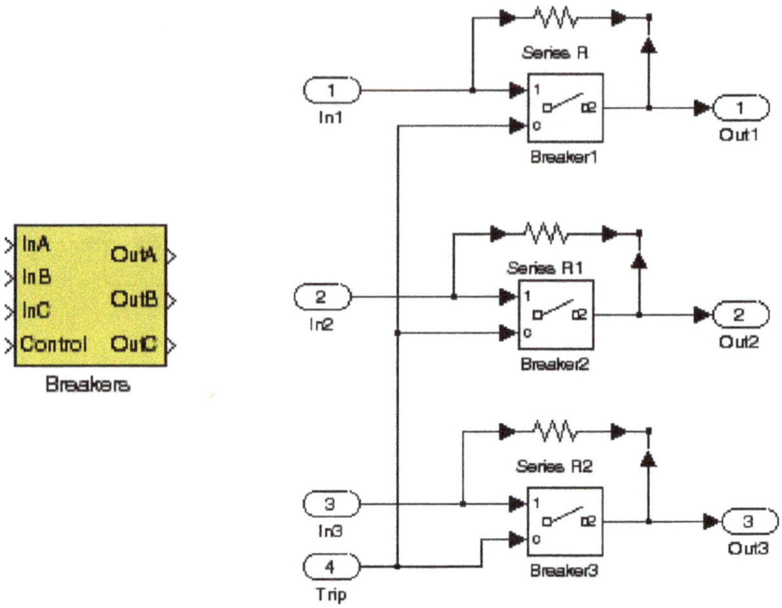

Fig. 6.13 Icon of the breaker subsystem and models in mask view

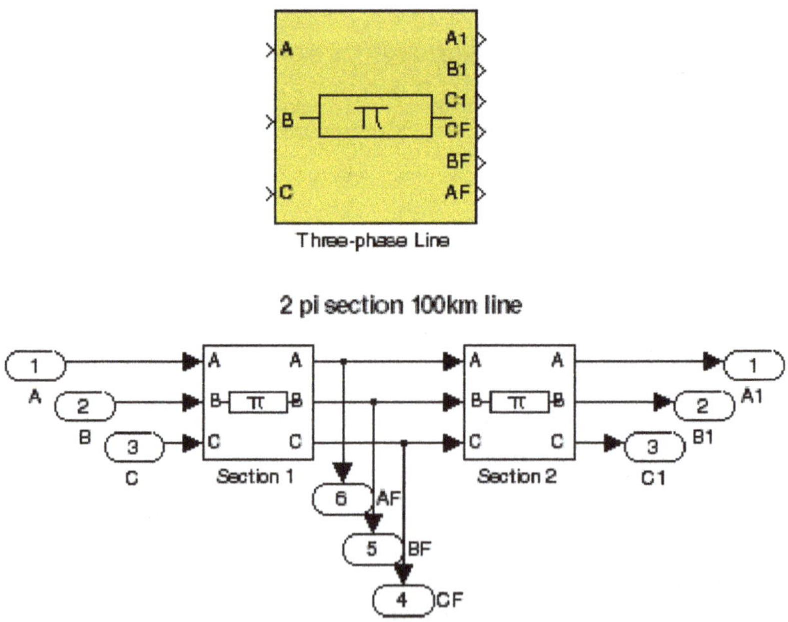

Fig. 6.14 Icon of the transmission line subsystem and models in mask view

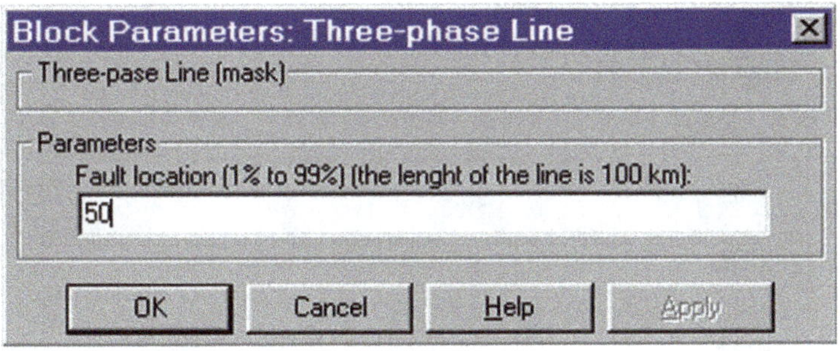

Fig. 6.15 Dialog box of the transmission line subsystem

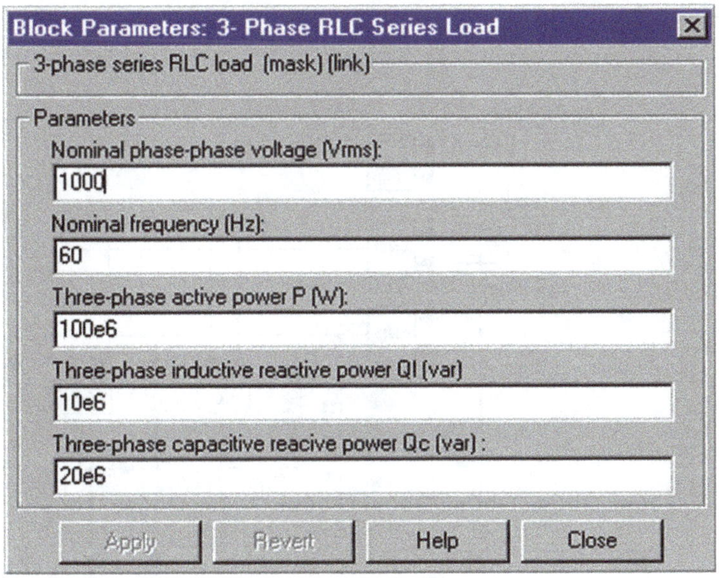

Fig. 6.16 Dialog box of the three-phase RLC series load subsystem

6.2.3.7 Faults

This block is intended to represent a fault somewhere in the transmission line. This block allows defining the type of fault to be staged, the time of occurrence, and the fault resistance (if needed, otherwise define it as a very low value other than zero) (Fig. 6.17).

The inputs of this subsystem are named AF, BF, and CF as can be seen in Fig. 6.18.

The proper operation of switches S1–S4 determines the type of fault. Notice that the initial status of the switches is open -0- and will change according to Table 6.2 at the time defined as *Fault time*.

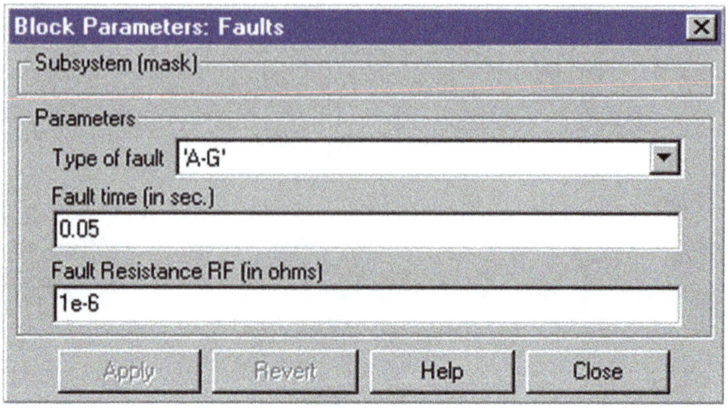

Fig. 6.17 Dialog box of the faults subsystem

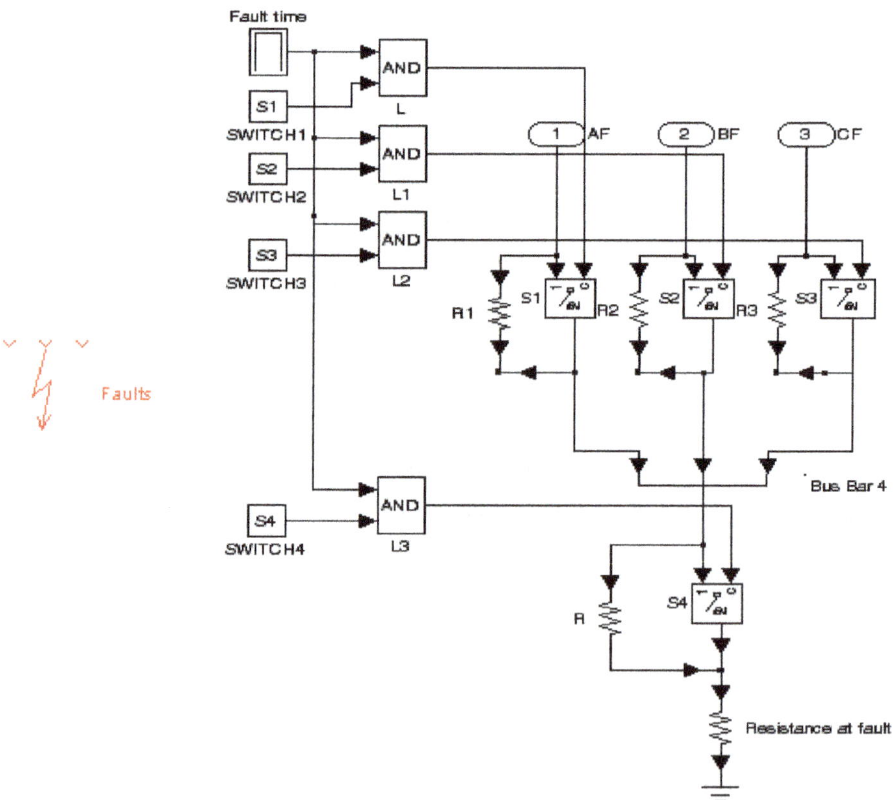

Fig. 6.18 Icon view of the faults subsystem and models in mask view

Table 6.2 Fault time for switches S1–S4

Type of fault	S1	S2	S3	S4
Normal operation	0	0	0	0
A-G	1	0	0	1
B-G	0	1	0	1
C-G	0	0	1	1
B-C	0	1	1	0
A-C	1	0	1	0
A-B	1	1	0	0
A-B-C	1	1	1	0
B-C-G	0	1	1	1
A-C-G	1	0	1	1
A-B-G	1	1	0	1
A-B-C-G	1	1	1	1

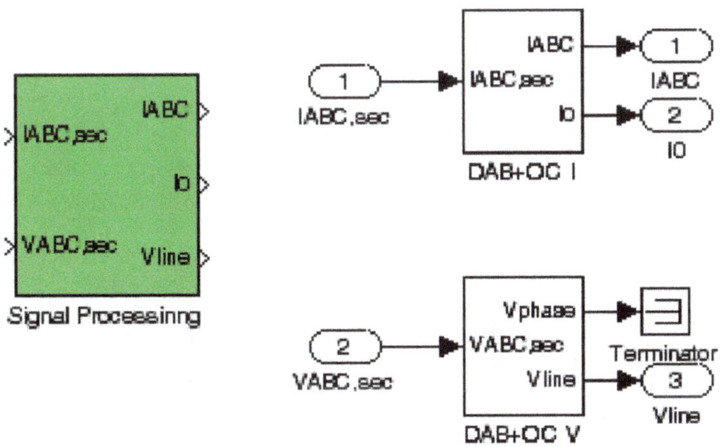

Fig. 6.19 Icon of the signal-processing subsystem and models in mask view

There is a small initialization program (in the mask) that generates the states for S1–S4 according to the input of *Type of fault*.

6.2.3.8 Signal Processing

This block is composed of two subsystems, one for voltages and one for currents. Each subsystem converts all input quantities from analog to digital format, then obtain the magnitude and phase angle of each one. The inputs come from the CT and VT subsystems (Fig. 6.19).

The subsystem for currents is shown in Fig. 6.20, where DAB stands for Data Acquisition Board and OC stands for Orthogonal Components. These two blocks belong to the library named *Elements*.

The outputs of this subsystem are the magnitude and phase of the line currents I_A, I_B, and I_C as well as the zero sequence current, I_0.

Figure 6.21 shows the subsystem for voltages. It can be noticed that the inputs of this subsystem are phase voltages (in analog format) and the outputs are phase

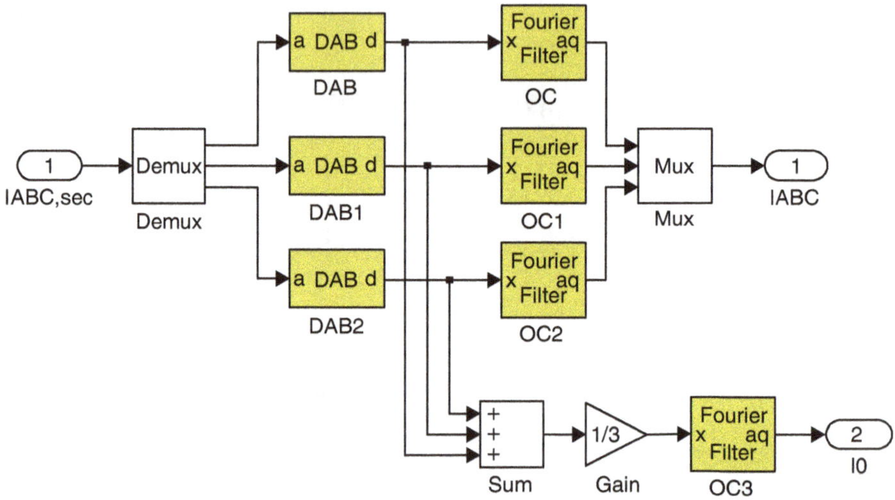

Fig. 6.20 DAB + OC *I* subsystem

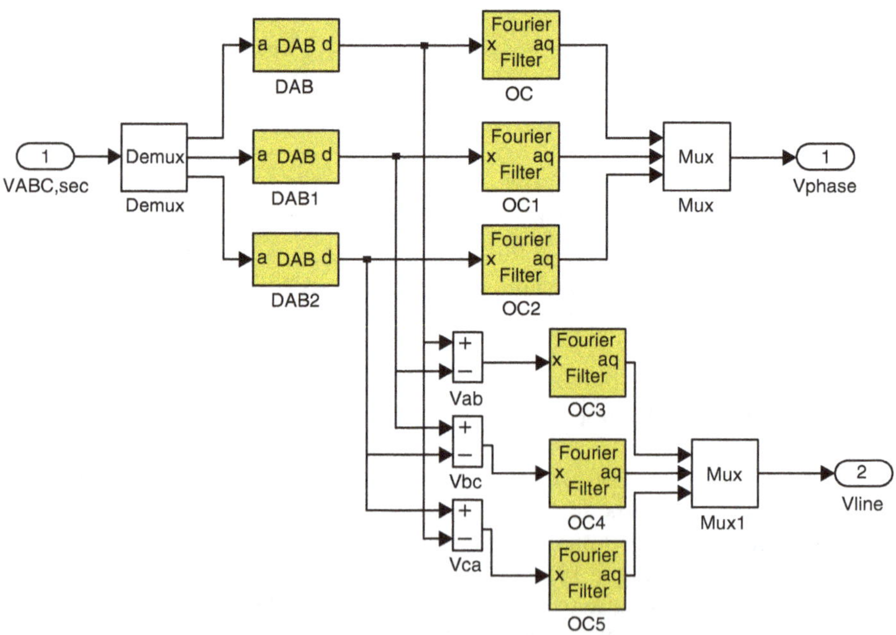

Fig. 6.21 DAB + OC *V* subsystem

voltages (in a vector representation) plus line voltages. Phase voltages are not required in our modeling, but are included for completeness.

6.2.3.9 Description of Protection Schemes

There are three different schemes implemented: A Directional Overcurrent Relay as a primary protection, Residual Overcurrent Relay, and Time Overcurrent Relay as backup protection. They are inside the subsystem named protective relays shown in Fig. 6.22.

After processing the input information, the relays should issue a trip signal if there is a fault. In fact, there are three trip signals, each one from each relay. Those trip signals are entered to a logic subsystem that will deliver only one trip signal. There are scopes (measurement elements) at every trip signal to help further understanding of each scheme. The logic subsystem is just an OR gate.

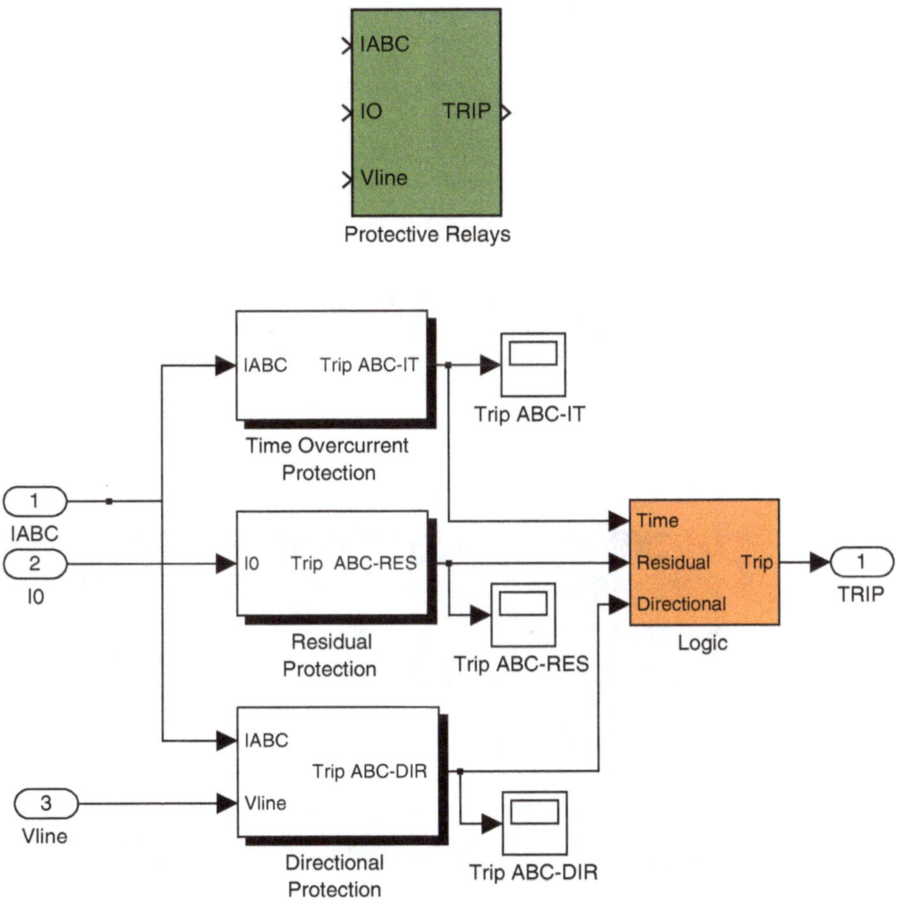

Fig. 6.22 Icon of the protective relays subsystem and models in mask view

6.2.3.10 Implementation of Directional Overcurrent Relay

The principle of operation for this relay is based on the description of Fig. 6.3. It uses the information from Fig. 6.1a and the information of Table 6.1 connection chart for phase-fault directional sensing, specifically connection 5. To perform this protection, a block named directional element is used as can be seen in Fig. 6.23.

Each directional element compares a line voltage (as reference) versus a line current. Each block is marked properly to know which quantities are being compared. To achieve this comparison, two parameters must be entered as shown in Fig. 6.24.

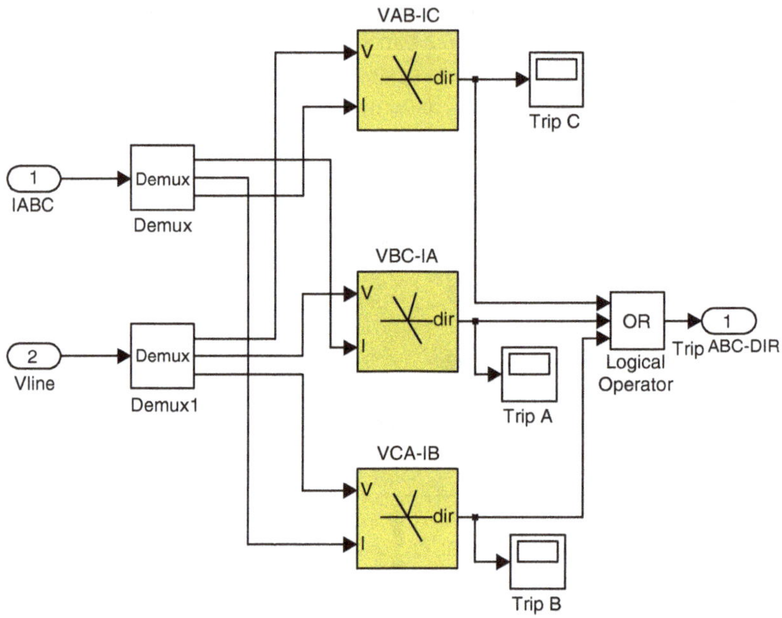

Fig. 6.23 View of the directional overcurrent relay subsystem

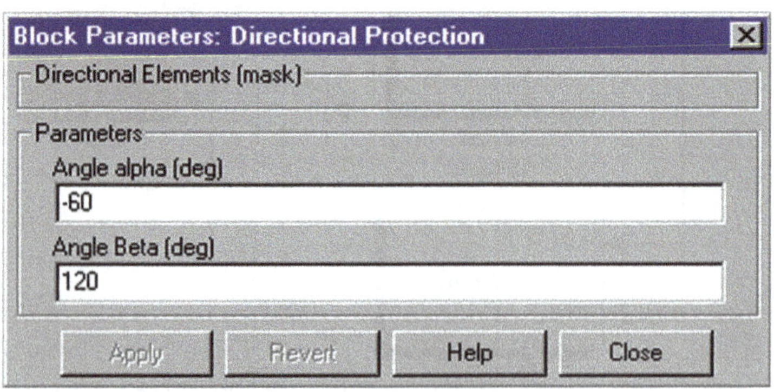

Fig. 6.24 Dialog box of the directional overcurrent relay subsystem

The values for alpha and beta come up from Fig. 6.1a as mentioned earlier. If the phase angle between the two input quantities (V_{line} and I_{line}) is between alpha and beta, then the output will be one and zero otherwise. A scope (measurement element) is placed with each directional element to know which phase(s) is (are) under fault.

6.2.3.11 Implementation of Time Overcurrent Relay

This relay uses the line currents as its only input. It compares this current against a threshold named pickup. When the input value is greater than the pickup value, the block named Universal Comparator starts a comparison against a predetermined curve and eventually will issue a trip signal (Fig. 6.25).

The input S2 of each Universal Comparator is connected as shown to avoid warning messages during simulation. The main parameters required for this subsystem are shown in Fig. 6.26.

6.2.3.12 Implementation of Residual Overcurrent Relay

This relay works using the same idea as the time overcurrent relay but instead of using the line currents, it uses the zero sequence current of the system (Fig. 6.27).

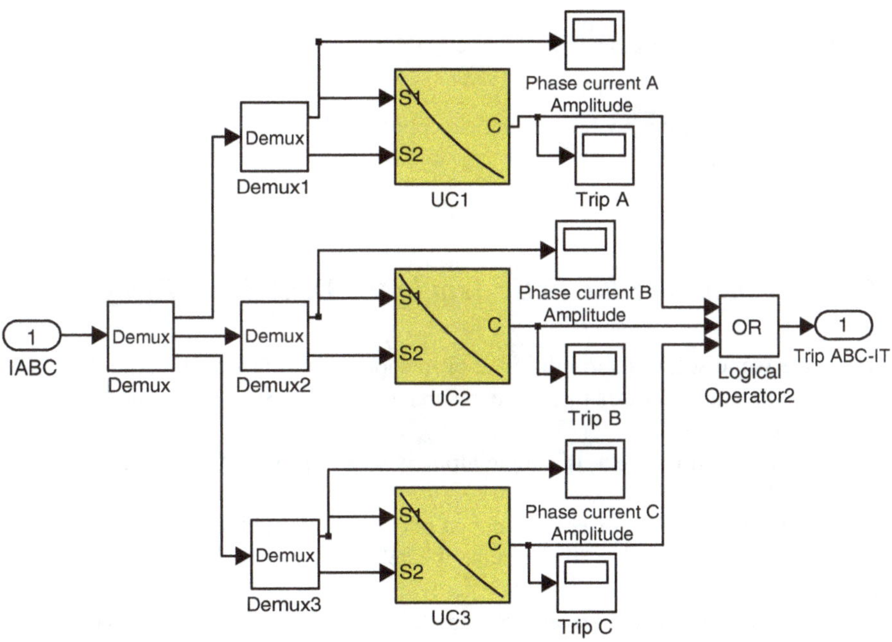

Fig. 6.25 View of the time overcurrent relay subsystem

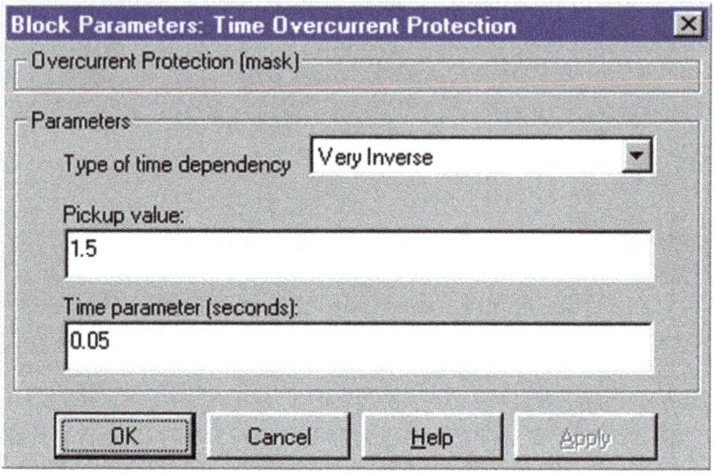

Fig. 6.26 Dialog box of the time overcurrent relay subsystem

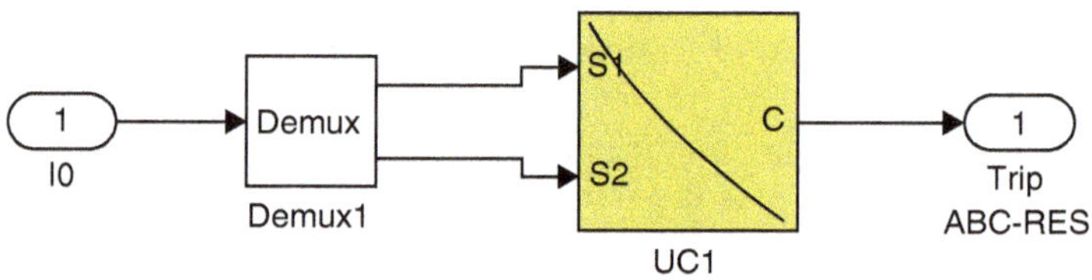

Fig. 6.27 View of the residual overcurrent relay subsystem

All the parameters to be defined are contained in the mask of the Universal Comparator, UC.

6.2.4 *Minimizing the False Trip in the Directional Relay*

There are some solutions to minimize or to suppress the false trip in the directional relay. In the next figure, there are results for this trip using alpha = −60° and beta = 120° (Fig. 6.28).

To minimize this effect, the angle alpha is reduced to −50° with the next results (Fig. 6.29).

As it can be seen, the effect is reduced but not eliminated. It is not recommended to reduce alpha beyond −50° because the relay's response becomes as shown in Fig. 6.30.

There is another way to overcome this false trip: using in the block named logic an array as is shown in Fig. 6.31.

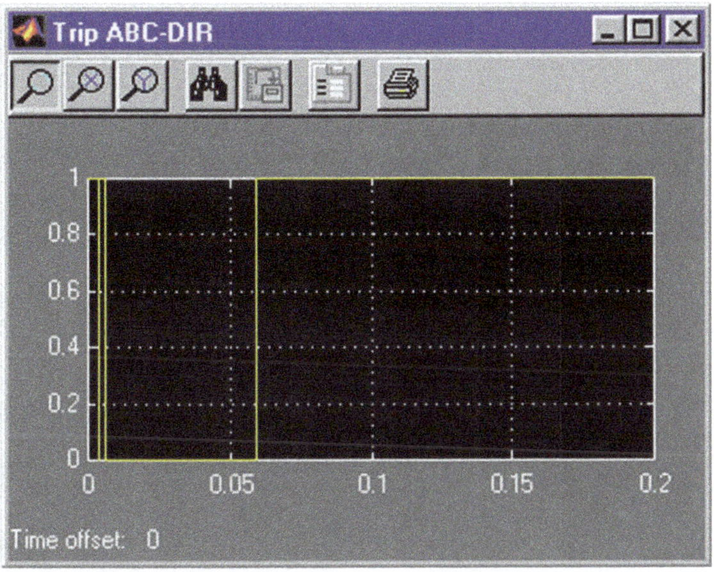

Fig. 6.28 Trip for directional relay using alpha = −60° and beta = 120°

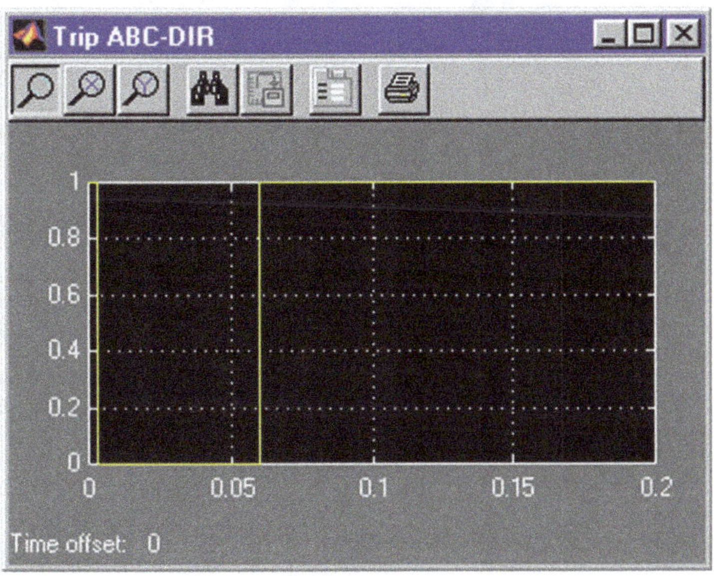

Fig. 6.29 Trip for directional relay using alpha = −50° and beta = 120°

In this approach the connection command gives a signal (high or 1) to close the breakers and connect the power source to the transmission line. When a fault appears and is detected by any relay, there will be a trip signal, which will reset

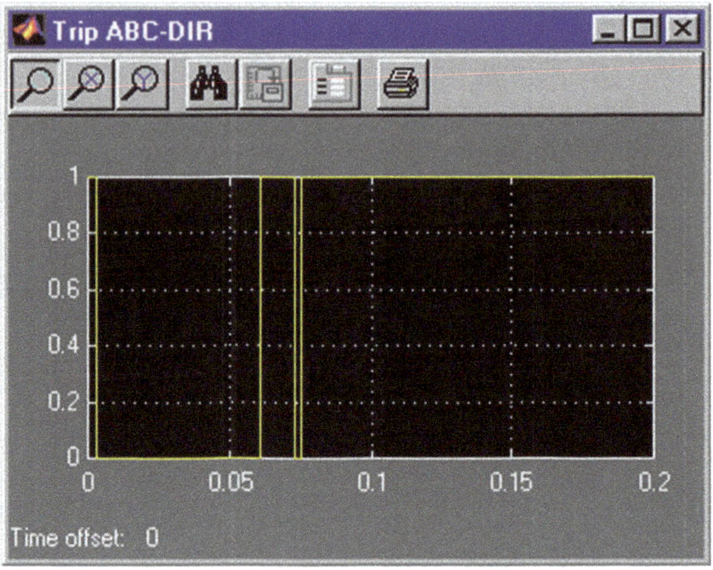

Fig. 6.30 Trip for directional relay using alpha = −40° and beta = 120°

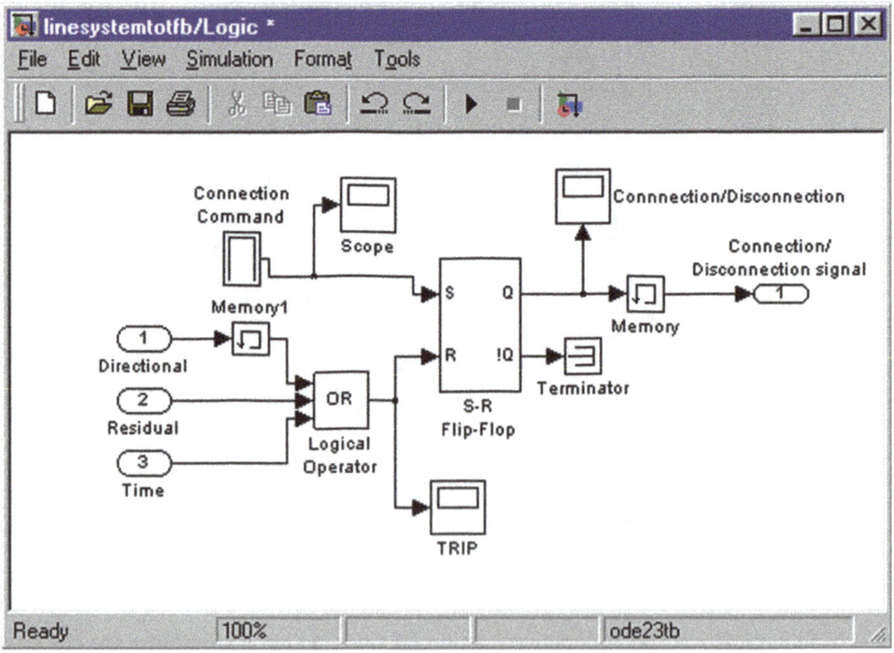

Fig. 6.31 Logic array to overcome the false trip of the directional relay

the flip-flop giving a disconnection command (low or 0) to open the breakers. This system is working as a "closed loop" system.

6.3 Line Protection System: Differential Relaying

6.3.1 Introduction

This section shows the design of a line protection system for two-terminal line with the following functions:

- Current differential element (phasors) with compensation for the line capacitance. Two-slope bias characteristic and harmonic restraint should be used.
- Six-element five-zone (four forward and one backward) impedance relay with the choice of three different zone shapes.

The measuring algorithms, operating principles, and logic will be selected, assuming the following signals are available: I_A, I_B, I_C—local currents, V_A, V_B, V_C—local voltages, I_{AR}, I_{BR}, I_{CR}—remote end currents (phasors) compensated for half of the line capacitances. Assume perfect sampling synchronization.

6.3.2 Theoretical Background

6.3.2.1 Power System Model

Consider the following simple power system (Fig. 6.32):

Two equivalent sources are connected at the two ends of the transmission line. For the transmission line, lumped parameter model will be used. Two protection systems are installed at two ends of the transmission line, the main protection is differential protection and the back-up protection is distance protection.

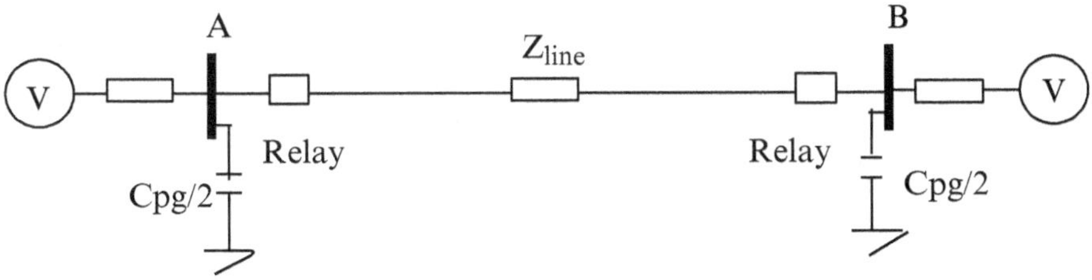

Fig. 6.32 Power system model

6.3.2.2 Differential Protection Scheme

The differential protection scheme for a transmission line is shown as Fig. 6.33. The relay operates on the sum of the currents flowing in the CTs' secondary side, $I_1 + I_2$. For normal operation or an external fault, the currents in the two CTs will be equal in magnitude and opposite in phase, thus there will be no current flowing in the relay operating coil and the relay will not operate.

If a fault occurs within the protected section between the two CTs, a current will flow through the operating circuit causing the relay to issue a trip signal.

To improve the selectivity and security of the current differential scheme, a bias characteristic is often been utilized. In a relay with bias characteristic, the operating current is the vector sum of the CTs' currents.

$$I_{\text{operate}} = |I_1 + I_2|$$

This operating current must be larger than the restraint current. This is a percentage of the sum of the magnitude of the individual CTs' currents, which is:

$$I_{\text{restraint}} = K[|I_1 - I_2|]$$

The operating characteristic of the bias current relay is shown in Fig. 6.34, which is a single slope characteristic.

In the scheme shown later in the simulation, a two-slope characteristic (Fig. 6.35) is used instead. The reason is that a two-slope bias characteristic improves the security of the current differential relay for external faults to the protected zone. This is particularly advantageous because the CTs may not accurately reproduce the primary fault currents under transient fault conditions.

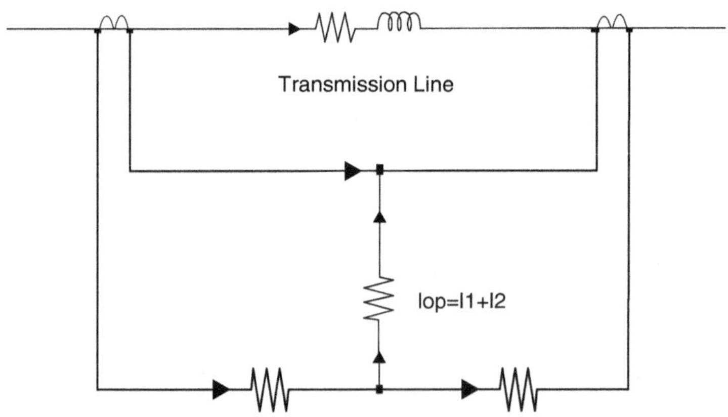

Fig. 6.33 Differential protection scheme

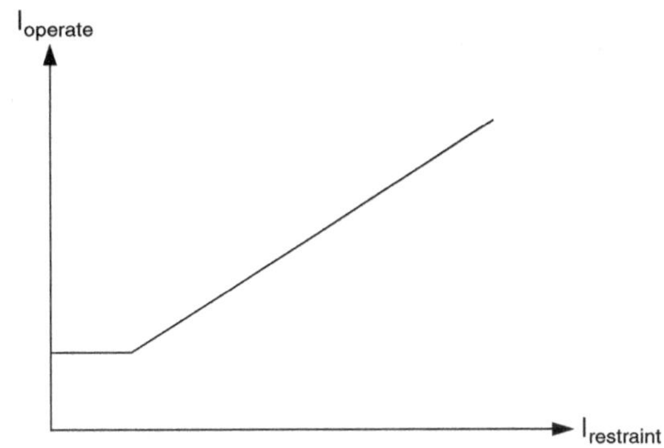

Fig. 6.34 Bias operating characteristic

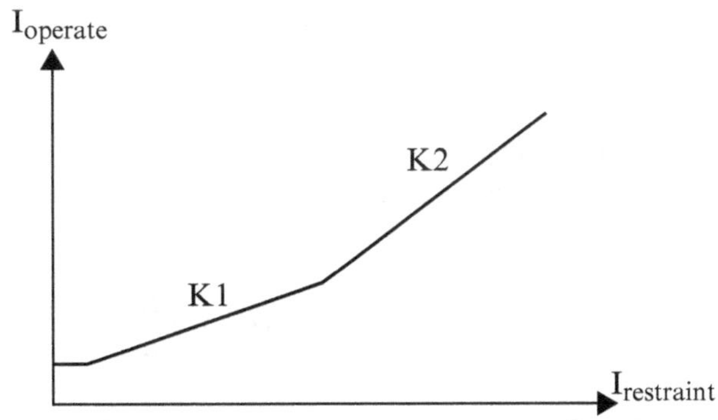

Fig. 6.35 Two-slope bias operating characteristic

The two-slope characteristic is a form of adaptive restraint in which the magnitude of the restraint quantity is increased for high current where CT accuracy is worse and its saturation becomes more probable.

The basic premise for the operation of differential protection scheme is that the sum of the currents entering the protection zone is zero when there is no fault or when an external fault occurs. But for a transmission line that may not be the actual case due to the capacitive charging current of the line. For short overhead transmission lines, the charging current can be treated as a small unknown error and is covered by the bias characteristic. For long transmission lines, the charging current will be too large to be treated as an unknown error and must be compensated.

In paper [1], an adaptive approach for charging current compensation was developed by subtracting a $C \times \mathrm{d}v/\mathrm{d}t$ term from the measured current at each terminal of the protected line.

The three-phase model at a terminal for charge compensation is shown in Fig. 6.36. Both phase-to-phase and phase-to-ground capacitance are shown. The system is assumed to be balanced so that the capacitance is the same for each phase.

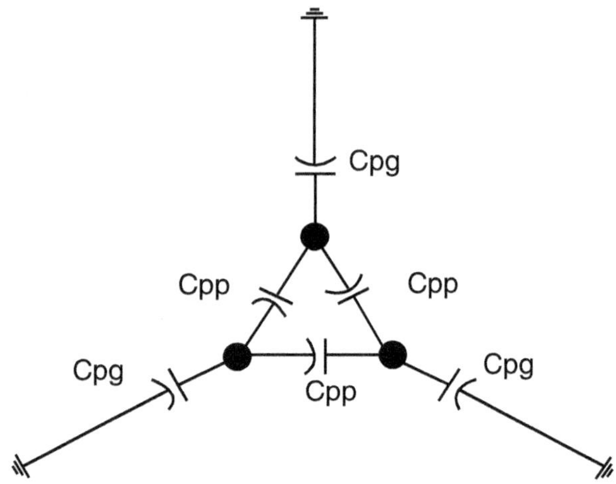

Fig. 6.36 Model for compensation

C_{pg} is the phase-to-ground capacitance, while C_{pp} is the phase-to-phase capacitance. The compensation for each phase involves all three phases. For example, the compensation for phase A is given by

$$C_{pg}\frac{dv_a}{dt}+C_{pp}\left(2\frac{dv_a}{dt}-\frac{dv_b}{dt}-\frac{dv_c}{dt}\right)$$

Since only local voltage signals are available, the charging current compensation will be conducted at the local relay as follows:

$$\begin{aligned}
I_{compensated,a} &= I_a - C_{pg}\frac{dv_a}{dt}+C_{pp}\left(2\frac{dv_a}{dt}-\frac{dv_b}{dt}-\frac{dv_c}{dt}\right)\\
I_{compensated,b} &= I_b - C_{pg}\frac{dv_b}{dt}+C_{pp}\left(2\frac{dv_b}{dt}-\frac{dv_a}{dt}-\frac{dv_c}{dt}\right)\\
I_{compensated,c} &= I_c - C_{pg}\frac{dv_c}{dt}+C_{pp}\left(2\frac{dv_c}{dt}-\frac{dv_a}{dt}-\frac{dv_b}{dt}\right)
\end{aligned}$$

6.3.2.3 Distance Protection Scheme

The distance protection is implemented as a five-zone (four forward and one backward zones) impedance relay (Fig. 6.37).

If the computed distance falls inside zone 1, which is usually set to 80 % of the protected line, the relay should operate immediately, but since the relay works in backup mode, it should operate with a time delay Δt. If the computed distance falls inside zone 2, which is usually set to 120–125 % of the protected line, the relay should operate with a time delay $\Delta t + t_2$. For a fault located in zone 3, which is up to the next adjacent bus, the relay should operate with a delay $\Delta t + t_3$. For a fault located inside zone 4, which extends even farther, the relay should operate with a

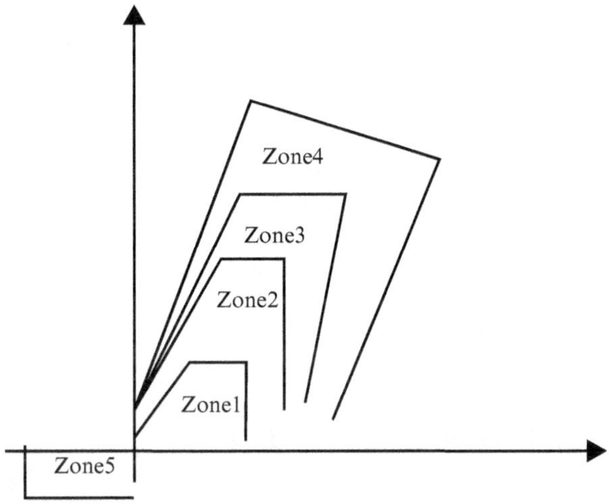

Fig. 6.37 Five protection zones

delay of $\Delta t + t_4$. The backward fault detection is set to detect faults that occur "behind the bus," which is in the direction pointing to the rest of the system.

The method used to calculate the impedance is based on the differential equation. The Fourier filter is used to remove some unwanted frequencies from the input signal. The form of Fourier filter is:

$$y(n) = \sum_{k=0}^{p} a_k \times x(n-k)$$

where $a_k = \sin\left[\omega_0 T_s\left(\frac{p}{2} - k\right)\right]$, p is the number of samples per cycle and T_s is the sample time.

The equation to solve is the following:

$$Ri + L\frac{\mathrm{d}i}{\mathrm{d}t} = V$$

Using Least Square Method, the equation has the following form:

$$\int_{n-d+1}^{n} \left(Ri + L\frac{\mathrm{d}i}{\mathrm{d}t} - V\right)^2 = 0$$

where d is the number of samples in the data window.

$$S = \sum_{u=n-d+1}^{n} \left[Ri(u) + \frac{L}{T_s}(i(u) - i(u-1) - v(u)) \right]^2$$

From the conditions $\frac{\partial S}{\partial R} = 0$ and $\frac{\partial S}{\partial L} = 0$ the expressions for R and L can be obtained. To solve the equation, Euler method is used.

In this document, a line protection system using current differential relays with compensation for the line capacitance and a six-element five zone back-up protection is described.

6.3.3 Simulation Models

6.3.3.1 Differential Relay

The relay is shown in Fig. 6.38, and its parameter dialog box is shown in Figs. 6.39. The relay consists of two units: current differential protection (primary protection) and distance protection (back-up protection). The two units are shown in Fig. 6.40.

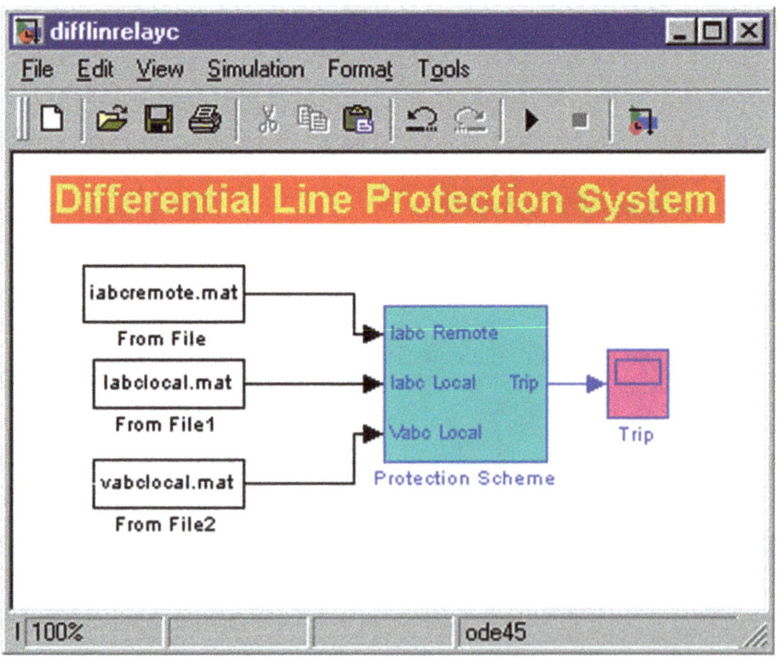

Fig. 6.38 Differential line protection relay

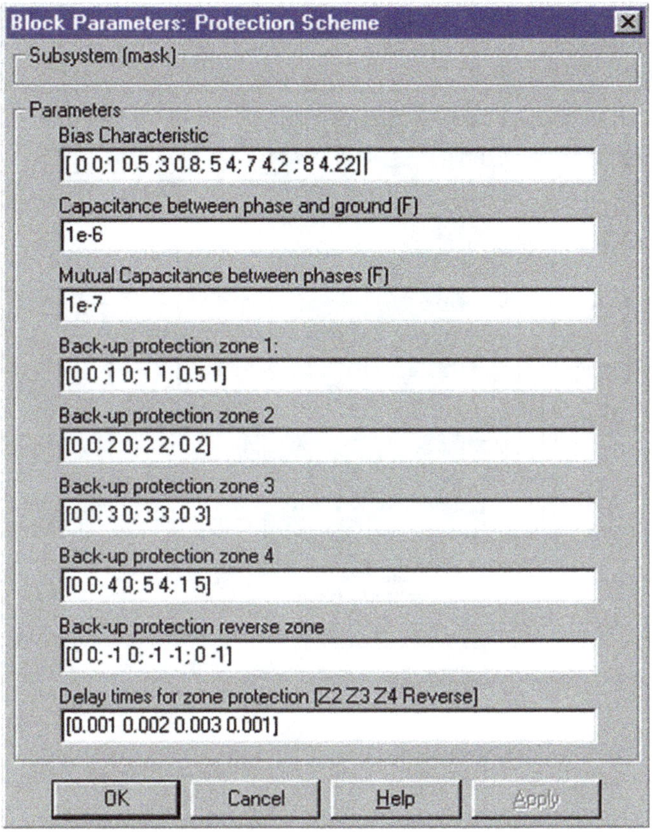

Fig. 6.39 Differential relay parameters dialog box

6.3.3.2 Current Differential Protection

The differential protection scheme is shown in Fig. 6.41.

The internal structure of the phase differential protection is shown in Fig. 6.42.

The charging compensation model is shown in Fig. 6.43.

This model is designed to compute a current to compensate the effects of the capacitance charging current:

$$i_{\text{compensation, a}} = C_{\text{pg}} \frac{\mathrm{d}v_{\text{a}}}{\mathrm{d}t} + C_{\text{pp}} \left(2 \frac{\mathrm{d}v_{\text{a}}}{\mathrm{d}t} - \frac{\mathrm{d}v_{\text{b}}}{\mathrm{d}t} - \frac{\mathrm{d}v_{\text{c}}}{\mathrm{d}t} \right)$$

After subtracting the compensating current from the local current, the operating and restraining signal will be formed and input into the bias characteristic.

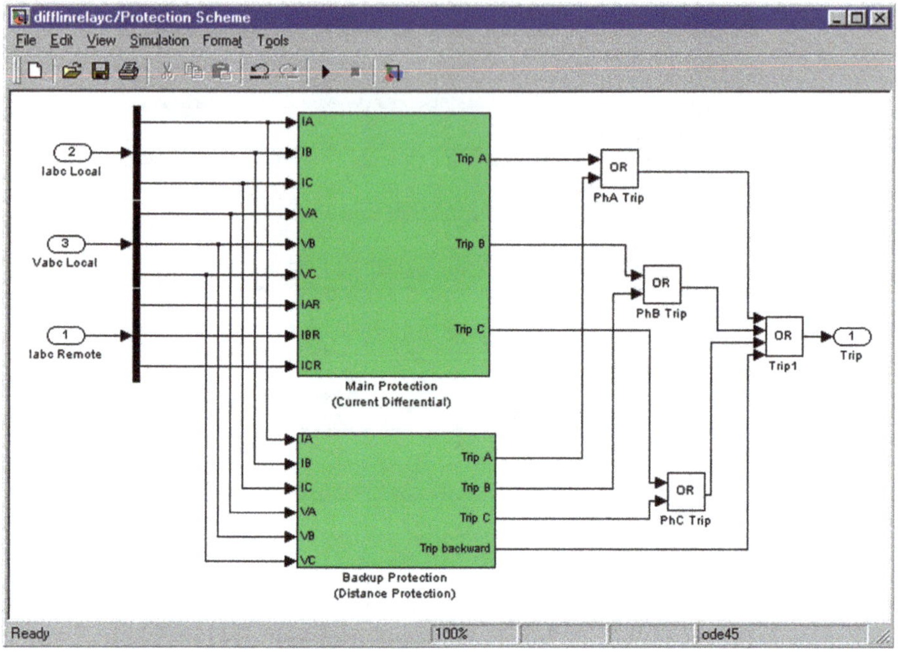

Fig. 6.40 Differential relay components

6.3.3.3 Distance Protection

The distance protection scheme is shown in Fig. 6.44. If the fault falls in zone 1, no time delay will be added (since differential relay inherently works faster than the distance relay). Otherwise some kind of time delay will be applied.

Though the pickup signal is issued by individual phases, the tripping signal will operate on the three breakers.

6.4 Line Protection System: Zone Protection

6.4.1 Introduction

In this section, a transmission line protection system for two-terminal lines is described. The design considers the following functions:

- Six-element five-zone (four forward and one backward) impedance relay with the choice of three different zone shapes.
- Power swing blocking element.
- Directional element (memory polarization).
- Logic for selected communication scheme (both sending and receiving signals).

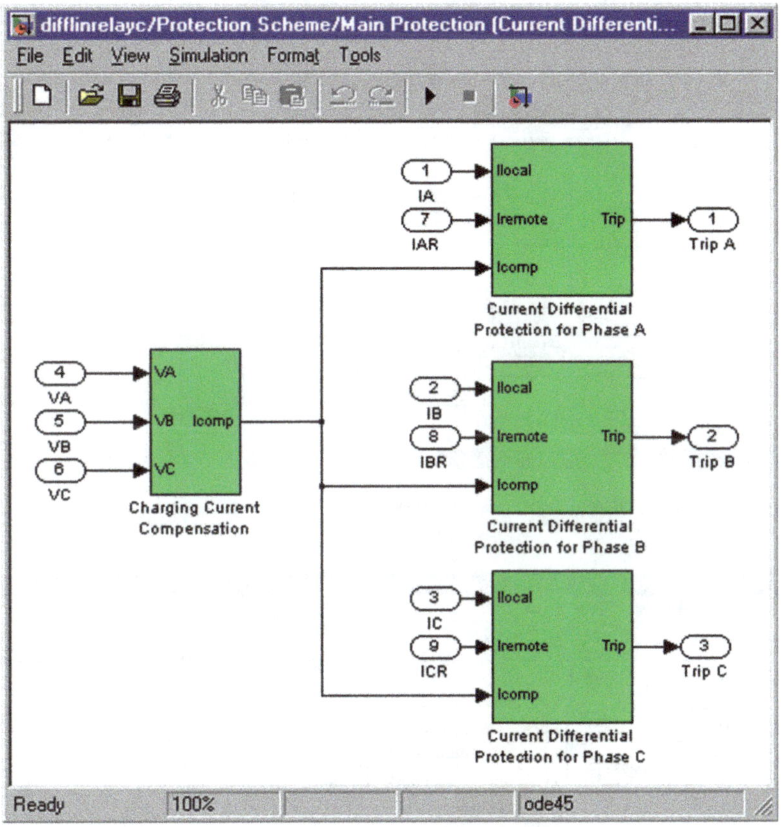

Fig. 6.41 Protection scheme

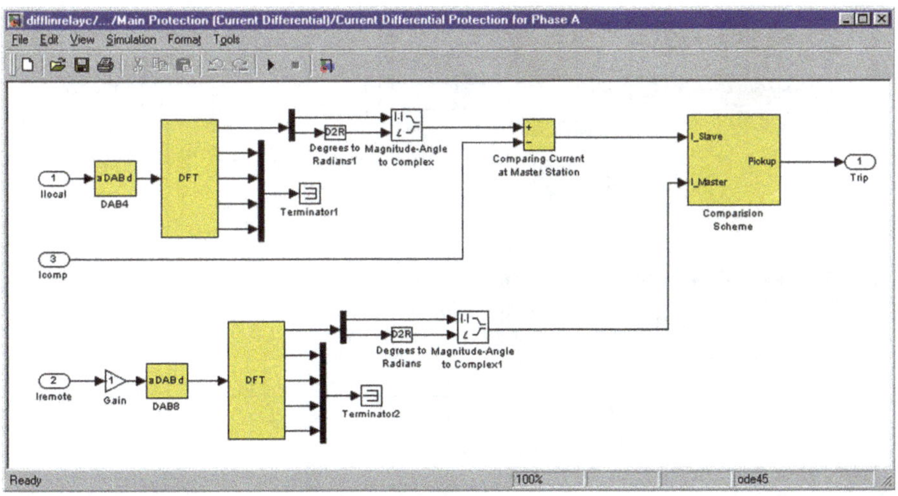

Fig. 6.42 Internal structure of differential protection

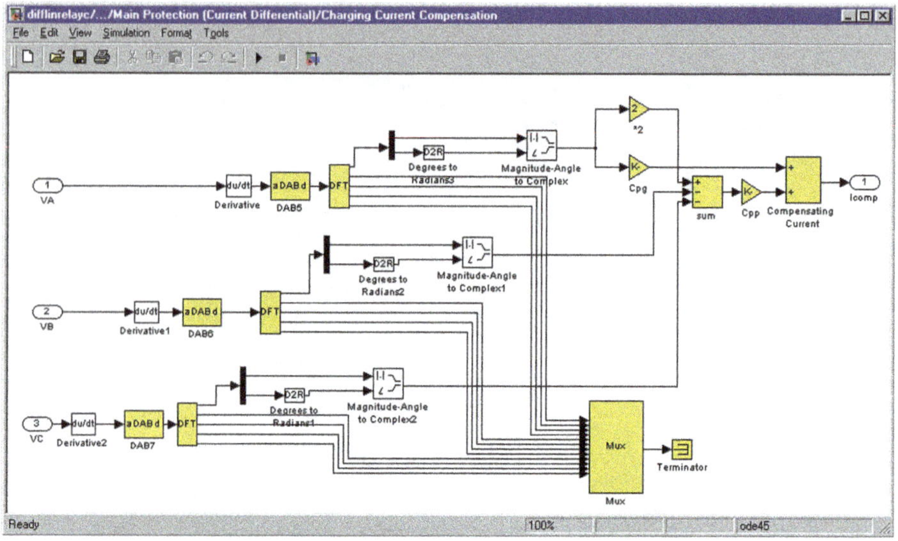

Fig. 6.43 Charging current compensation model

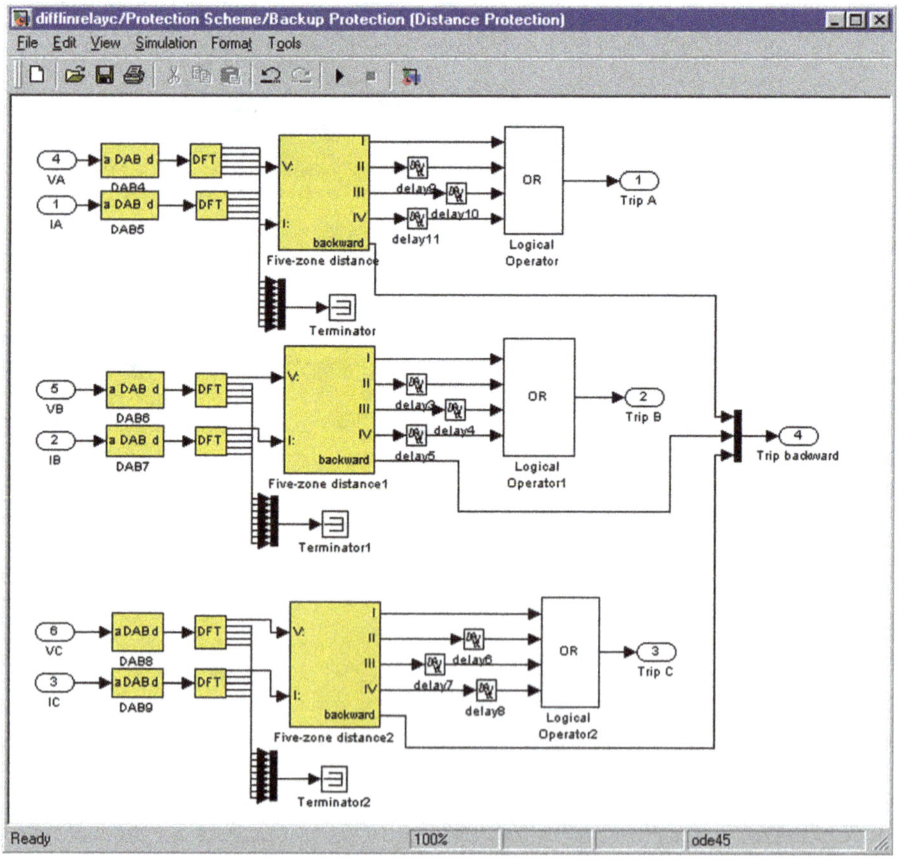

Fig. 6.44 Distance protection scheme

- Three-phase inverse-time directional overcurrent relays for the back-up protection.

Next, a line protection system using impedance relays with power swing blocking element and directional element as well as inverse-time overcurrent protection for the back-up protection is described.

6.4.2 Simulation Models

6.4.2.1 Relay Design

This protection system is composed of three main elements (see Fig. 6.45):

- Measurements
- Relay logic
- Trip logic

6.4.2.2 Measurement Block

The measurement block collects and processes data from the Test Set model. The inputs are the three-phase currents and voltage signals in time domain. The outputs are the single phase and phase-to-phase current and voltage phasors (Fig. 6.46). The other blocks of the protection system will use these phasors.

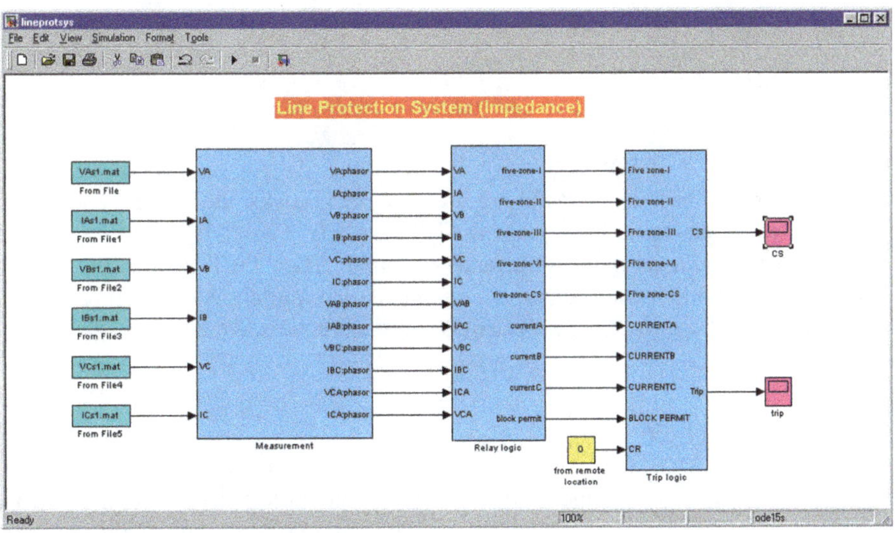

Fig. 6.45 Protection system overall structure

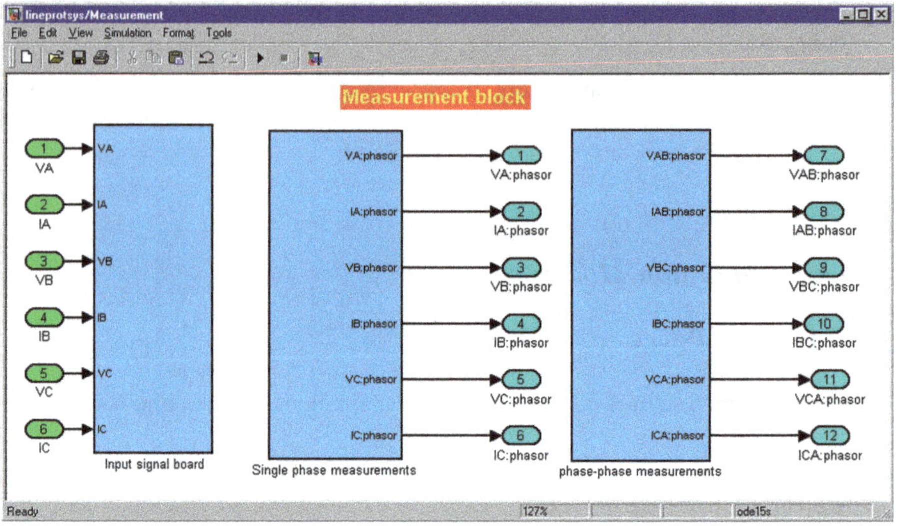

Fig. 6.46 The measurement block

The following blocks compose the measurement block: input signal board, single-phase measurements, and phase-phase measurements. In the input signal board (Fig. 6.47), phase currents and voltages are collected and the zero compensation components of the phase current measurements are calculated. The quantities used in relay block are:

For single-phase faults:

$I_{\text{measure}} = I_{\text{phase}} + (k \times I_{\text{zero}})$.
$V_{\text{measure}} = V_{\text{phase}}$. k is a coefficient.

For phase–phase faults:

$I_{\text{measure}} = I_{\text{phase1}} - I_{\text{phase2}}$
$V_{\text{measure}} = V_{\text{phase1}} - V_{\text{phase2}}$.

In the single phase and multiphase measurement blocks, the "DAB" and "DFT" from the relay elements library are used (Fig. 6.47). The "DAB" block performs A/D conversion, analog filtering, and sampling. The "DFT" block obtains the fundamental and harmonic phasor components of input signals. For this application, only the phasor of the fundamental component is used (see Fig. 6.48).

6.4.2.3 Relay Logic

The relay logic block includes all the relay protection function elements listed above. These elements are listed below. The Internal structure of this block is shown in Fig. 6.49.

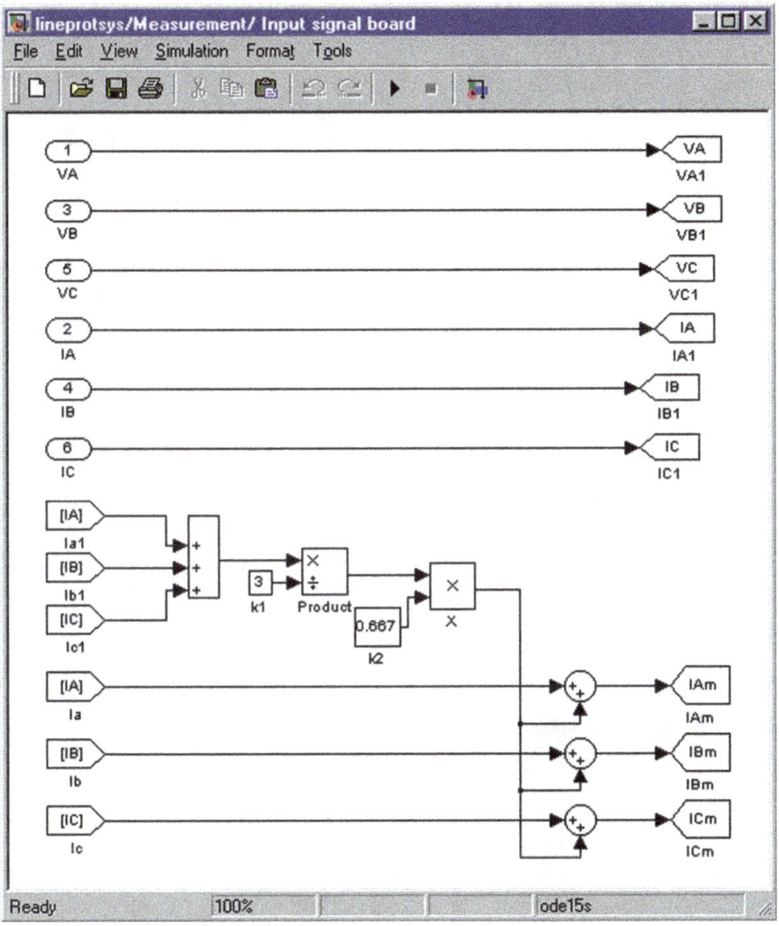

Fig. 6.47 Internal structure of the input signal board

6.4.2.4 Six-Element Five-Zone Distance Relay

The internal structure of the six-element five-zone relay is shown in Fig. 6.50.

6.4.2.5 Five-Zone Distance Block

The Five-zone distance block (Fig. 6.51) calculates the phasor of the impedance by using the "BM" block. Then the resistance and reactance components of the impedance are extracted. The judgment whether an impedance is located in a given zone is done by the zone comparison (ZC) block. For example, if the calculated impedance is within the first zone, then the first output corresponding the first zone trip will issue a "1" signal.

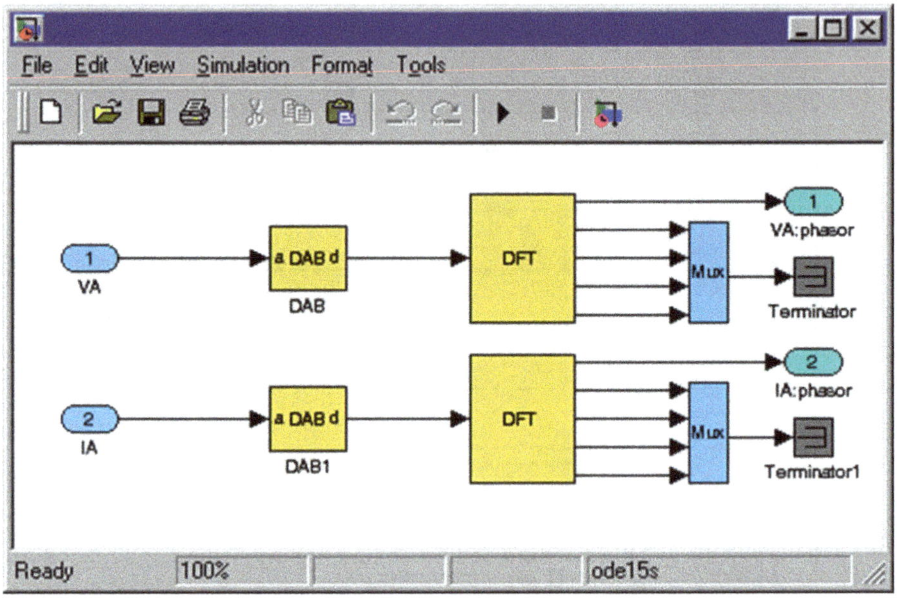

Fig. 6.48 Phase a voltage and current measurements

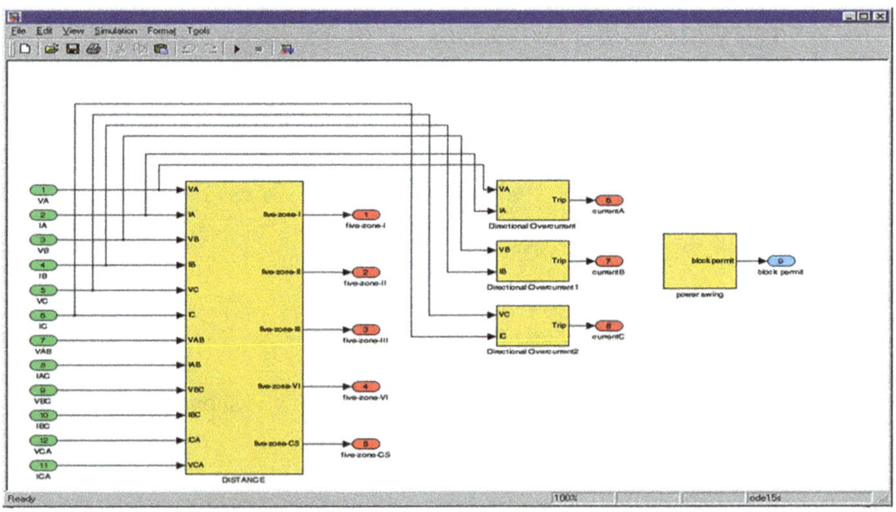

Fig. 6.49 Internal structure of the relay logic block

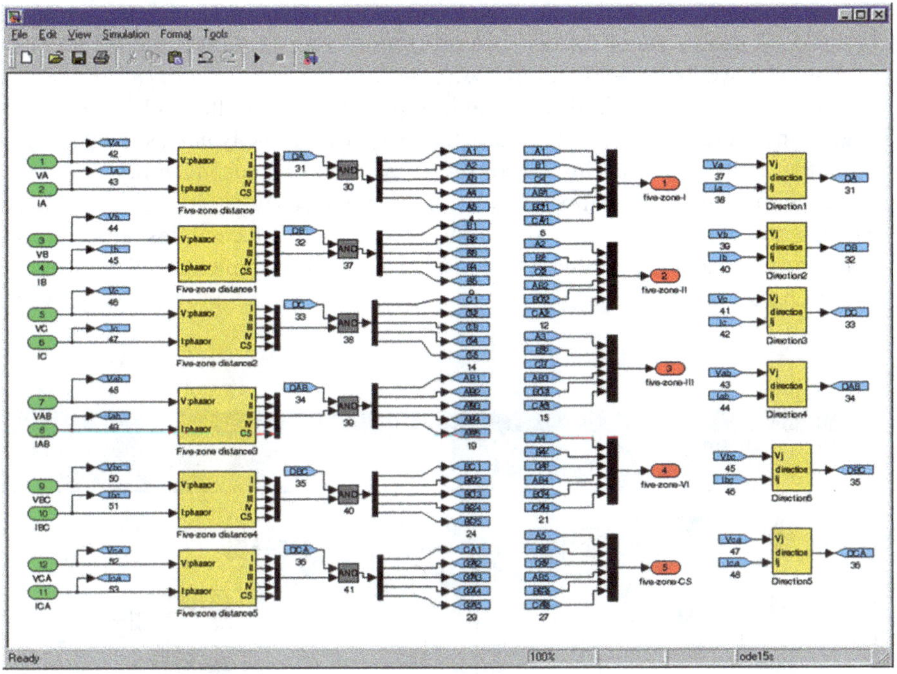

Fig. 6.50 The five zone elements and directional elements

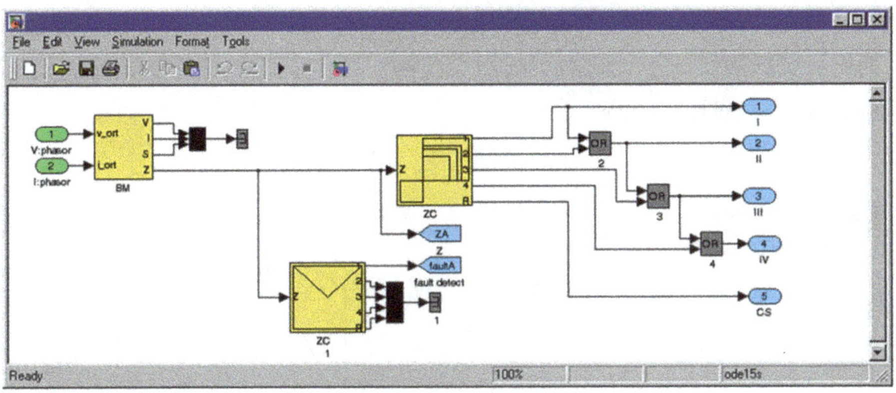

Fig. 6.51 Internal structure of the five-zone distance element

6.4.2.6 Directional Element

When a fault occurs in a location very close to the relay measurements, the voltage that will be measured by the relay can be almost zero. This may cause the relay to block a trip. In this case, the memory polarization block memorize the pre-fault voltage and use it to calculate the impedance so its calculated value is not too small. In this way, dead zone of the distance protection is avoided. See Fig. 6.52.

Normally we have: $-90 < \arg[(I_j \times Z_{set} - U_j)/U_j] < 90$.

We can see in Fig. 6.53 that if a fault impedance is zero, the fault is just on the edge of the protection zone. Therefore, the relay will probably not issue a trip signal when that fault happens. This causes the relay to lose its dependability.

In the memory polarization scheme, U_j is replaced by U_p in the denominator, which is the pre-fault voltage. So the impedance will be located inside the protection zone when a fault happens. The equation to do this is implemented in the model shown in Fig. 6.52.

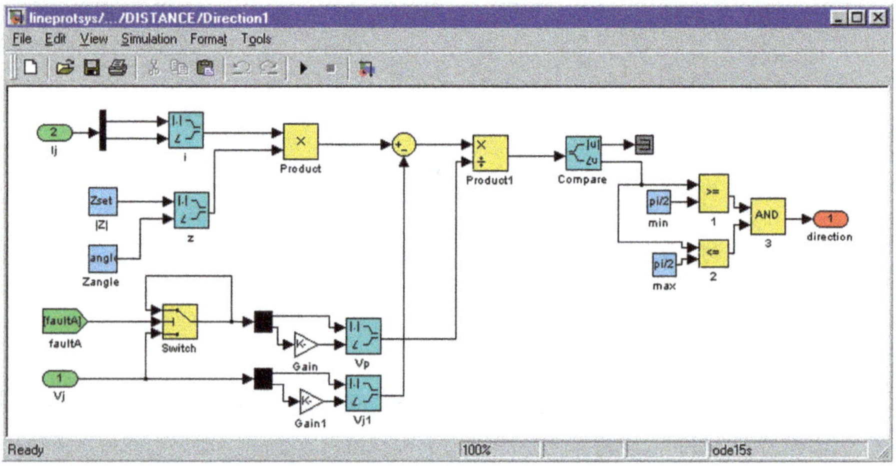

Fig. 6.52 Directional element

Fig. 6.53 Impedance characteristic

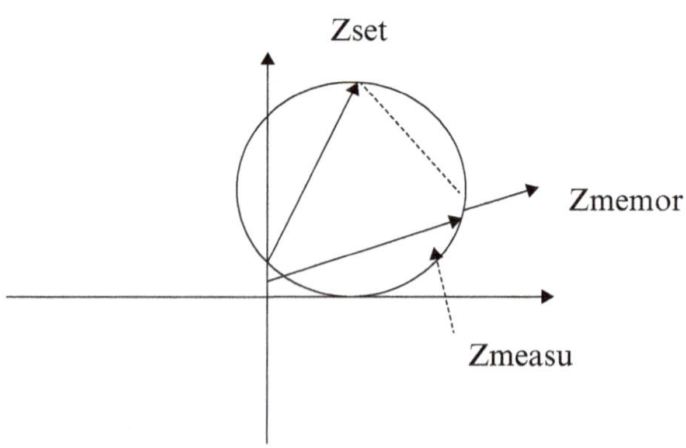

6.4.2.7 Power Swing Block

The internal structure of the power swing block is shown in Fig. 6.54.

The power swing is a big problem in real power systems. When a power swing happens, the impedance that the relay "senses" will travel through the protected zones, but it is not a real fault. If the impedance goes through the first zone, the first-zone distance protection will trip the breaker causing an unexpected trip. In order to avoid this misoperation, a power swing block scheme must be applied. This block can detect the power swing and block the distance relay from tripping. Normally, the power swing has no sudden changes in electric vectors. So we can realize the detection according to the changing time of the impedance values.

When a power swing happens, the impedance that a relay senses may go through the path shown in the Fig. 6.56 above. It travels through the protection zones slowly, from the outer zone to the inner zone as time passes. Normally the time elapsed is much bigger than when a real fault occurs. So we can calculate the elapsed time t and compare it with a threshold. If t is bigger than a threshold, then a power swing is supposed to happen and a power swing block signal is issued. Although some zone elements will send trip signals, the overall trip signal is not sent out.

The model shown in Fig. 6.55 memorizes the time t_1 (time instant in which the impedance begins to enter the first zone) and time t_2 (time in which the impedance begins to enter the second zone). Then it compares the difference of the two times with the threshold, and then the power swing will be detected.

The time sequence is shown in Fig. 6.57.

Furthermore, the relay has a communication scheme, which works as follows: if a fault is located in the backward zone, the relay will send a CS signal to the remote

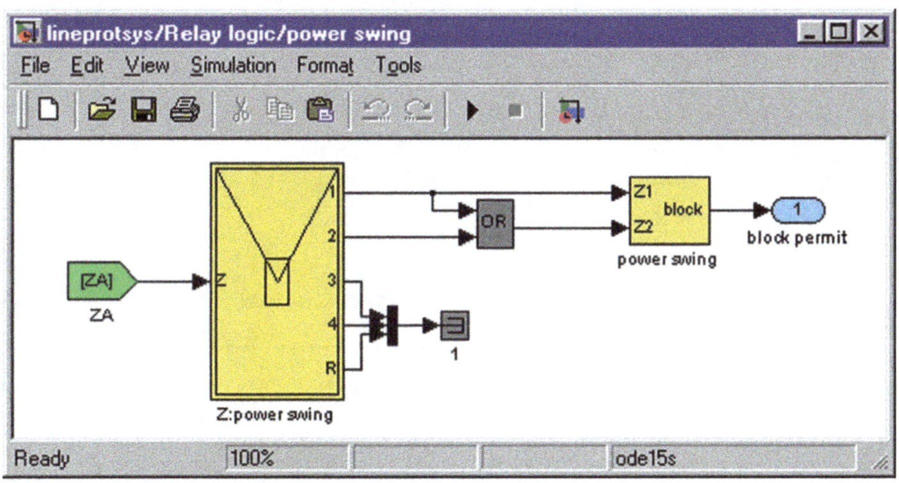

Fig. 6.54 Detection of power swing

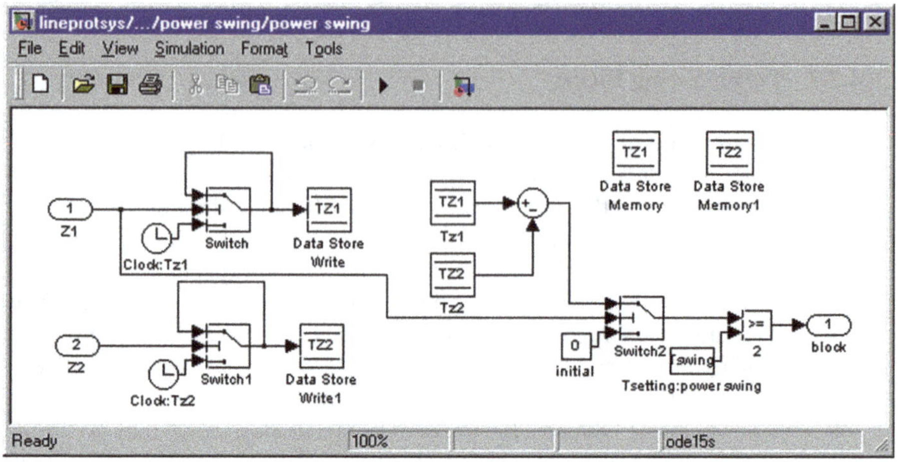

Fig. 6.55 Calculation of the difference of T1 and T2

Fig. 6.56 Protection zones and apparent impedance path

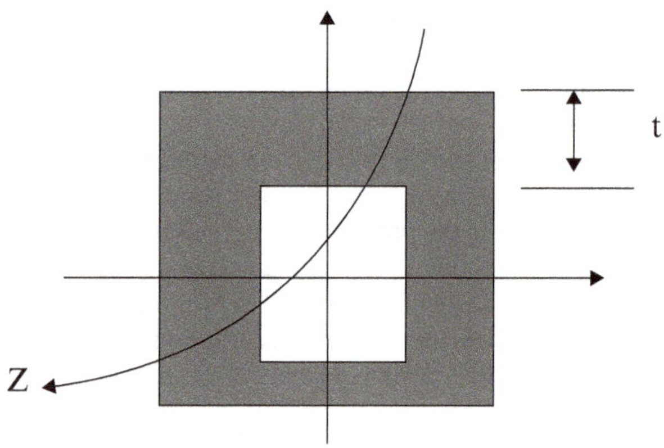

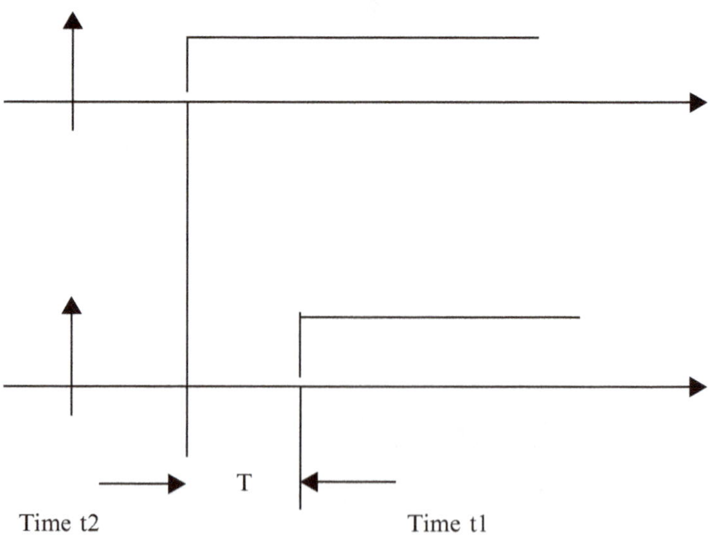

Fig. 6.57 Operation time sequence for impedance relaying

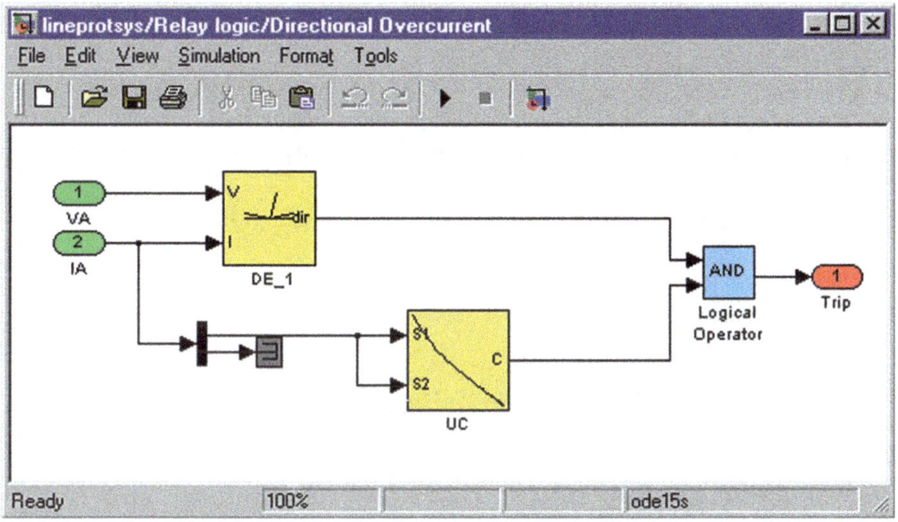

Fig. 6.58 The directional currents

relay in order to block it, because it is an external fault. And it also receives the same signal from the remote relay, in order to judge whether to trip.

6.4.2.8 Directional Overcurrent Blocks

First, the input voltage and current phasors are used to detect the direction of a fault, to see whether it is in the forward direction. The output is "one" when the current phasor falls into the operating characteristic and "zero" otherwise. Second, the "UC" block and its function "signal-to-time" are used. It has time inverse characteristic. The bigger the current is, the quicker the time of response will be (Fig. 6.58).

6.5 Line Protection System: Pilot Protection

6.5.1 Introduction

Next, a transmission line protection system for two-terminal lines is described. The design considers the following functions:

- Pilot logic for typical communication schemes
- All the measuring functions needed by the pilot logic (measuring zones, undervoltage elements, etc.).

The distance relay scheme offers the following functions:

- Three separate MHO forward-sensing zones for multi-phase faults
- Three separate "quadrangular" forward-sensing zones for phase-to-ground faults
- One MHO reverse-sensing zone for multi-phase faults
- One reverse-sensing zone with Quadrilateral (Quad) characteristic for phase-to-ground faults
- Six separate MHO starters, one for each fault-measured loop,
- Undervoltage element
- Permissive Underreaching (PUTT) and blocking overreaching (BLOV + TB) pilot logics.
- CR and CS logic signals.

In this example a distance protection relay that uses pilot logic is described.

6.5.2 Relay Design

The main block of the relay is shown at Fig. 6.59. The relay has the following inputs:

- Phase currents signals coming from CT's
- Phase voltages signals coming from VT's
- CR (carrier receive) signal from the relay at the other end of the protected transmission line.

The main outputs of the relay are:

- TRIP signal for the Circuit Breaker
- CS (carrier send) signal

Additionally, other outputs allow the user to follow what is happening inside the relay. These signals are:

- Fault type—indicates which phases are faulted
- Faulted zone—indicates the zone where the fault occurred
- Logic signals—internal signal from tripping logic module
- R_f, X_f—curves of measured impedance components of the faulted-loop: R_{fault} and X_{fault}
- V_f, I_f—voltage and current signals after analog preprocessing

The settings of the relay are given in Fig. 6.60. User can set up the following:

- Reach of each zone, including reverse zone
- Operating time in each of the zones, including reverse zone
- Reach of starter and its operating time (backup, works as fourth zone)
- Characteristic angle of the line
- k_o—coefficient, for current compensation in case of phase-ground faults
- Threshold for undervoltage element.

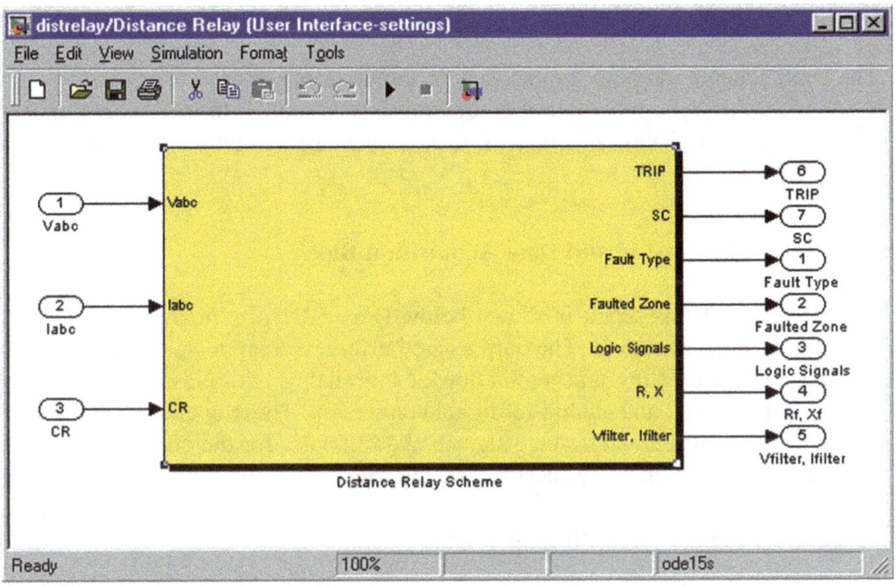

Fig. 6.59 Distance relay inputs and outputs

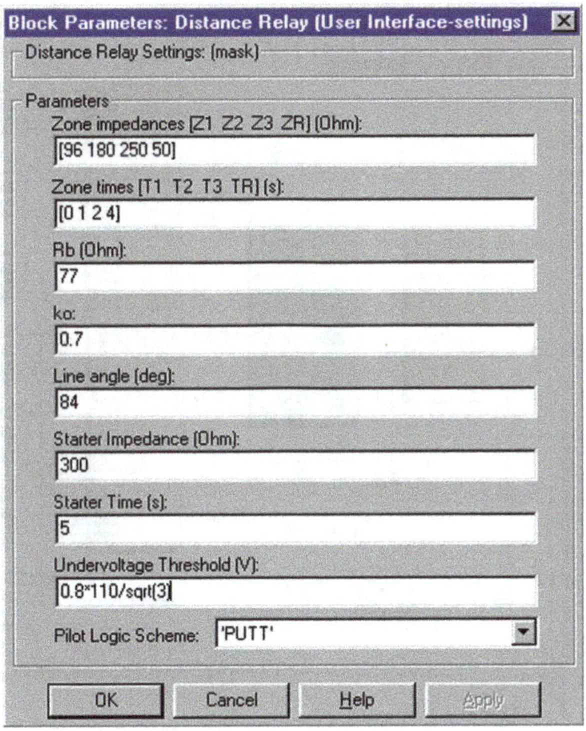

Fig. 6.60 Distance relay settings

6.5.2.1 Scheme Structure

The main scheme structure of the relay is shown in Fig. 6.61.

As it can be seen, the whole structure consists of smaller blocks, which perform specific functions. These blocks will be described separately.

6.5.2.2 Preprocessing and Data Acquisition Block

The structure of this block is shown below (Fig. 6.62). Its inputs are the analog voltage and current signals. The outputs are buffers with input signal samples. This block enables modeling and simulation of the analog anti-aliasing filter, analog signal conditioning, and analog-to-digital conversion. These operations are applied to each phase current and voltage signals. Additionally for the current signals, the coefficient k_o multiplied by the residual current is computed based on the phase current signals.

The default settings for this block are:

- Cut-off frequency for analog low-pass filter: 250 Hz
- Analog low-pass filter: second-order Butterworth approximation

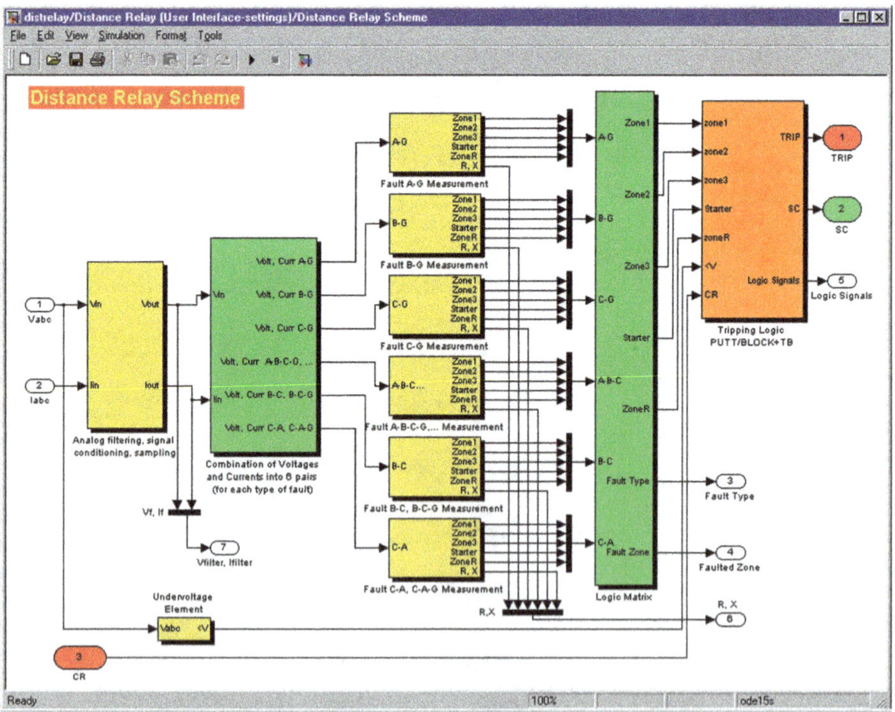

Fig. 6.61 Relay main scheme structure

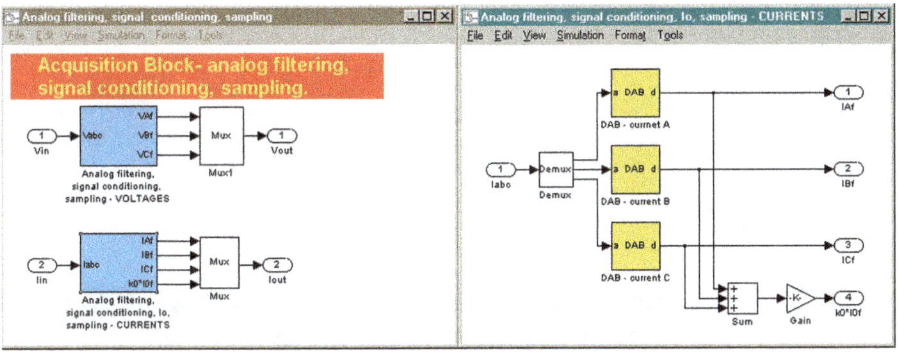

Fig. 6.62 Preprocessing block

- Sampling frequency: 1920 Hz (32 samples/cycle)
- Window size: 10 (The window contains the ten previous samples of the signal)
- Vertical resolution of A/D: 16 bit
- Current vertical range: 80 A ($I_n = 5$ A)
- Voltage vertical range: 80 V, ($V_n = 69$ V)

6.5.2.3 Voltage and Current Combination Block

The function of this block is to provide all appropriate combinations of phase voltages and currents needed by impedance-measuring elements block (next block). The inputs for this block are the phase voltages and currents, as well as the residual current kI_o. The outputs are the six combined voltage–current pairs (Table 6.3 and Fig. 6.63).

6.5.2.4 Impedance Measurement and Zone Comparison Blocks

The purpose of these six similar blocks (three for phase-to-ground fault cases, and three for multi-phase fault cases) is primarily to measure the impedance of the faulted-loops and then perform the zone comparison function. Each individual block is designed to measure impedance in different loops according to Table 6.4. The input for each of these blocks (Fig. 6.64) is a unique pair of combined voltages and currents as described above. Each output corresponds to a zone, and if the measured impedance is within limits, it is set to high. Also, the values of R and X impedance components are taken out for monitoring purposes.

The measure of impedance is based on a differential equation model (DEIM), using Least Square Method and trapezoidal solution rule with a data window of ten samples. The basic equations of this method are:

Table 6.3 Output of the combination block

Pair #	Voltage	Current
1	V_a	$I_a + kI_o$
2	V_b	$I_b + kI_o$
3	V_c	$I_c + kI_o$
4	$V_a - V_b$	$I_a - I_b$
5	$V_b - V_c$	$I_a - I_b$
6	$V_c - V_a$	$I_a - I_b$

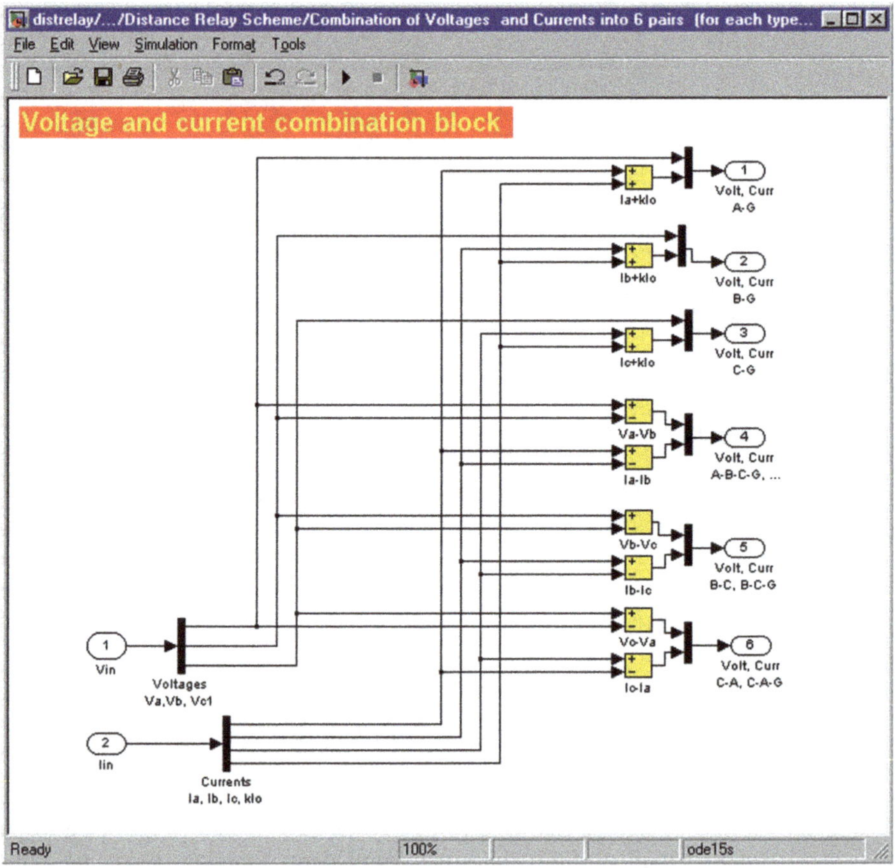

Fig. 6.63 Voltage and current combination block

Table 6.4 Fault type and quantities used for impedance measurement

Type of fault	Voltage	Current
A-G	V_a	$I_a + kI_o$
B-G	V_b	$I_b + kI_o$
C-G	V_c	$I_c + kI_o$
A-B-C-G, A-B-C, A-B	$V_a - V_b$	$I_a - I_b$
B-C, B-C-G	$V_b - V_c$	$I_a - I_b$
C-A, C-A-G	$V_c - V_a$	$I_a - I_b$

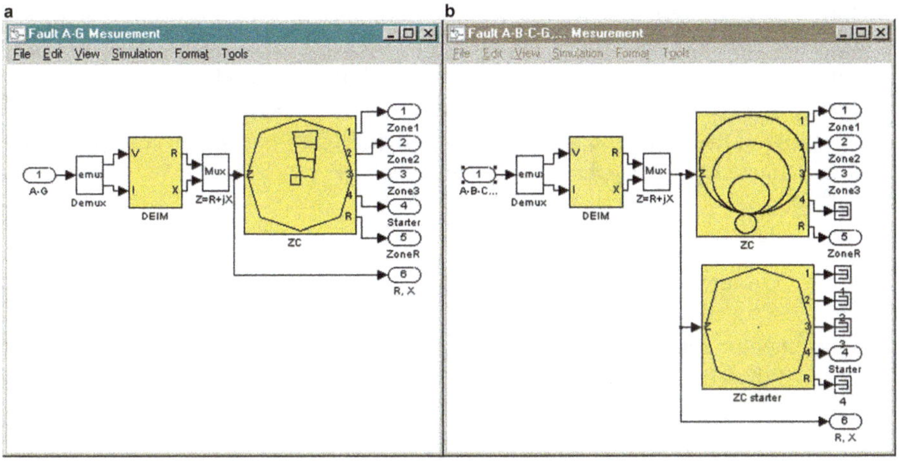

Fig. 6.64 Impedance components measurement elements with zone comparison function, (**a**) for phase-ground phase elements and (**b**) multi-phase elements

$$\int_{n-d+1}^{n} \left(R + L\frac{\mathrm{d}i}{\mathrm{d}t} - e\right)^2 = 0$$

where n is the sample index and d is the window size.

$$S = \sum_{u=n-d+1}^{u=n} \left[R\frac{T_s}{2}(i(u) + i(u-1)) + L(i(u) - i(u-1)) - \frac{T_s}{2}(v(u) + i(v-1))\right]^2$$

From the conditions $\frac{\partial S}{\partial R} = 0$, and $\frac{\partial S}{\partial L} = 0$, the expressions for R and L can be obtained, and subsequently the impedance to the fault point. This method is quite accurate and at the same time is not sensitive for DC components in input signals and also needs a relatively low number of computations.

The outputs from DEIM are used for zone comparison subblock. As we can observe in Fig. 6.64a, for the case of one phase-to-ground fault, there is only one comparison block. It determines if the measured impedance is within each zone. There are three forward Quad zones, one reverse zone, and one starter ("4th zone").

In case of multi-phase fault measurements, the situation is slightly different. There are three MHO forward zones, one MHO reverse zone, and one Quad starter element.

The six measurement elements work in parallel, giving six sets of outputs (six sets of zones), which have to be combined.

6.5.2.5 Logic Matrix Block

As it was mentioned above, it is needed to combine all six sets of outputs from the measurement elements into only one common set: zone 1, zone 2, zone 3, zone *R*, and Starter. This is the job of the block called "logic matrix". It creates the final faulted-zone and faulted-phase indicator signals (Fig. 6.65).

6.5.2.6 Undervoltage Element

This element compares phase voltages against a threshold, and if any of the phase voltages fall below this threshold, the "<V" output signal is activated (Fig. 6.66).

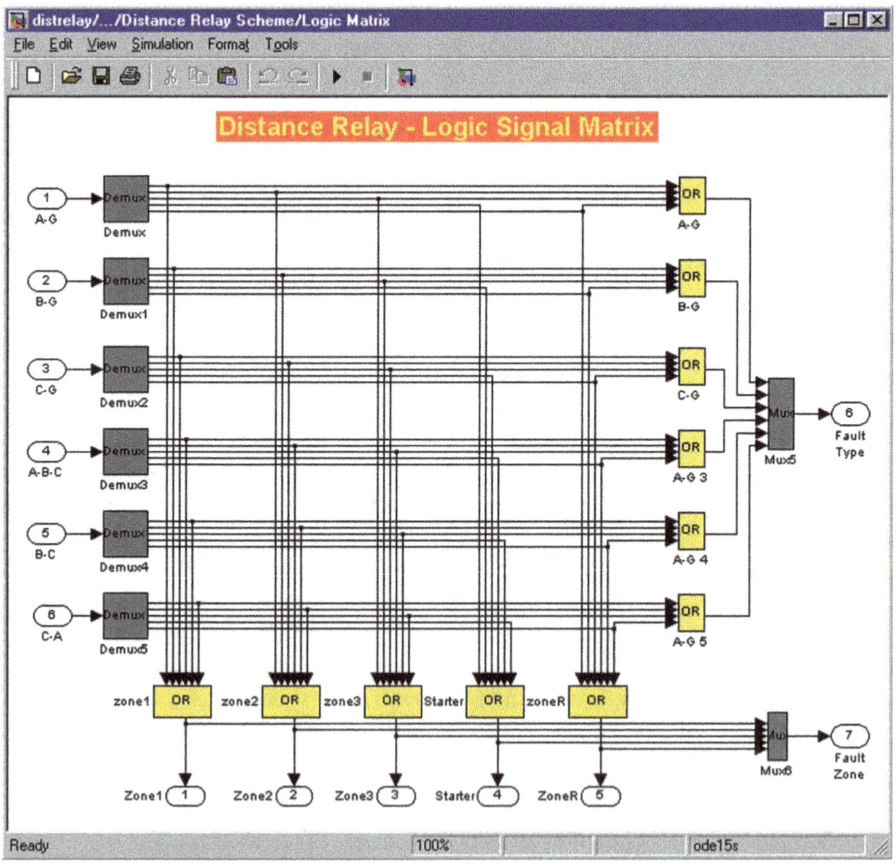

Fig. 6.65 Logic matrix block

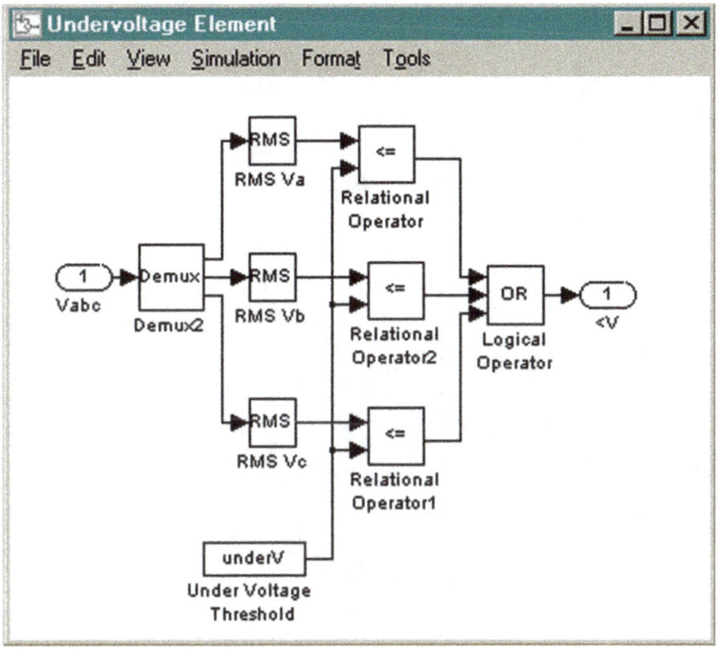

Fig. 6.66 Undervoltage element

6.5.2.7 Pilot Logic

In this relay two communication schemes are implemented. These are PUTT (Permissive Underreaching Transfer Tripping) and BLOV + TB (blocking overreaching with current reversal logic). The pilot tripping logic scheme is presented in Fig. 6.67.

PUTT

Zone 1 has an underreaching setting. It trips independently of the signaling channel and sends an enabling signal to the opposite end of the line. Upon receiving an enabling signal, the relay at the opposite end of the line trips if either the starters or the undervoltage relay has picked up.

This is a relatively simple scheme, which provides fast protection for the whole length of the protected line. Most of the line is still properly protected even if the signaling channel fails. Also in the case of a weak infeed from one end, tripping along 85 % of the line is fast and takes place at both ends.

$$\mathrm{TRIP} = \mathrm{ZM}_1 + (\mathrm{S} + \mathrm{U} <) \times \mathrm{CR} + \mathrm{ZM}_2 \times t_2 + \mathrm{ZM}_3 \times t_3 + \mathrm{S} \times t_S$$

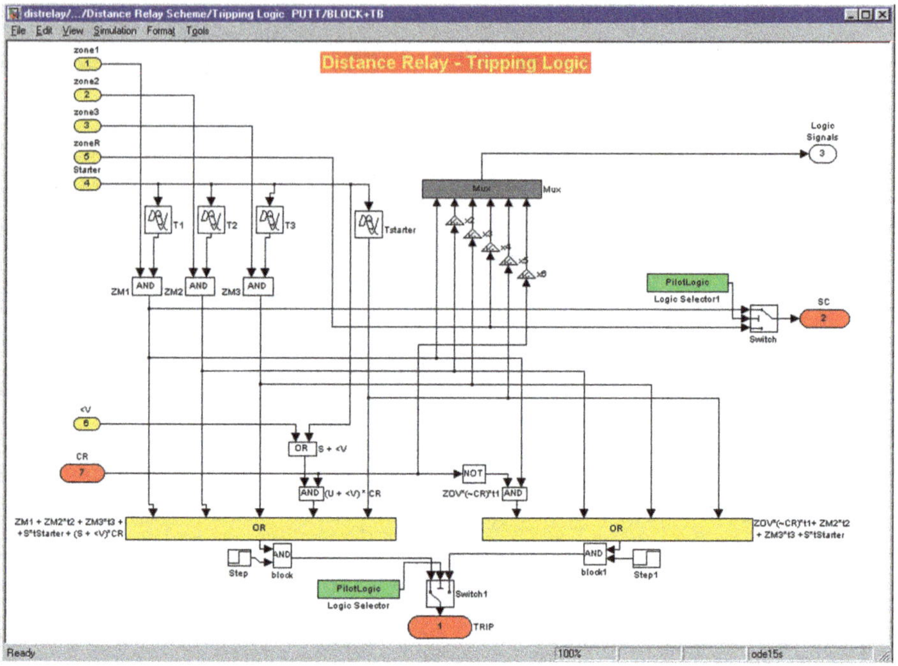

Fig. 6.67 Pilot tripping logic

BLOV ± TB

Zone 1 has an overreaching setting. Upon picking up, the starters also enabled a time delay Thf (setting T1). Provided no blocking signal is received during this time from the opposite end of the line, zone 1 will trip. Should the fault be behind the relay at the opposite end of the line, i.e., not on the protected line, its reverse measuring unit picks up and sends a blocking signal to the opposite end of the line.

This scheme provides protection for the whole length of the protected line. When the fault is fed from one side by a weak source, tripping will take place at both ends in most cases, but in sequence. Since the delay Thf must take the longest signaling time into account, trip times will frequently be correspondingly long. The failure of signaling channel during a fault on an adjacent line may lead to maloperation.

$$\mathrm{TRIP} = \mathrm{ZOV} \times (\tilde{\ }\mathrm{CR}) \times t_1 + \mathrm{ZM}_2 \times t_2 + \mathrm{ZM}_3 \times t_3 + \mathrm{S} \times t_4$$

6.6 Summary

This section introduces the transmission line protection using overcurrent, pilot, and distance protection schemes. The principles for transmission line protection are firstly described. Then the software models for each protection schemes are provided. Details of the power system models such as the voltage source, instrumentation, circuit breaker, and line are provided. The protection schemes and their specific modules are provided as well. After this section, the readers should have a deep understanding of the line protection systems and the available model resources for building a protection scheme.

Reference

1. Anderson PM (1995) Analysis of faulted power systems. IEEE Press, Piscataway

Chapter 7
Design and Implementation of Transformer and Busbar Differential Protection

7.1 Introduction

This chapter discusses protection systems for transformers and busbars using differential schemes.

7.2 Transformer Protection Systems

7.2.1 Introduction

This section contains the description of a transformer protection system for two-winding transformers with at least the following protection functions:

- Biased differential relay including the two-slope bias characteristics, instantaneous unbiased differential overcurrent element, second and fifth harmonic restraints, and differential and restraining currents formed digitally. It considers the wye-delta 11 (330° phase shift) connection.
- Restricted earth protection for the wye-connected winding. It provides the choice of the zero sequence (neutral point current) overcurrent protection and the zero sequence differential protection.
- Three-phase inverse-time overcurrent relays for the back-up protection of the high-voltage winding.
- Six-element three-zone (two forward and one backward) impedance relays for the back-up protection of the high-voltage winding.
- The following signals are available: **IHA**, **IHB**, **IHC**—high-voltage side currents, **IXA**, **IXB**, **IXC**—low-voltage currents, **IN**—neutral current of the wye-connected winding, **VHA**, **VHB**, **VHC**—high-voltage side voltage. Use the suggested variable names.

© Springer International Publishing Switzerland 2016

M. Kezunovic et al., *Design, Modeling and Evaluation of Protective Relays for Power Systems*, DOI 10.1007/978-3-319-20919-7_7

A transformer protection system using differential relays is described. After the following examples, the reader will have an understanding of this type of relay system.

7.2.2 Theoretical Background

Small transformers are usually protected by fuses or overcurrent relays. Large transformers (2.5 MVA or greater) are usually protected by biased current differential relays. Consider a two-winding single-phase transformer illustrated in Fig. 7.1.

When there is no faults within the zone defined by the two CTs, we have,

$$I_1N_1 = -I_2N_2T \tag{7.1}$$

Equation (7.1) is an approximation, because it does not take into account the magnetizing current. N_1 and N_2 are the nominal turns of the two winding, and T is the ratio of the tap changer. If the two current transformers have turns ratios of $1{:}n_1$ and $1{:}n_2$, respectively, then

$$\begin{aligned} I_1 &= n_1 i_1 \\ I_2 &= n_2 i_2 \end{aligned} \tag{7.2}$$

The CT secondary currents i_1 and i_2 may be made equal in magnitude by choosing n_1 and n_2 such that

$$N_1 n_1 = N_2 n_2 \tag{7.3}$$

In general, there are always CT errors, therefore,

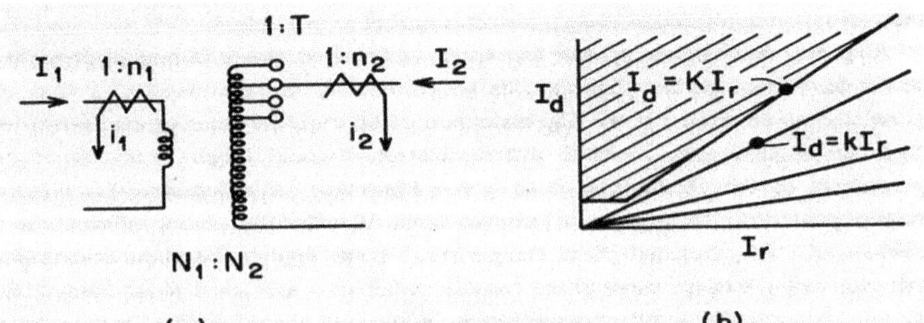

Fig. 7.1 Analysis of differential relay for transformer

$$i_1 + i_2 = k[(i_1 - i_2)/2] \tag{7.4}$$

or

$$I_d = kI_r \tag{7.5}$$

The algebraic sum $(i_1 + i_2)$ is the differential current I_d, and $(i_1 - i_2)/2$ is the average value of the two winding currents referred to the CTs' secondaries, which is known as restraining current I_r. In order for the differential relay to refrain from tripping, it becomes necessary to shape the relay characteristic as shown in Fig. 7.1b. The biased slope K is made greater than the k of (7.4), in order to allow for some safety margin. Typical settings available for biased differential relays are 10, 20, or 40 %.

During the energization of a transformer, abnormal current may flow in the winding that is being energized. These are known as the magnetizing inrush currents, caused by the saturation of the transformer core. To block the relay trip under conditions of inrush, the second harmonic current is used as another restraining signal.

In addition, an overexcited transformer has a significant fifth harmonic component in its magnetizing current, and hence it is desirable to create a composite restraint function with the second and fifth harmonics.

7.2.3 *Simulation Models*

7.2.3.1 Overall Design

The overall design diagram of the relay system is shown as Fig. 7.2.

There are six main parts of the design, which are explained below.

7.2.3.2 Measurements

The transformer protection system needs measurements of the three phase currents and voltages, and neutral current. These signals are essential to the operation of the relay. The voltage and current signals are generated with a model of the power system with a transformer as shown in Fig. 7.2. The signals are stored in .mat files. Then the protection system uses them as its inputs. The names of the input files and their contents are explained in Table 7.1.

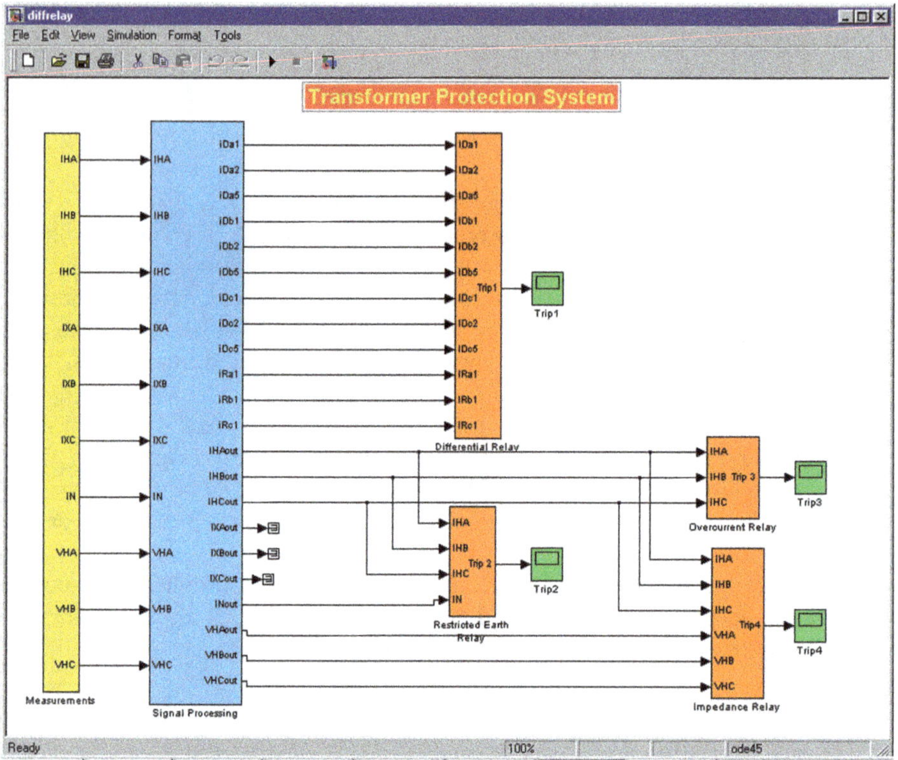

Fig. 7.2 Design diagram of transformer protection system

Table 7.1 Input files and their contents

File	Contents
IHA*.mat	Phase A current at the high voltage side
IHB*.mat	Phase B current at the high voltage side
IHC*.mat	Phase C current at the high voltage side
IXA*.mat	Phase A current at the low voltage side
IXB*.mat	Phase B current at the low voltage side
IXC*.mat	Phase C current at the low voltage side
IN*.mat	Neutral current
VHA*.mat	Phase A voltage at the high voltage side
VHB*.mat	Phase B voltage at the high voltage side
VHC*.mat	Phase C voltage at the high voltage side

where symbol * represents a text used to distinguish the simulation cases

7.2.3.3 Signal Processing

The structure of this block under its mask is shown in Figs. 7.3, 7.4, and 7.5.

The DAB block is used to perform analog filtering, signal conditioning, and sampling, while the DFT block calculates the phasors of the selected harmonics of the input signal. In the subsystem, we use a block named VG-2 to get the instantaneous values of the three-phase differential and restraining currents.

The input parameters for the signal-processing block are shown in Fig. 7.6.

Note that the transformer connection is wye-delta-11.

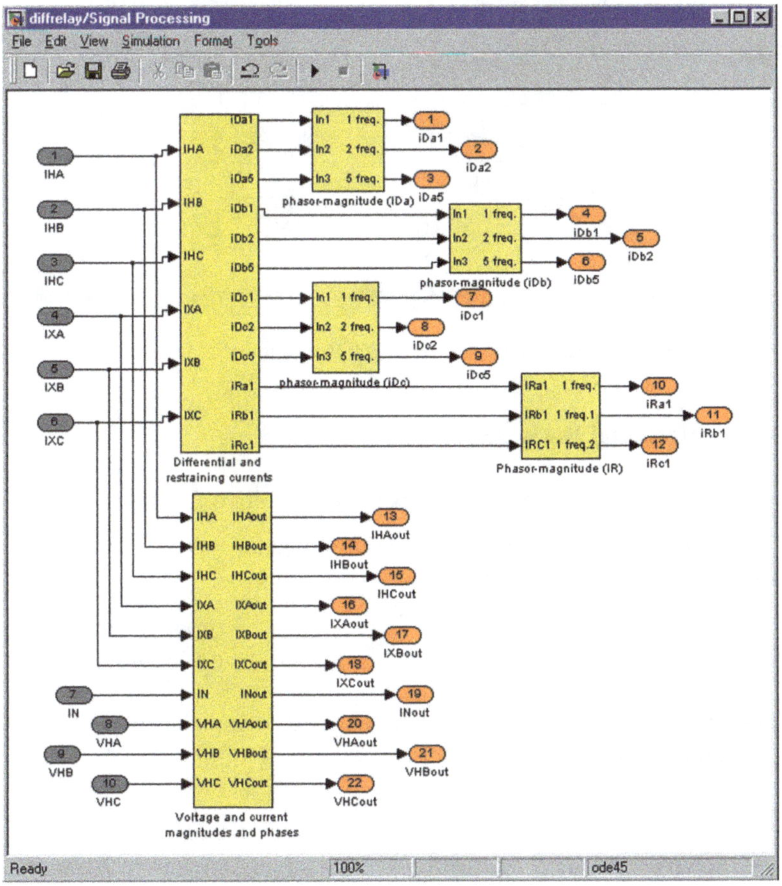

Fig. 7.3 Signal processing block

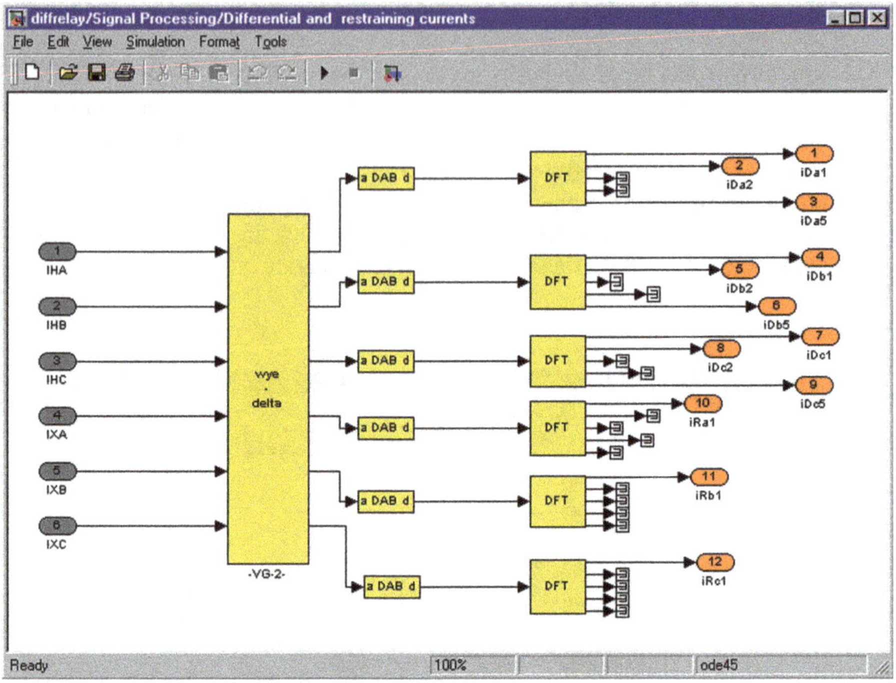

Fig. 7.4 Differential and restraining currents block (for differential relays)

7.2.3.4 Differential Relay

The designed differential relay provides three protective functions:

- Two-slope bias characteristic differential protection.
- Instantaneous unbiased differential overcurrent protection.
- Second and fifth harmonic restraints.

Figures 7.7 and 7.8 show the design structure of the differential relay (Figs. 7.9 and 7.10).

The pick-up value is used to eliminate errors due to imbalance and assure that the relay would not act under the normal conditions.

7.2.3.5 Restricted Earth Protection for the Wye-Connected Winding

The zero-sequence differential relay is similar to the one mentioned above, but in this case a one-slope bias characteristic is used (Figs. 7.11, 7.12, and 7.13).

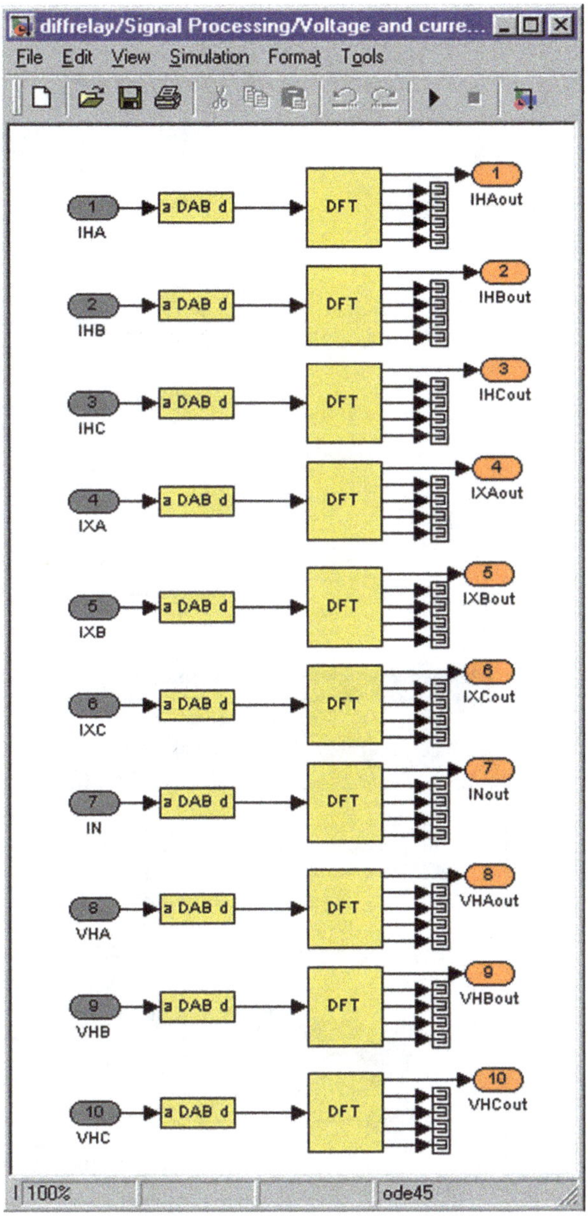

Fig. 7.5 Voltage and current magnitudes block (for other relays)

Note that the block named RC ("Ratio Mismatch Compensation" in Fig. 7.14) is to handle possible mismatches of the CT ratios between the two sides. There are two options to choose from in the type of protection parameter: Overcurrent or Differential protection.

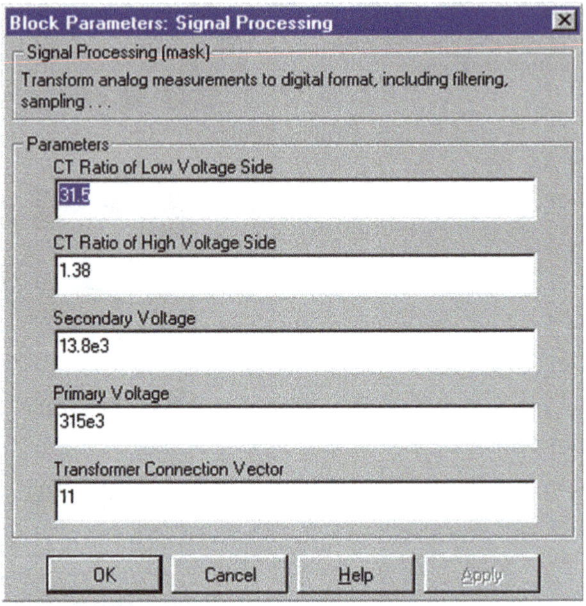

Fig. 7.6 Interface of signal processing block

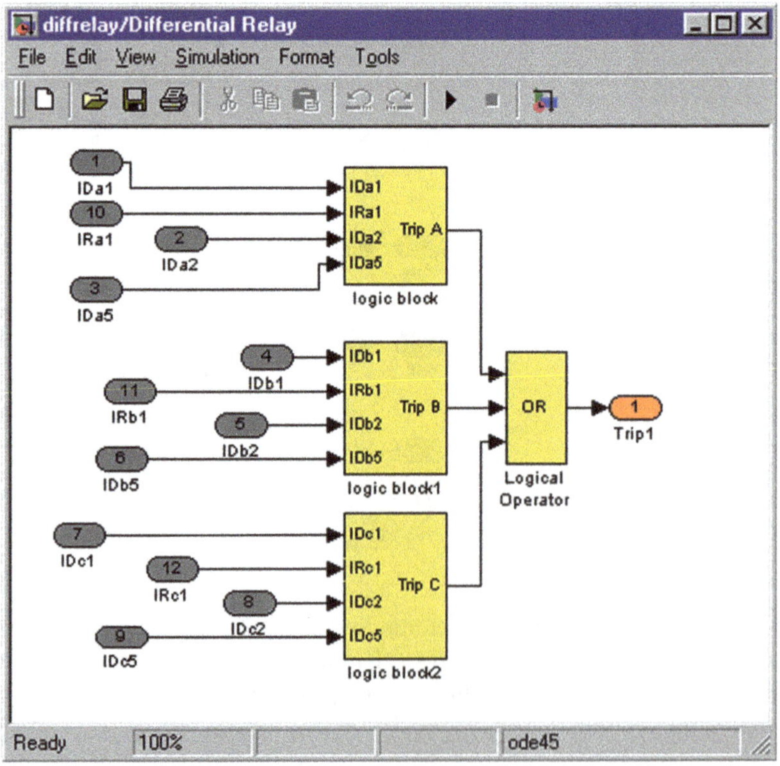

Fig. 7.7 Differential relay

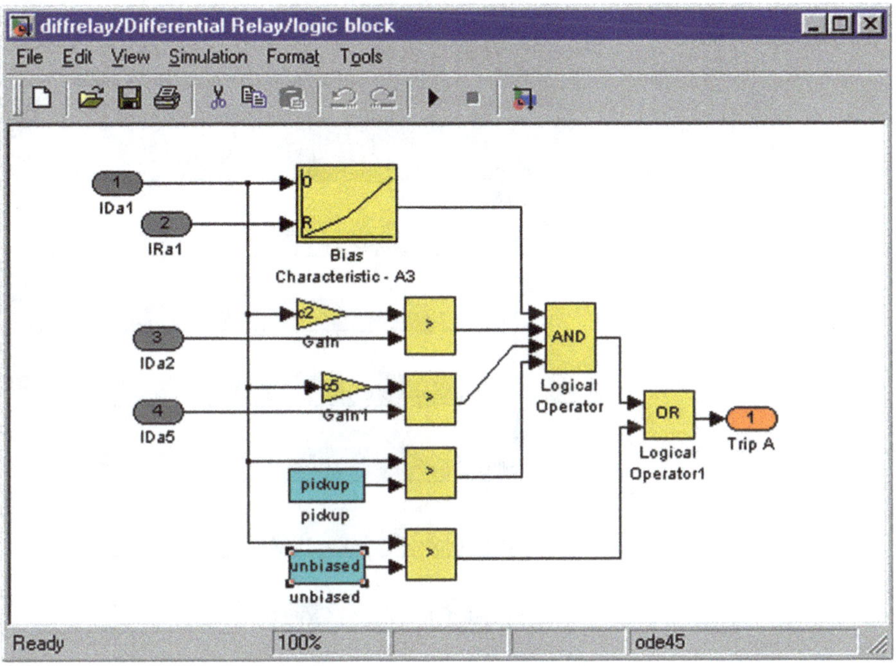

Fig. 7.8 Differential relay (logic block)

7.2.3.6 Three-Phase Inverse-Time Overcurrent Relay

Using the *UC* element in the Relay Library, a three-phase inverse-time overcurrent relay was designed. This relay is shown in Fig. 7.15.

"Very inverse" in "Signal-to-time comparison" option was chosen in the UC block. In this type of comparison, for the input signal S1, the operating (t_{op}) time is calculated. The operating time is then compared against the elapsed time since the signal became greater than the pick-up value. If t is greater than t_{op}, then the output of the block becomes one. The elapsed time is reset every time the signal goes below the pick-up value.

$$t_{op} = (13.5k)/(I_n - 1)$$

t_{op} = operation time
I_n = normalized input signal. I_n = S1/(pick-up value)
k = time parameter

The user interface for this relay is shown in Fig. 7.16.

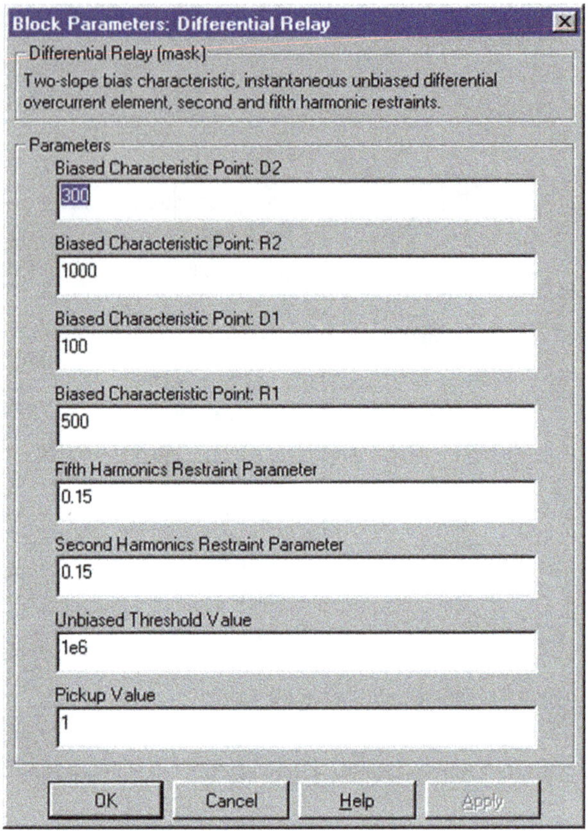

Fig. 7.9 User interface of the differential relay

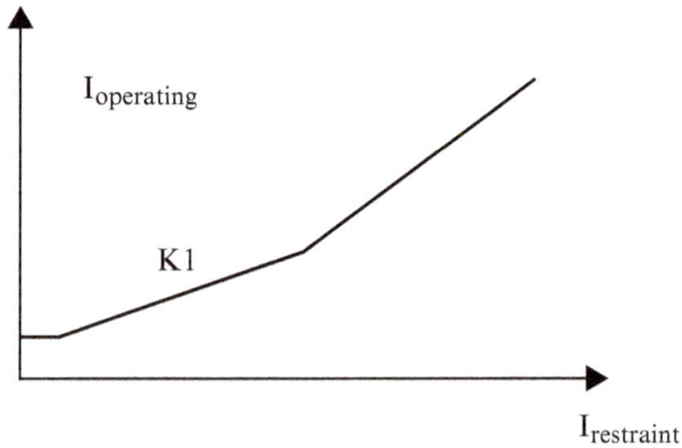

Fig. 7.10 Differential relay (two-slope bias characteristic)

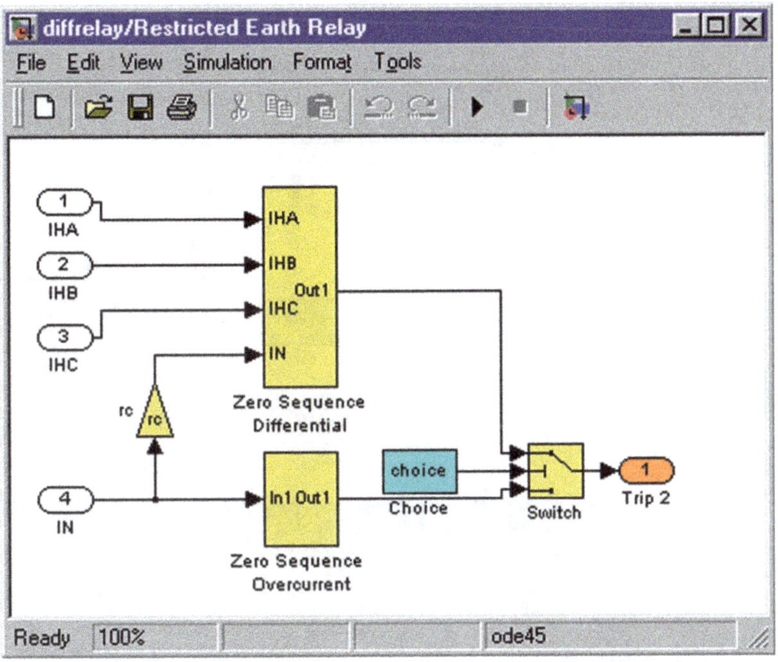

Fig. 7.11 Restricted earth protection

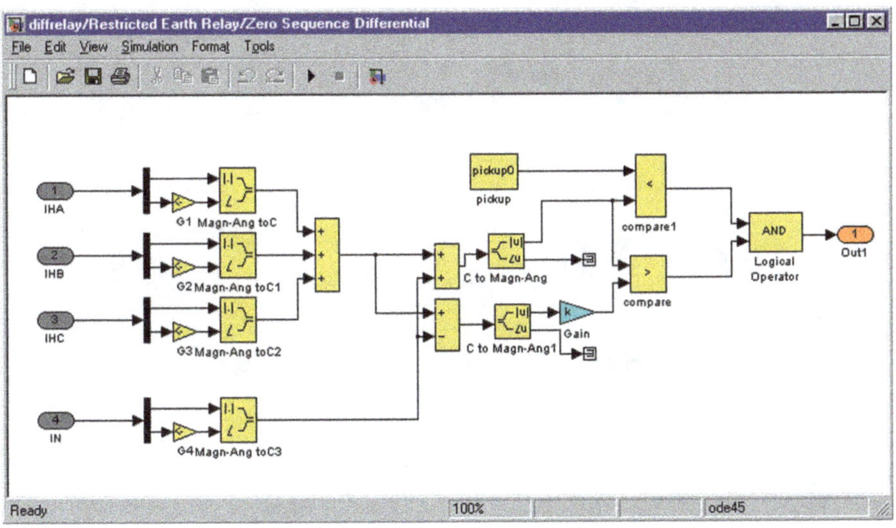

Fig. 7.12 Zero-sequence differential block

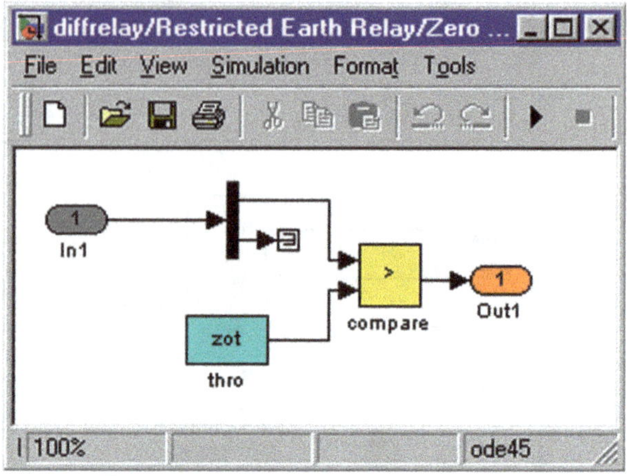

Fig. 7.13 Zero-sequence overcurrent block

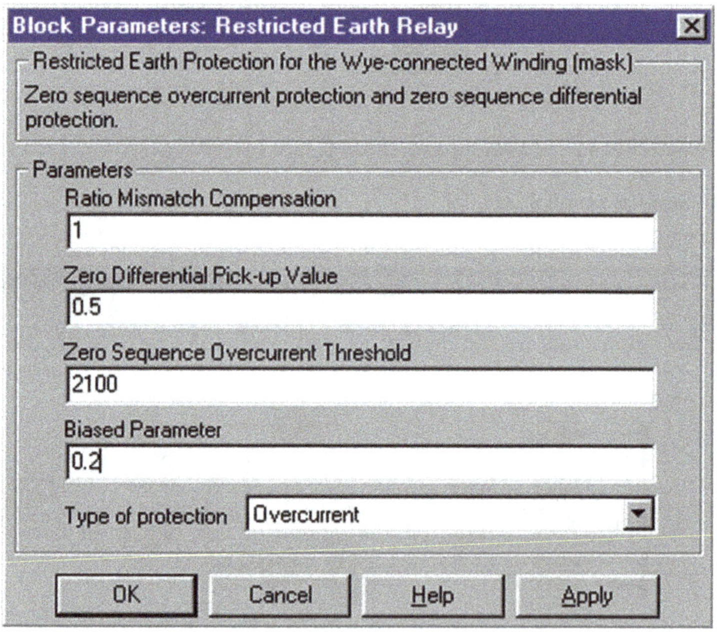

Fig. 7.14 User interface of the restricted earth relay

7.2.3.7 Six-Element Three-Zone Impedance Relay

The inputs of the six elements are:

- V_A, $I_A + 3I_0$
- V_B, $I_B + 3I_0$
- V_C, $I_C + 3I_0$

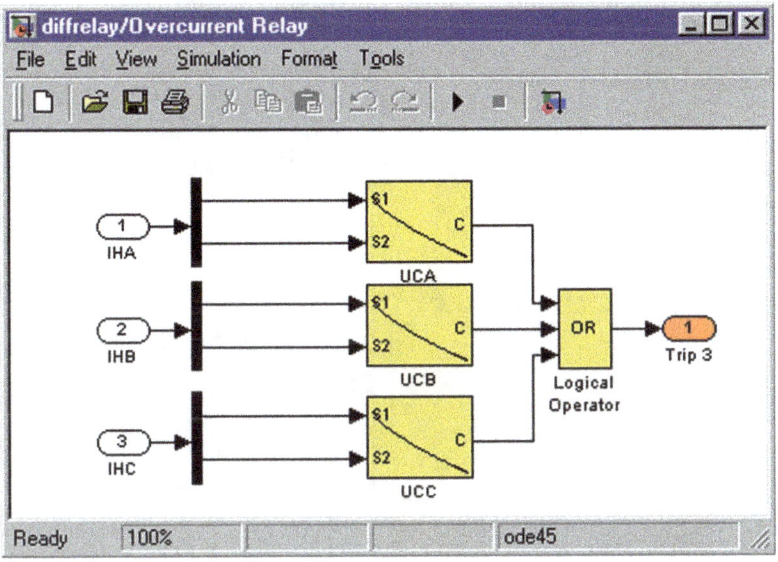

Fig. 7.15 Three-phase inverse-time overcurrent relay

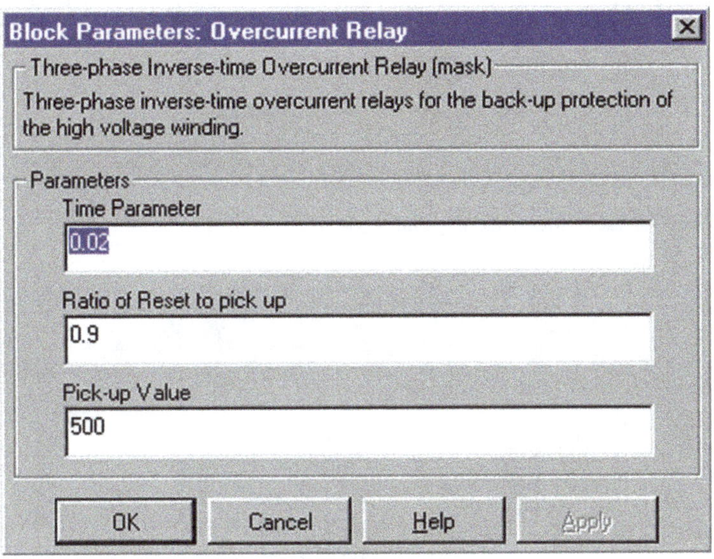

Fig. 7.16 User interface

- $V_A - V_B$, $I_A - I_B$
- $V_B - V_C$, $I_B - I_C$
- $V_C - V_A$, $I_C - I_A$ (Figs. 7.17, 7.18, and 7.19)

Note that we use a free expression (polygon) to represent the operation zones.

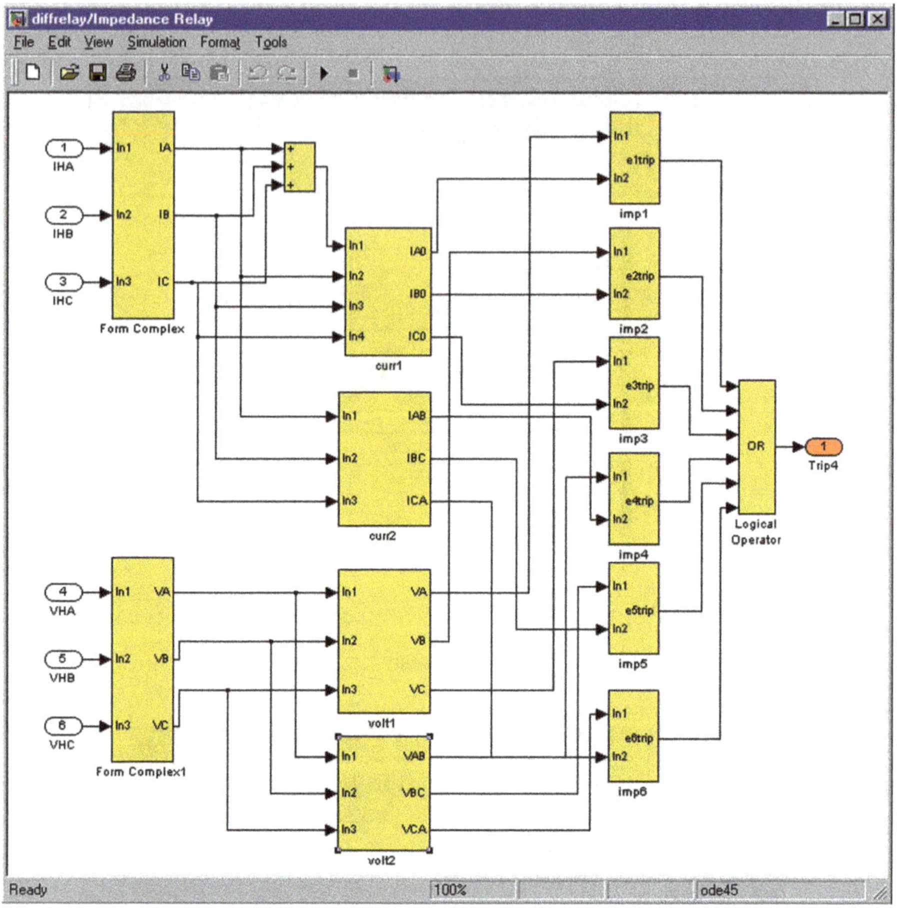

Fig. 7.17 Six-element three-zone impedance relay

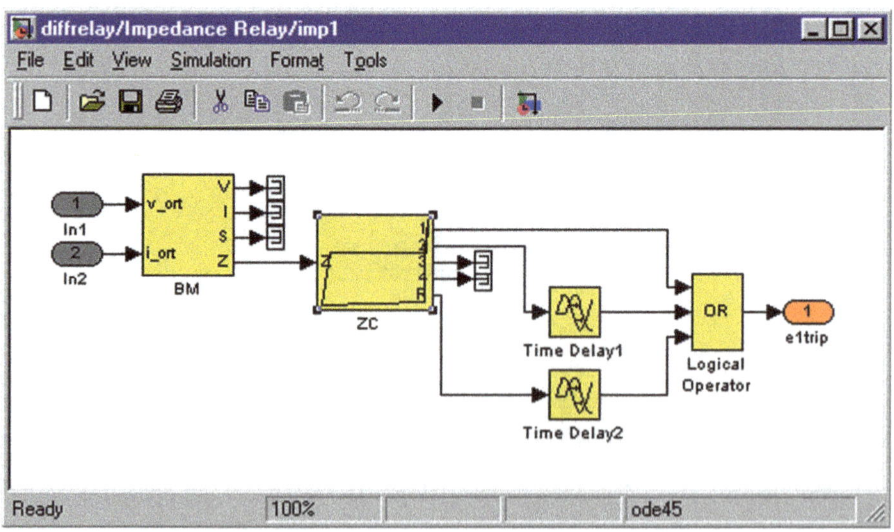

Fig. 7.18 Six-element three-zone impedance relay (creation for trip signal)

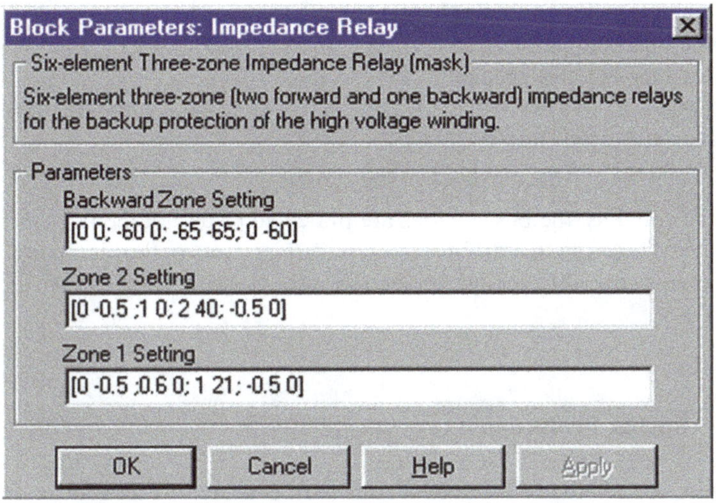

Fig. 7.19 User interface

7.3 Busbar Protection Systems

7.3.1 Introduction

In this section, a typical bus arrangements and the busbar protection scheme are introduced. The design of the busbar protection system for N-terminal busbars contains the following protection functions:

- Biased differential relay with a three-slope bias characteristic and harmonic restraint should be used.
- Detector of CT saturation should be implemented to control the bias adaptively.
- Trip suppresser should be used to distinguish between the saturation of CT under internal and external faults and block the relay in the latter case.

In this document, a busbar protection system using differential relays is described. After studying the following examples, the reader will have an understanding of this type of relay system.

7.3.2 Theoretical Background

7.3.2.1 Typical Bus Arrangements

The different bus arrangements are the following:

- Single bus—single breaker
- Double bus with bus tie—single breaker

- Main and transfer bus—single breaker
- Double bus—single breaker
- Double bus—double breaker
- Ring bus
- Breaker-and-a-half bus
- Bus and transformer—single breaker

For the ring bus, the bus sections are protected as part of the lines or connected equipment. Therefore, the differential protection is not applicable. In the rest of the arrangements, the differential protection should be considered as the primary protection.

7.3.2.2 Busbar Protection

Bus faults are relatively infrequent, but they can be very disruptive to a power system. Therefore, fast and sensitive protection using the various differential schemes is required.

The typical relay protection schemes are shown below:

7.3.2.3 Time-Overcurrent Differential

The secondary of all the CTs is paralleled and connected to an inverse-time-overcurrent relay. There is no restraint, so the relay must be set above the maximum CT magnetizing difference current for the external fault.

Advantages

- The inverse characteristic is an advantage to override unequal CT saturation, particularly on the dc component.
- It is inexpensive.

Drawbacks

- Difficult to set, except through long experience
- Typical operating time: 15–20 cycles (60 Hz base) for internal faults

Application

- Small low-voltage buses

7.3.2.4 High-Impedance Voltage Differential

This scheme loads the CTs with a high impedance to force the error differential current through the CTs instead of the relay operating coil.

Advantages

- Discriminates between external and internal faults by overcoming the current transformer saturation.

Drawbacks

- All the CTs should have the same ratio.
- Typical operating time: 20–30 ms or 8–16 ms (instantaneous unit used)

Application

- Not widely used now.

7.3.2.5 Air-Core Transformers Differential

This system uses linear couplers instead of conventional iron core bushing-type CTs.

Advantages

- Very flexible. Circuits can be added or subtracted with minimum problems

Drawbacks

- Existing and conventional CTs cannot be used
- Linear coupler CTs cannot be used for other applications

Typical Operating Time

- $<$16 ms, with sensitivity of 2–50 mA

7.3.2.6 Low-Impedance Differential

This kind of scheme is one of the most popular protection schemes, since it has no special requirement on instrument transformers and it can discriminate between internal and external faults efficiently.

This scheme is also named *biased differential* or *percentage differential*. The restraint current makes the relay insensitive to the effects of CT saturation on external faults. This scheme was the one selected for the protection system discussed later on.

7.3.2.7 Directional Comparison Differential

A directional sensing unit is connected to each bus circuit "looking into" the bus, with the trip contacts in series. For internal faults, all contacts should close to trip the bus. Normally closed contacts are necessary for feeder circuits with no feed to the fault.

Advantages

- Almost complete independence of CT performance, characteristic, and ratio

Disadvantages

- Relative high cost (relays required for each circuit)
- Contacts in series are difficult to coordinate

Application

- Not in general use in recent years

7.3.2.8 Partial Differential

This scheme is used for: (1) feeders supplying negligible current to bus faults, or (2) buses that do not have adequate or suitable CTs for a complete differential application

Application

- Buses in industrial and lower-voltage distribution substations

7.3.2.9 Zero-Sequence Current/Voltage

The bus is grounded at one point through an overcurrent relay.

Application

- Not used in the USA because of relatively higher costs, difficulties of construction, and difficulties in protecting personnel.

7.3.3 Simulation Models

7.3.3.1 Overall Relay Design

The low-impedance differential scheme will be used as an example. The main difficulty of busbar relay protection is the saturation of CTs. When a CT saturates, there will be some harmonics produced in the secondary side of the CT, especially the second and the fifth harmonic. This phenomenon can be used to detect the CT saturation.

The restraining current is the sum of the magnitudes of the currents, which is higher during external faults than during internal faults, while the differential current is lower during external faults than during internal faults. By comparing the differential current against the restraining current using a biased characteristic, the internal faults and external faults can also be distinguished.

Figure 7.20 shows the common modules of the differential current relay which consists of several common components, such as Low-pass filter, Sampling/Holding, and A/D converter. The relaying algorithm is executed in the processor. The whole protection system designed in MATLAB is shown in Figs. 7.21 and 7.22.

7.3.3.2 Data Acquisition System

In the data acquisition system, the signals measured by CTs are passed through DAB blocks. The cut-off frequency is selected as 360 Hz and the sampling frequency is selected as 60×20 Hz. As aforementioned, the interested harmonics are the second and the fifth harmonic, so such cut-off frequency is selected accordingly.

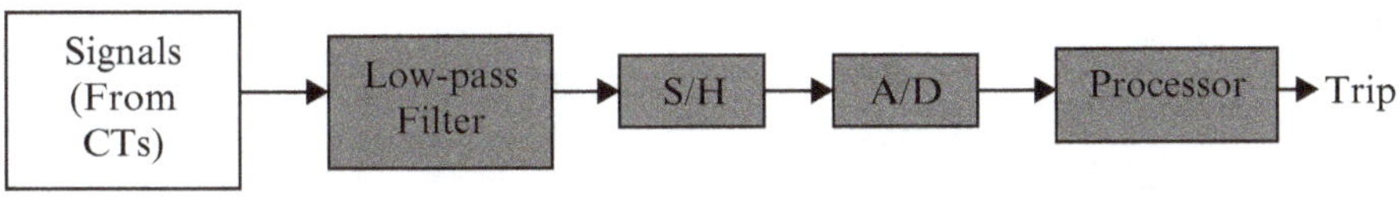

Fig. 7.20 Busbar protection system components

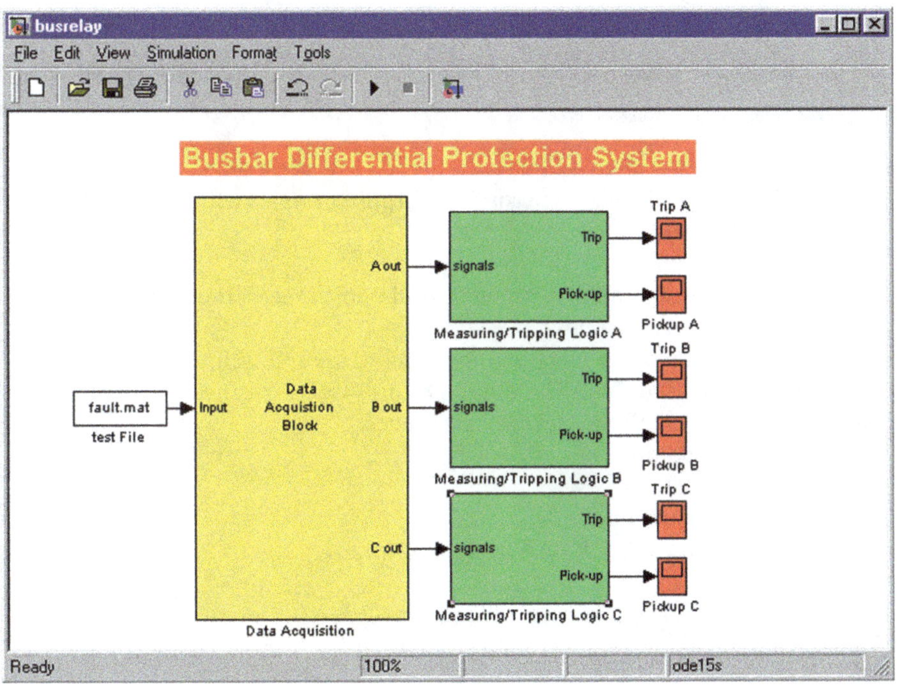

Fig. 7.21 Bus differential protection system

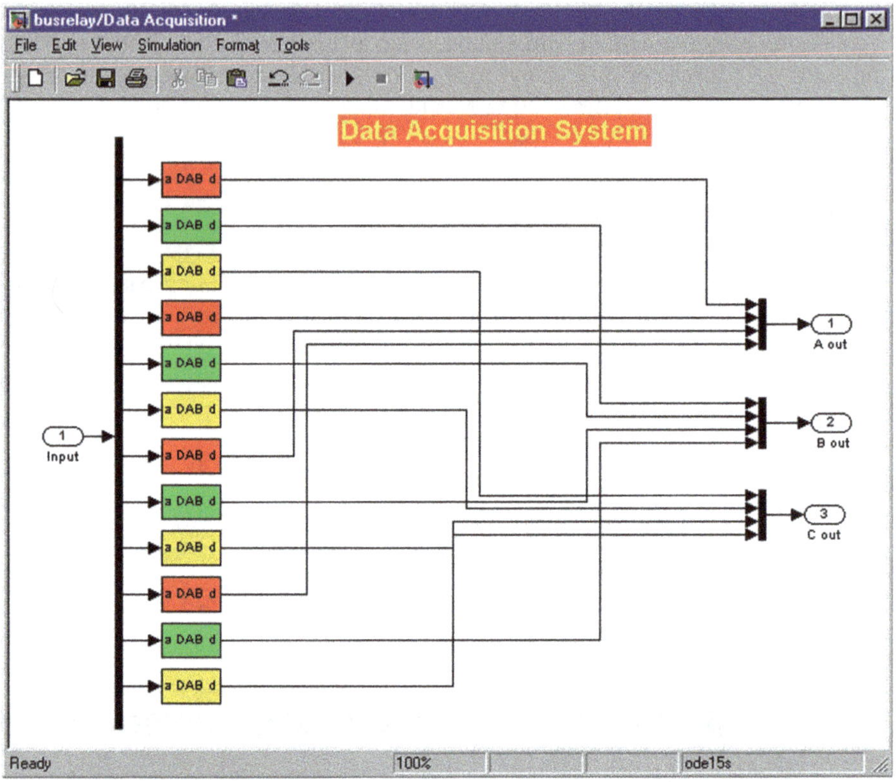

Fig. 7.22 Data acquisition system

7.3.3.3 Differential Relay Measuring/Tripping Logic (with Harmonic Restraint)

Figure 7.23 shows the internal structure of the relay measuring algorithm and the trip logic.

The fundamental component and the second, third, fourth, and fifth harmonics are measured, and the Total Harmonic Distort (THD) is calculated. The restraint current is the sum of all the current magnitudes.

The pickup condition is:

$I_{op} > I_{nom} \times A0$

The tripping condition is:

pickup **AND** ($I_{op} > I_r \times K$) **AND** (THD $<$ C1 $\times I_{op}$)

where:

I_{op}—differential current

I_{nom}—nominal load current

I_r—restraint current

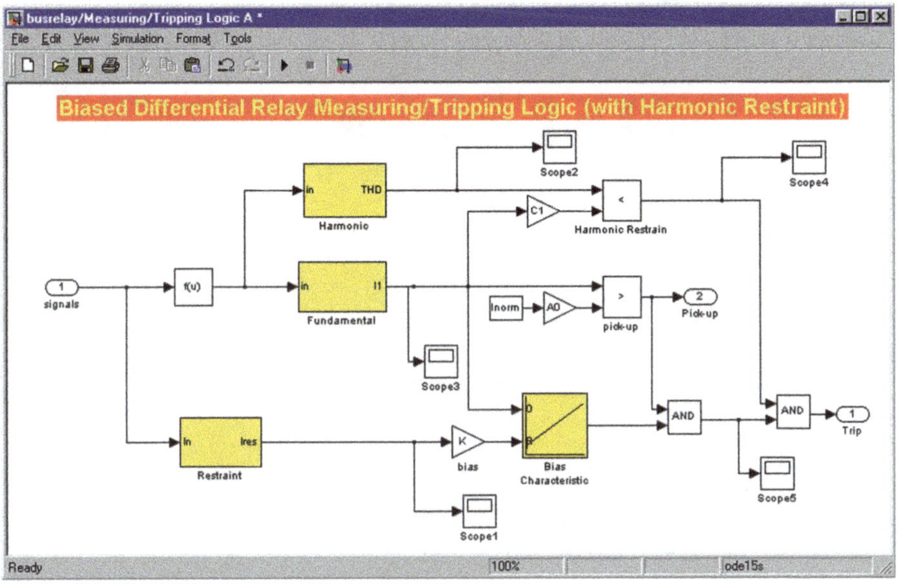

Fig. 7.23 Inside the relay measuring/tripping logic

7.4 Summary

This chapter introduces the protection for transformers and busbars using differential schemes. For transformer protection, the biased differential relaying and restricted earth protection are described. The three-phase inverse-time overcurrent protection is also described as back-up protection. For busbar protection, the biased differential relaying scheme is described. The issues need to be considered such as CT saturation and trip suppression are covered as well. After this chapter, the readers should have a better understanding of the transformer and busbar protection and the available model resources for building a protection scheme.

Chapter 8
Testing of Digital Protective Relays

8.1 Introduction

This chapter discusses the online (closed-loop) and off-line (open-loop) testing of the digital relay models using developed software modules and physical relays using the low-voltage digital simulator. The software examples are developed in MATLAB/SIMULINK as described in above sections. The first subsection provides protective relay models including overcurrent relay and distance relay. The digital simulator-based low-voltage relay test platform is described in the second subsection. Examples of laboratory exercise for students are given at the end of each subsection.

- Models of three-phase overcurrent relay and impedance relay are developed using SimPowerSystems in MATLAB/SIMULINK
- 230 kV three-phase power network models for performing relay tests are developed using SimPowerSystems in MATLAB/SIMULINK
- Low-voltage simulator-based protective relay test system is described
- Circuit breakers (CBs), voltage transformers (VTs), and current transformers (CTs) are modeled as ideal elements
- Examples of closed-loop and open-loop protective relay tests are presented

The configuration of the models can be changed by using the cut-and-paste and drag-and-drop operations on the elements in the power system. In addition, the parameters of each module can be changed by double clicking the icon corresponding to the element of interest.

The test plan provided in this chapter requires the following activities:

- Perform closed-loop and open-loop protective relay tests with various power system disturbance including fault types, locations, and resistance.
- Determine appropriate protection scheme for a given power system and configure relays with correct settings.
- Verify various protection functions including back-up protection scheme.
- Analyze test results and characterize relay operation performance.

© Springer International Publishing Switzerland 2016

M. Kezunovic et al., *Design, Modeling and Evaluation of Protective Relays for Power Systems*, DOI 10.1007/978-3-319-20919-7_8

8.2 Modeling and Testing Digital Relays

In this section, digital relay models (overcurrent and impedance) and three-phase power system models developed in MATLAB/SIMULINK are described. Compared to commercial relays, these relays are modeled with only the basic functions. Examples of lab exercise to perform online relay testing are provided at the end.

8.2.1 Modeling and Testing Overcurrent Relay

This exercise shows how overcurrent relaying algorithm and overcurrent definite-time relays can be used to protect a radial electric power network. Rules for protecting a network using overcurrent relays, relaying principle and algorithm, requirements for instrumentation, and the principle of calculating settings can be referred to Sect. 8.3.

Note the following:

- In order to protect a given element, a current transformer (CT) needs to be used to measure the current. The CT should be installed at the element's terminal that is closest to the supplying source.
- A circuit breaker (CB) needs to be installed at the terminal closest to the supplying source to disconnect the element in the case of a fault.
- An overcurrent relay must measure the current magnitude and compare it against the threshold.
- The fault clearing time corresponds to the sum of the operating time of the relay and the operating time of the CB.

Figure 8.1 shows a radial network consisting of an equivalent system source and three transmission lines. This radial network can be protected against faults using the overcurrent relaying (OR) principle. In order to protect Line 2, for example,

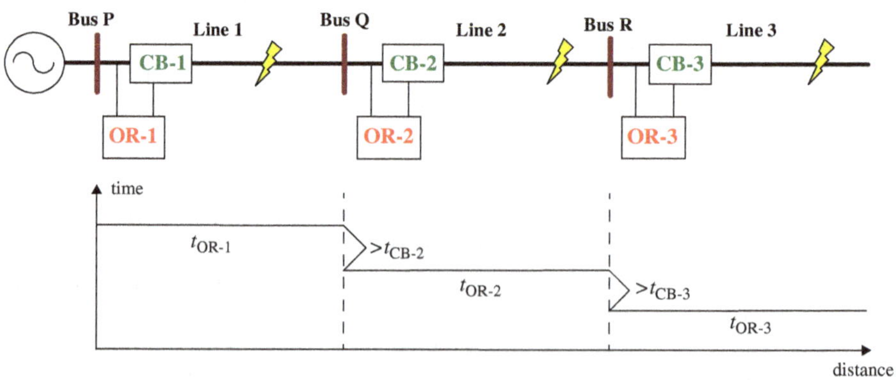

Fig. 8.1 Radial network and illustration of the definite-time overcurrent protection

OR-2, CT-2, and CB-2 should be installed at Bus Q. OR-2 must be coordinated with the relay downstream (in this case OR-3) to ensure selectivity of operation. The definite-time overcurrent relaying principle is illustrated in Fig. 8.1, which shows the time delays of the relays in the protection system of the network.

The protection system shown in Fig. 8.1 operates as follows:

- If a fault occurs on one of the transmission line (feeder) segments, i.e., a fault occurs on Line 2, some or all relays upstream (i.e. OR-1) between the fault and the supply system will pick-up (active) and begin timing-out. However, only the relay closest to the fault (OR-2 in this case) is supposed to trip since it has the shortest time delay. After the relay trips, the remaining upstream relays will reset and none will trip. And the relays (OR-3) installed downstream from the fault will not pick-up because there is no fault current flowing through. The part of the network downstream from the operating relay will be cut-off, while the upstream part will operate normally after the fault is cleared.
- If a fault occurs on Line 3 and the OR-3 or CB-3 fail to operate, the fault will be cleared by the next upstream relay, i.e., OR-2 in this case. In this particular condition, the upstream relay OR-2 will be functioning as a back-up protection and the fault will be tripped by OR-2. In this case, more elements will be removed from system than desired.

It should be noted that to protect a radial power network, overcurrent relays and CBs should be installed between each power system element and the supply system. The closer a fault is to the source, the higher the fault current and the longer the clearing time. The combination of these two issues is a major disadvantage of this overcurrent protection. To shorten the average clearing time, relay may be designed in such a way that their delay time depends on the fault current magnitude.

The following activities should be done for designing protection scheme:

- Analysis of the load and fault conditions
- Selecting CT and CB
- Setting relay: pick-up current and time delay
- Sensitivity check

An online overcurrent relay testing program developed in SIMULINK is shown in Fig. 8.2, in which three definite-time overcurrent relays, OR-1, OR-2, and OR-3 are used as primary protection for feeder segments designated as Line 1, Line 2, and Line 3, respectively. OR-1 is used as back-up protection for Line 2 and Line 3, and OR-2 is used as back-up protection for Line 3. Elements used in the model are described in following subsections.

8.2.1.1 Overcurrent Relay Model

The model icon is shown in Fig. 8.3. Double click on it to open parameters dialog frame in which one can input settings for the relay model. Relay's parameters include Pickup Current, Time Delay, and Residual Current Threshold, as

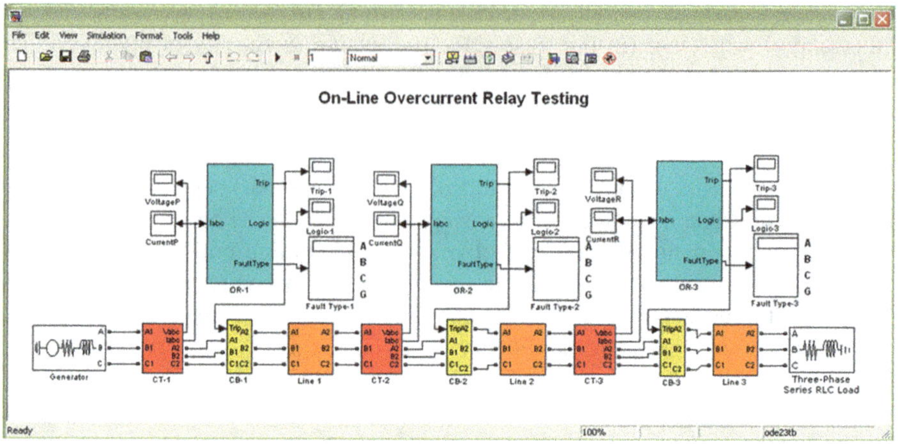

Fig. 8.2 Online overcurrent relay testing model

Fig. 8.3 Overcurrent relay model

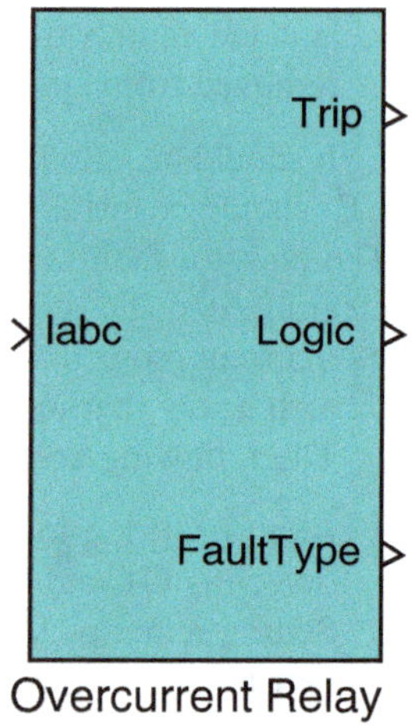

shown in Fig. 8.4. The output terminal "Trip" should be fed back to the CB to clear faults.

8.2.1.2 Transmission Line Model

Figure 8.5 shows the three-phase transmission line model. One should note that this model integrates the fault creation function. Double click on the icon to open the parameter dialog frame, as shown in Fig. 8.6; it includes the following parameters:

- Nominal frequency
- Resistance, inductance, capacitance in positive sequence and zero sequence

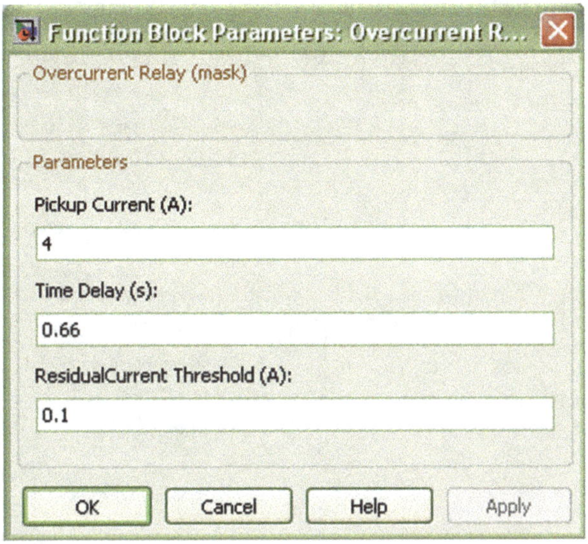

Fig. 8.4 Overcurrent relay model parameter dialog frame

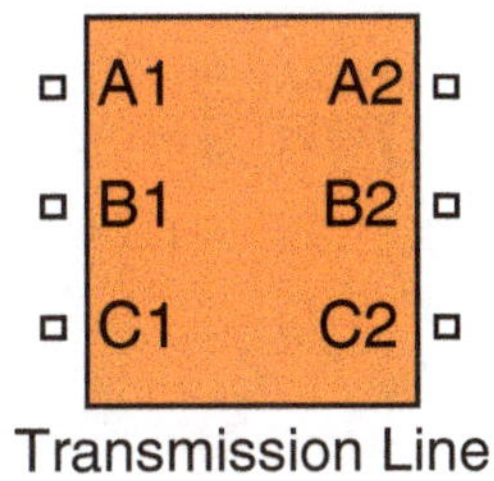

Fig. 8.5 Transmission line model

- Line length
- Fault location in p.u.
- Fault type including A-G, B-G, C-G, B-C, B-C-G, C-A, C-A-G, A-B-C, A-B-C-G
- Fault resistance and fault time sequence

8.2.1.3 Generator Model

The icon of a three-phase generator model is shown Fig. 8.7. This model consists of an ideal voltage source and a series RL model. Double click on it to open parameters dialog frame in which one can set up the voltage source. Parameters include voltage magnitude, phase angle, frequency, source resistance, and inductance, as shown in Fig. 8.8.

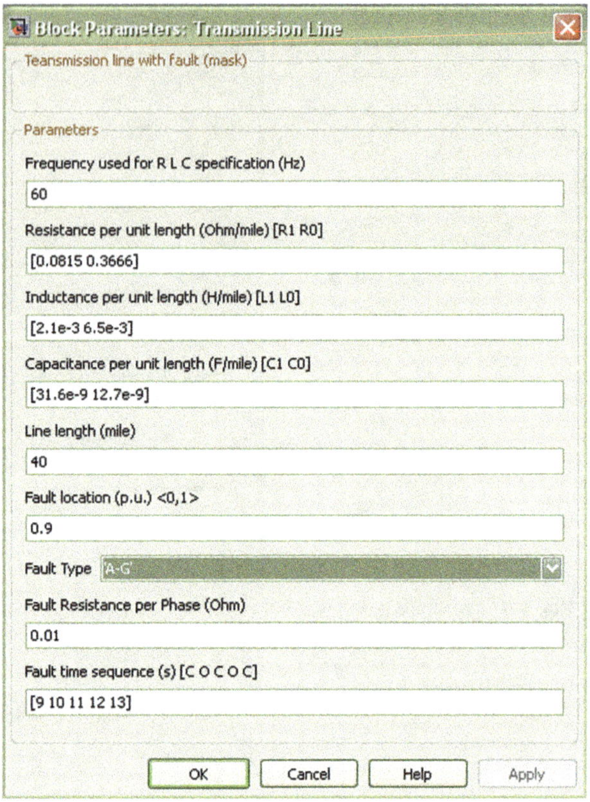

Fig. 8.6 Transmission line model parameter dialog frame

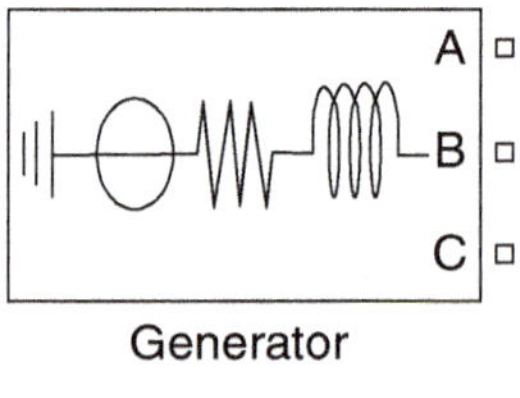

Fig. 8.7 Generator model

8.2.1.4 CT and VT Model

Figure 8.9 shows the three-phase CT and VT model. Double click on the icon to change the CT ration and VT ration through parameter dialog in Fig. 8.10.

8.2.1.5 Breaker Model

Figure 8.11 shows the three-phase circuit breaker model. Double click on the icon to change the operating time through parameter dialog frame as shown in Fig. 8.12.

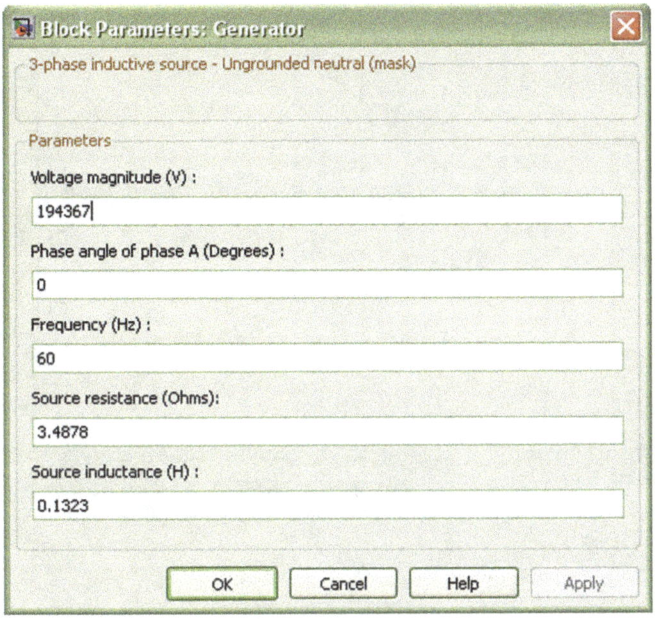

Fig. 8.8 Generator model parameter dialog frame

Fig. 8.9 CT and VT model

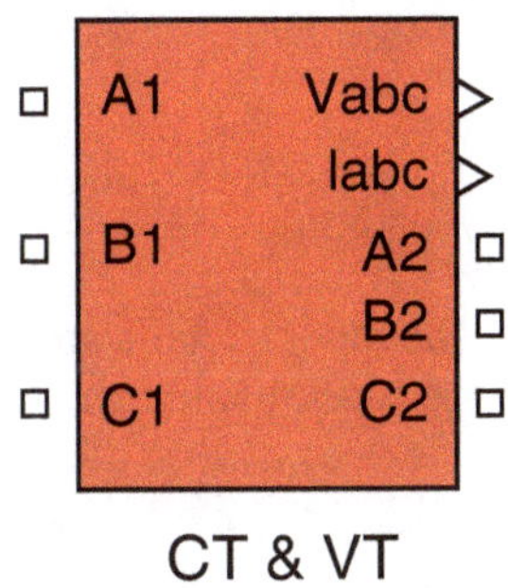

Fig. 8.10 CT and VT model dialog frame

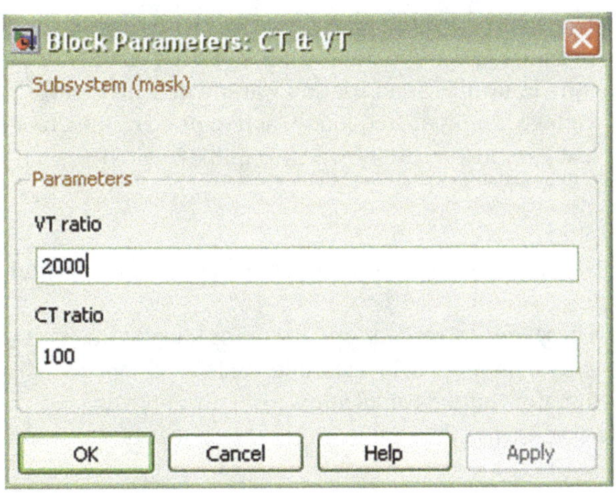

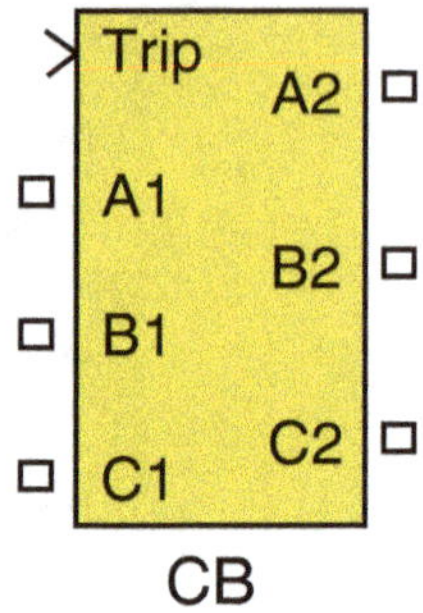

Fig. 8.11 Circuit breaker model icon

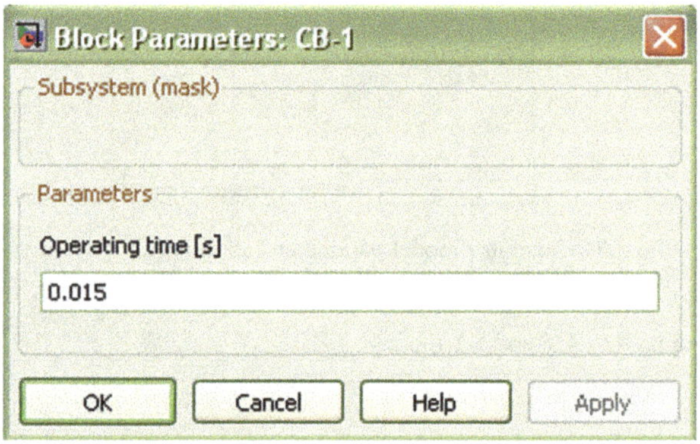

Fig. 8.12 Circuit breaker parameter dialog frame

The model details are given in Fig. 8.13. Double click on the breaker icon to open the block parameter box as shown in Fig. 8.14.

8.2.1.6 Load Model

The icon of a three-phase load model is shown Fig. 8.15. This model consists of a series RLC model. Double click on it to open parameters dialog frame in which one can set up this load model. Parameters include grounding configuration, nominal voltage, nominal frequency, active power, inductive reactive power, and capacitive reactive power, as shown in Fig. 8.16.

8.2.1.7 Voltage and Current Scopes (Measurement Elements)

These two elements are used for displaying voltage and current waveforms as shown in Fig. 8.17. Double click will open a display window which contains function buttons that allow user to zoom in and out on different sections of the waveforms.

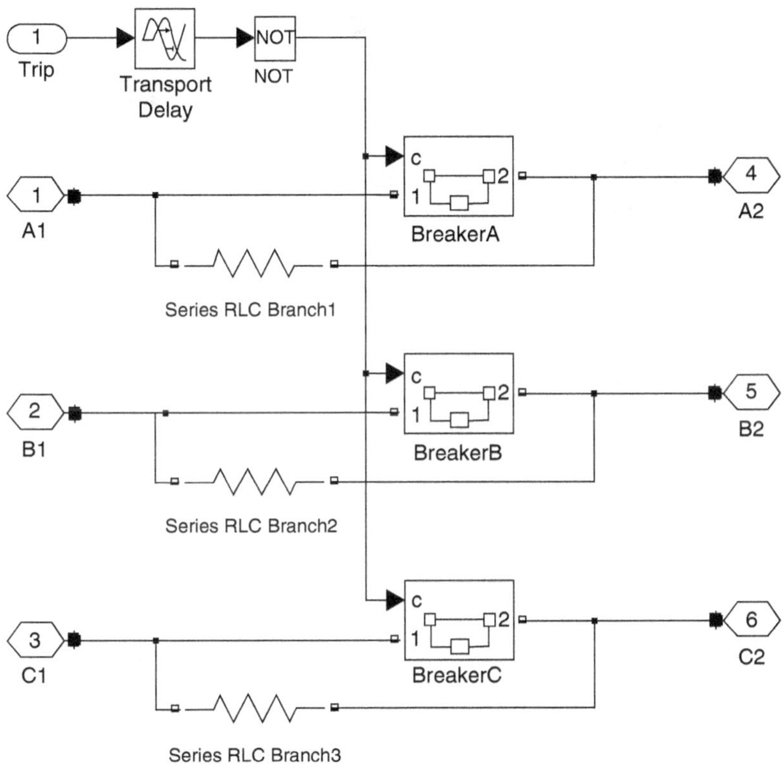

Fig. 8.13 Circuit breaker model detail

8.2.1.8 Trip and Logic Scopes (Measurement Elements)

These two elements are used for displaying trip and logic signals as shown in Fig. 8.18. Double click will open a display window which contains function buttons that allow user to zoom in and out on different sections of the waveforms.

8.2.1.9 Fault Type Display

This element is used for displaying the result of phase selection, which is one of the overcurrent relay functions. As shown in Fig. 8.19, Digits "1" and "0" indicate the corresponding character in the same row effective or ineffective respectively. For example, the digit "1" in first and fourth row indicates that the fault type is phase A-to-Ground "A-G".

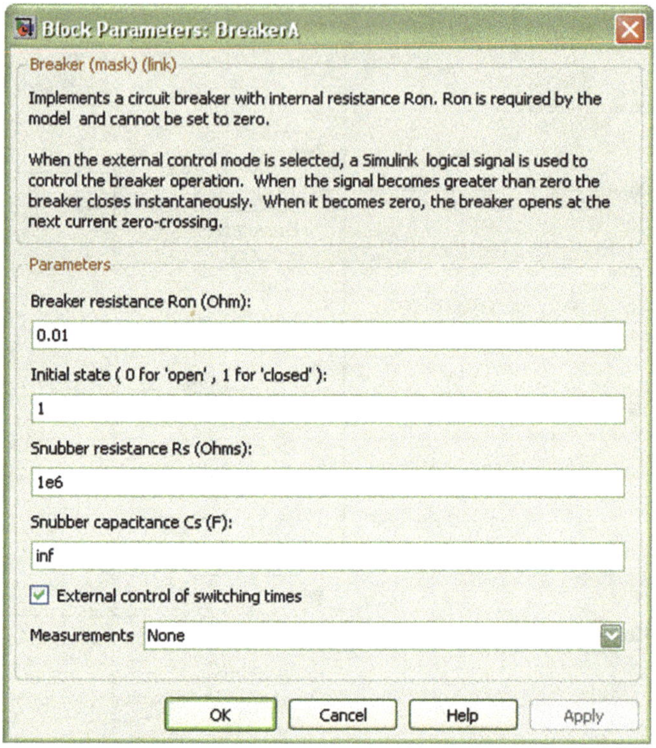

Fig. 8.14 Circuit breaker model dialog frame

Fig. 8.15 Load model

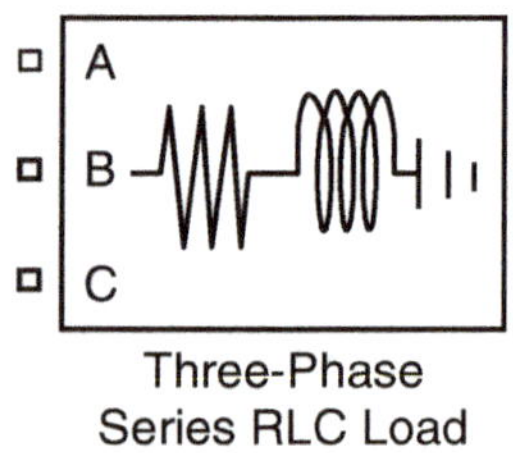

8.2.1.10 Default System Data

Generator

- Voltage magnitude: 194,367 V (230 kV system)
- Phase angle: 0°
- Frequency: 60 Hz
- Source impedance: $3.4878 + j0.1323 \times 2\pi \times 60\,\Omega$

Transmission Line 1, Line 2, and Line 3

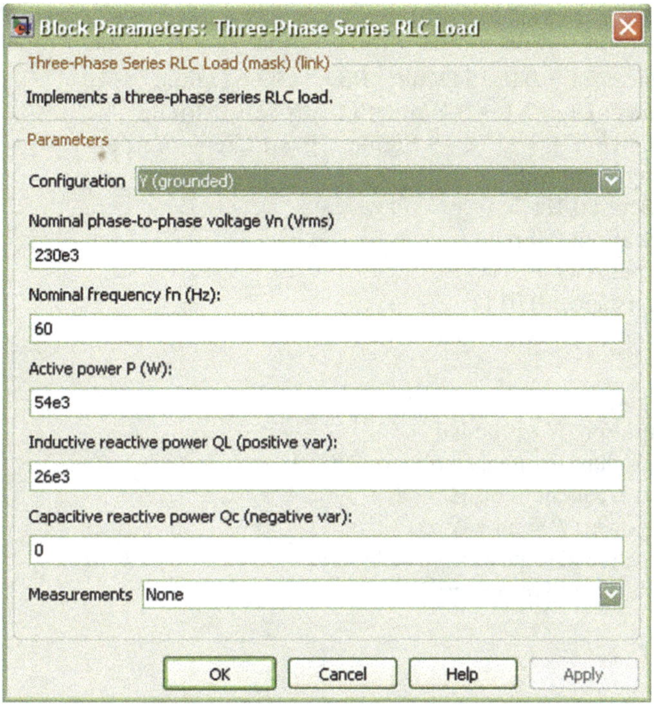

Fig. 8.16 Load model dialog frame

Fig. 8.17 Voltage and current scopes (measurement elements)

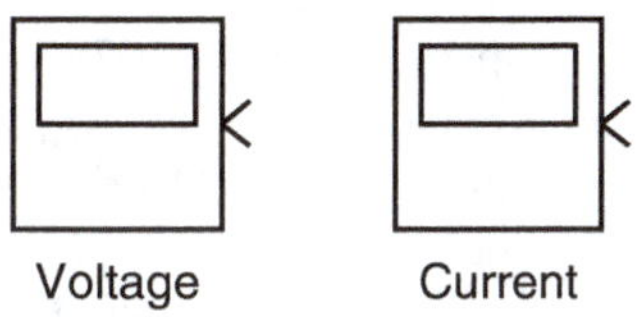

Fig. 8.18 Trip and logic scopes (measurement elements)

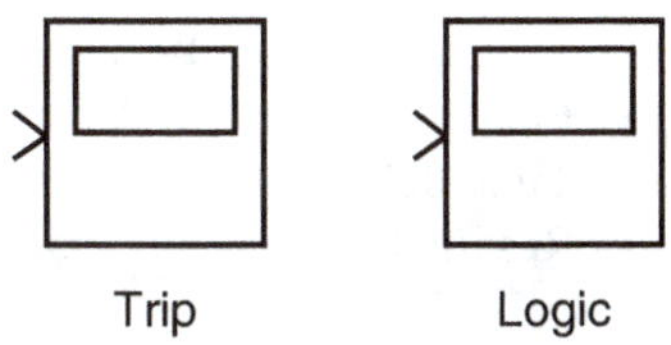

Fig. 8.19 Fault type display

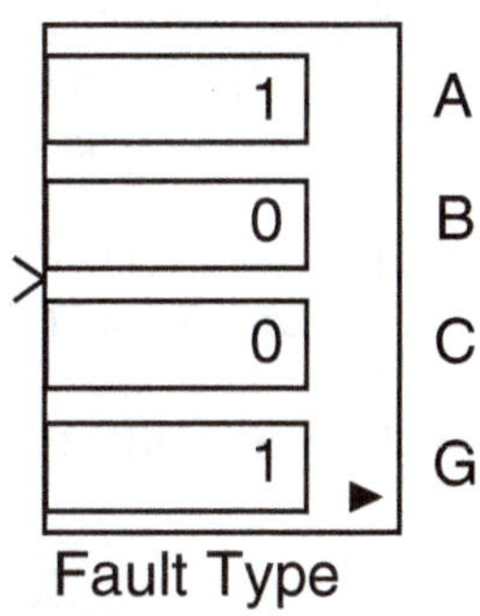

- Resistance: $R1 = 0.0815\ \Omega$/mile; $R0 = 0.3666\ \Omega$/mile
- Inductance: $L1 = 2.1\text{e}{-3}$ H/mile; $L0 = 6.5\text{e}{-3}$ H/mile
- Capacitance: $C1 = 31.6\text{e}{-9}$ F/mile; $C0 = 12.7\text{e}{-9}$ F/mile
- Line length: 40 mile
- Frequency: 60 Hz
- Fault location: 0.9 p.u.
- Fault type: A-G
- Fault resistance: 0.01 Ω
- Fault time sequence: [9 10 11 12 13] s

Load
- Configuration: Y grounded
- Nominal phase-to-phase voltage: 230 kV
- Nominal frequency: 60 Hz
- Active power: $P = 54$ kW
- Inductive reactive power: $Q_l = 26$ kvar
- Capacitive reactive power: $Q_c = 0$ var

CT and VT
- CT ratio: 100
- VT ratio: 2000

Circuit Breaker
- Operating time: 0.02 s
- Breaker resistance: $R_{on} = 0.01\ \Omega$
- Initial state: 1 (closed)
- Snubber resistance: $R_s = 1\text{e}6\ \Omega$
- Snubber capacitance: $C_s = \text{inf}$ F

Overcurrent Relay OR-1
- Pickup current: 3.8 A
- Time delay: 1.0 s
- Residual current threshold: 0.1 A

Overcurrent Relay OR-2
- Pickup current: 2.6 A
- Time delay: 0.6 s
- Residual current threshold: 0.1 A

Overcurrent Relay OR-3
- Pickup current: 1.4 A
- Time delay: 0 s
- Residual current threshold: 0.1 A

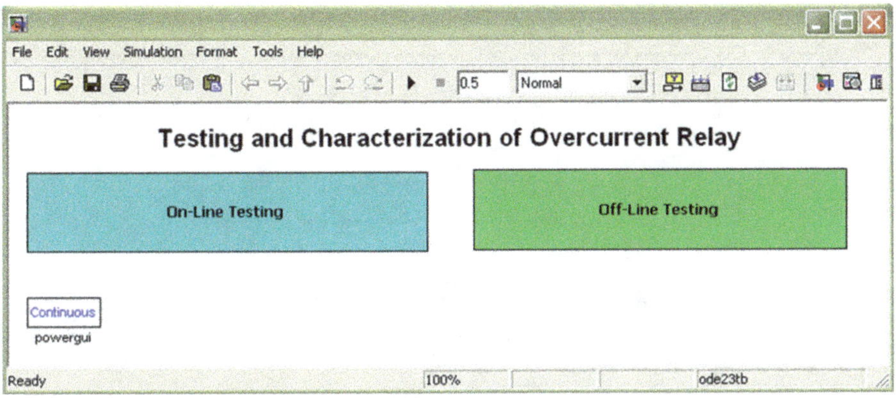

Fig. 8.20 Main menu of overcurrent relay testing model

8.2.1.11 Model Activation

The model is stored as a MATLAB file named "OORT.mdl." To use the model, firstly run MATLAB (R2008b or later version). Then search the Current Directory to find where this file is stored or copy the file into MATLAB's default working directory (typically MATLAB\R2008b\work) and type "OORT.mdl". The main menu will display in Fig. 8.20. Double clicking on "On-Line Testing" icon will open the model shown in Fig. 8.2.

8.2.2 Modeling and Testing Impedance Relay

This exercise shows how impedance relaying algorithm and four-zone protection scheme can be used to protect EHV electric power network. Rules for protecting a network using impedance relays, relaying principle and algorithm, requirements for instrumentation, and the principle of computing settings can be referred to previous chapters.

Note the following:

- In order to protect a given element, a voltage transformer (VT) and a current transformer (CT) need to be used to measure the voltage and current. The VT and CT should be installed at the element's terminal that is closest to the supplying source.
- A circuit breaker (CB) needs to be installed at the terminal closest to the supply source to disconnect the element in the case of a fault.
- An impedance relay must measure the magnitude phase angle of voltage and current, compute the apparent impedance, and compare it against the threshold.

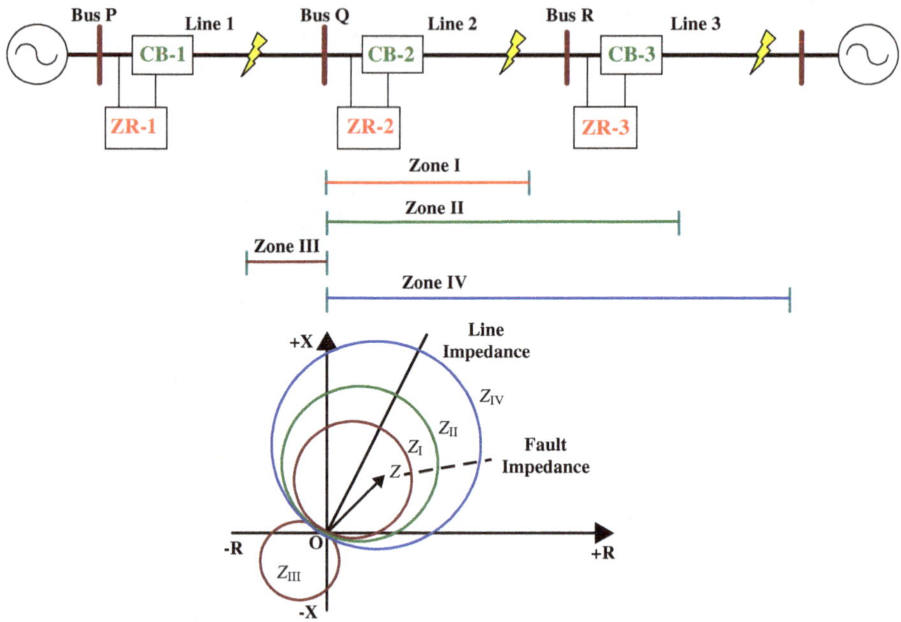

Fig. 8.21 An impedance relay with three forward zones and one reverse zone

- The fault-clearing time corresponds to the sum of the operating time of the relay and the operating time of the CB.

Figure 8.21 shows a power network consisting of two equivalent power sources and three transmission lines connecting three buses. This power network can be protected against faults using the impedance relaying (ZR) principle. In order to protect Line 2, for example, ZR-2, CB-2, and CT&VT-2 should be installed at Bus Q. ZR-2 must be coordinated with the relay downstream (in this case ZR-3) to ensure selectivity of operation. Four-zone protection scheme is used as illustrated in Fig. 8.21, which includes three forward zones (Zone I, II and IV) and one reverse zone (Zone III).

The first zone covers 75–80 % of the line and operates without any time delay. The second zone typically covers 120–150 % of the line and operates with a delay longer than the slowest relay for the far end bus elements. It provides primary protection for the remaining 20–25 % of the line and back-up protection for the elements adjacent to the far end bus. The third zone provides the reverse direction monitoring for some specific protection scheme, i.e., back-up protection for the local bus. The fourth zone provides back-up protection for certain distant elements in the system.

The protection system illustrated in Fig. 8.21 operates as follows:

- Primary protection for Line 1 is provided by ZR-1. For faults located 0–80 % from ZR-1, ZR-1 operates instantaneously. For fault located 80–100 % of the line, ZR-1 operates after Zone II time delay.

- Back-up protection for Line 1 is provided by the Zone III of ZR-2.
- Primary protection for Line 2 is provided by ZR-2. For faults located 0–80 % from ZR-2, ZR-2 operates instantaneously. For fault located 80–100 % of the line, ZR-2 operates after Zone II time delay.
- Back-up protection for Line 2 is provided by the Zone II and Zone IV of ZR-1 and the reverse Zone III of ZR-3.
- Primary protection for Line 3 is provided by ZR-3. For faults located 0–80 % from ZR-3, ZR-3 operates instantaneously. For fault located 80–100 % of the line, ZR-3 operates after Zone II time delay.
- Back-up protection for Line 3 is provided by the Zone III and Zone IV of ZR-2.

It should be noted that in reality each transmission line must be equipped with ZRs and CBs at two ends since the fault current is supplied from both ends of the faulted line. For the sake of simplifying the problem, impedance relays are only provided at one end to protect the transmission line.

The following activities should be done for designing protection scheme:

- Analysis of system impedance and fault conditions
- Selecting CT and CB
- Setting relay: pickup impedances and time delays for four protection zones
- Sensitivity check

An online impedance relay testing program developed in SIMULINK is given in Fig. 8.22, in which three impedance relays are used as primary protection for Line 1, Line 2, and Line 3. Elements used in the model are described in following subsections. The models of transmission line, generator, CT and VT, and breaker,

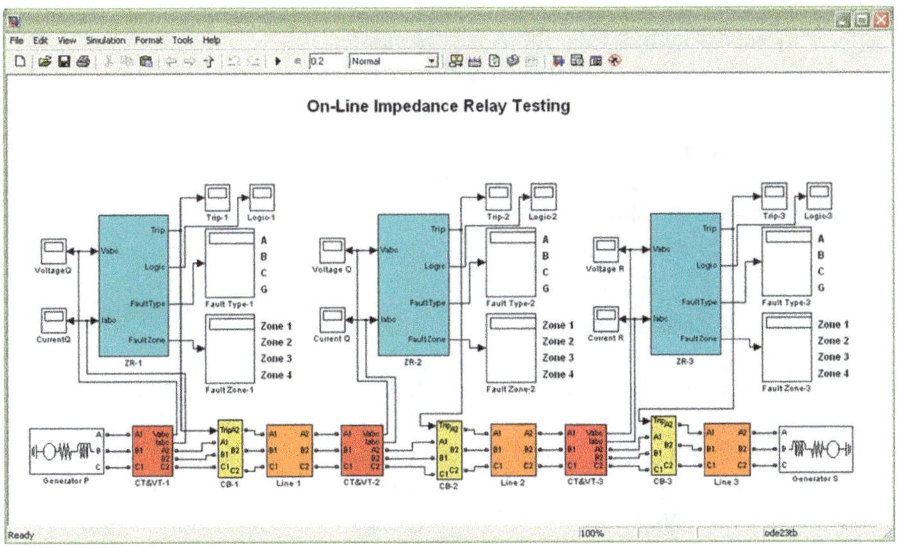

Fig. 8.22 Online impedance relay testing model

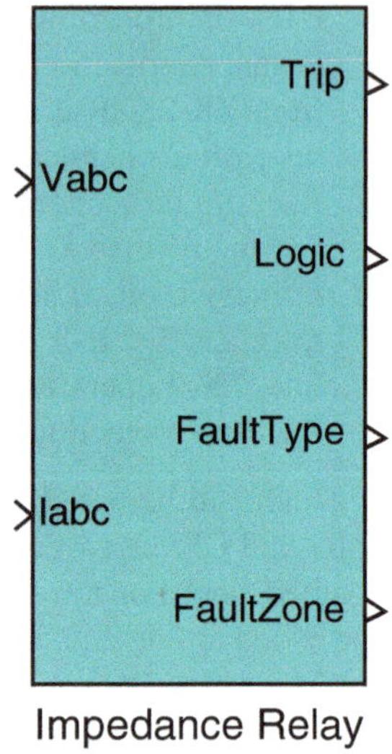

Fig. 8.23 Impedance relay model

and the display for voltage and current waveforms, trip logic, and result of fault type have been described in previous sections.

8.2.2.1 Impedance Relay Model

The model icon is shown in Fig. 8.23. Double click on it to open parameters dialog frame as shown in Fig. 8.24 in which one can input settings for the relay model. This relay model includes the following parameters:

- Zone pickup impedance: $Z1$, $Z2$, $Z3$, $Z4$
- Zone delay time: $T1$, $T2$, $T3$, $T4$
- Transmission line angle
- Zero sequence current compensator: k0M1 and K0A1
- Under voltage threshold

8.2.2.2 Fault Zone Display

This element is used for displaying the result of fault zone detection, which is one of the overcurrent relay functions as a starting element. As shown in Fig. 8.25, Digits "1" and "0" indicate the corresponding character in the same row effective or ineffective, respectively. For example, the digit "1" in first row indicates that the fault occurred within the range of Zone I.

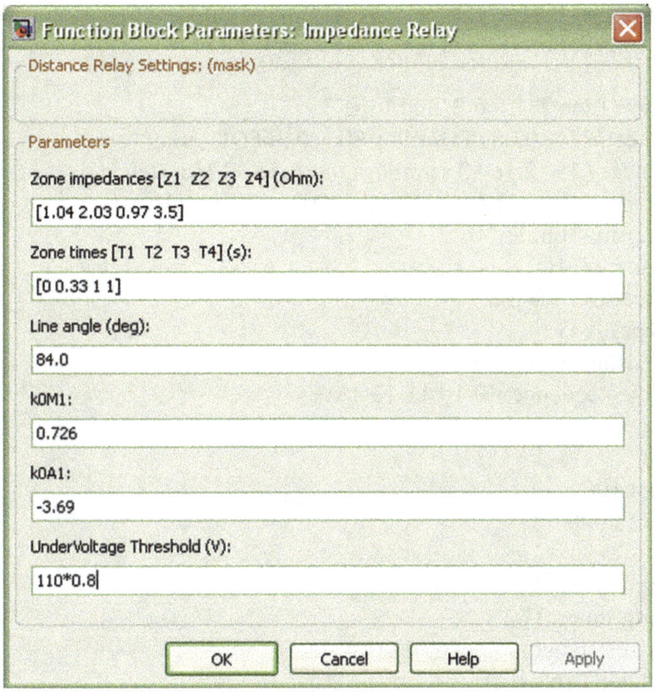

Fig. 8.24 Impedance relay model parameter dialog frame

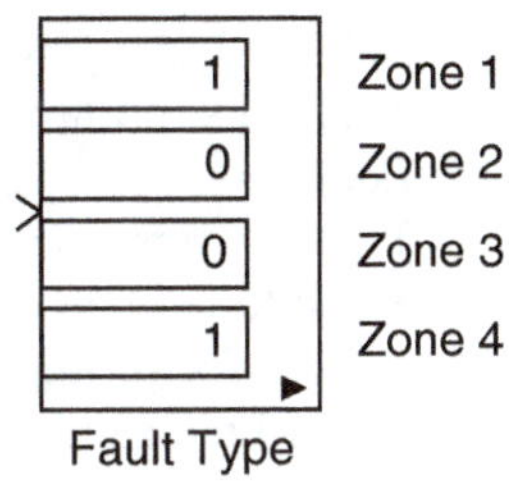

Fig. 8.25 Fault zone display

8.2.2.3 Default System Data

Generator P

- Voltage magnitude: 194,367 V (230 kV system)
- Phase angle: 10°
- Frequency: 60 Hz
- Source impedance: $3.4878 + j0.1323 \times 2\pi \times 60\,\Omega$

Generator S

- Voltage magnitude: 194,367 V (230 kV system)
- Phase angle: 0°

- Frequency: 60 Hz
- Source impedance: $3.4878 + j0.1323 \times 2\pi \times 60\,\Omega$

Transmission Line 1, Line 2, and Line 3

- Resistance: $R1 = 0.0815\,\Omega$/mile; $R0 = 0.3666\,\Omega$/mile
- Inductance: $L1 = 2.1\text{e}-3$ H/mile; $L0 = 6.5\text{e}-3$ H/mile
- Capacitance: $C1 = 31.6\text{e}-9$ F/mile; $C0 = 12.7\text{e}-9$ F/mile
- Line length: 40 mile
- Frequency: 60 Hz
- Fault location: 0.9 p.u.
- Fault type: A-G
- Fault resistance: $0.01\,\Omega$
- Fault time sequence: [9 10 11 12 13] s

CT and VT

- CT ratio: 100
- VT ratio: 2000

Circuit Breaker

- Operating time: 0.02 s
- Breaker resistance: $R_{on} = 0.01\,\Omega$
- Initial state: 1 (closed)
- Snubber resistance: $R_s = 1\text{e}6\,\Omega$
- Snubber capacitance: $C_s = \text{inf}$ F

Impedance Relay ZR-1

- Zone pickup impedance: $Z1 = 1.04\,\Omega$, $Z2 = 2.03\,\Omega$, $Z3 = 0.52\,\Omega$, $Z4 = 3.5\,\Omega$
- Zone time delay: $T1 = 0$ s, $T2 = 0.33$ s, $T3 = 1.0$ s, $T4 = 0.66$ s
- Transmission line angle: 84.0°
- Magnitude of zero-impedance current compensator: k0M1 = 0.726
- Angle of zero-impedance current compensator: k0A1 = −3.69
- Under voltage threshold: 110 * 0.8

Impedance Relay ZR-2

- Zone pickup impedance: $Z1 = 1.04\,\Omega$, $Z2 = 2.03\,\Omega$, $Z3 = 0.52\,\Omega$, $Z4 = 3.5\,\Omega$
- Zone time delay: $T1 = 0$ s, $T2 = 0.33$ s, $T3 = 1.0$ s, $T4 = 0.66$ s
- Transmission line angle: 84.0°
- Magnitude of zero-impedance current compensator: k0M1 = 0.726
- Angle of zero-impedance current compensator: k0A1 = −3.69
- Under voltage threshold: 110 * 0.8

Impedance Relay ZR-3

- Zone pickup impedance: $Z1 = 1.04\,\Omega$, $Z2 = 2.03\,\Omega$, $Z3 = 0.52\,\Omega$, $Z4 = 3.5\,\Omega$
- Zone time delay: $T1 = 0$ s, $T2 = 0.33$ s, $T3 = 1.0$ s, $T4 = 0.66$ s
- Transmission line angle: 84.0°

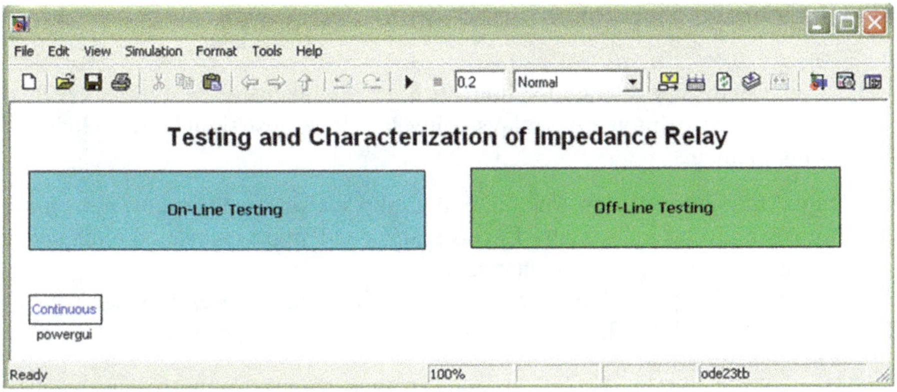

Fig. 8.26 Main menu of impedance relay testing model

- Magnitude of zero-impedance current compensator: kOM1 = 0.726
- Angle of zero-impedance current compensator: k0A1 = −3.69
- Under voltage threshold: 110 * 0.8

8.2.2.4 Model Activation

The model is stored as a MATLAB file named "OIRT.mdl." To use the model, firstly run MATLAB (R2008b or later version). Then search the Current Directory to find where this file is stored or copy the file into MATLAB's default working directory (typically MATLAB\R2008b\work) and type "OIRT.mdl". The main menu will be displayed in Fig. 8.26. Double clicking on "On-Line Testing" icon will open the model shown in Fig. 8.22.

8.3 Test Using Digital Simulator

In this section, the digital simulator-based protective relay testing platform including hardware and software is presented. The open-loop testing system for overcurrent and impedance relays and three-phase power system models developed in MATLAB/SIMULINK are described. Compared to the digital relay models described in Sect. 8.2, these relays feature multiple functions. A set of lab exercises has been designed for students to get familiar with these testing facilities. Test results can be compared with those exercises of Sect. 8.2.

8.3.1 Digital Simulator-Based Relay Test System

This test system is developed for students to study the characteristics and evaluate the performance of protective relays using low-level test signals. It includes the following elements:

- Digital Simulator: Reconstruct input digital waveforms and output four-channel voltage and current signals. Receive and send multi-channel digital signals through digital input/output channels
- Protective Relays: Include overcurrent relay and impedance/distance relay
- MATLAB/SIMULINK: Model power networks and generate test scenarios through simulation
- Relay Assistant™ software [1]: Commercial software creates test session for each test scenario and plays reconstructed waveforms through digital simulator to relays being tested
- Application software for relays: Commercial software provides user-friendly GUI to implement relay setting, status monitoring, and event report retrieving, which is developed by relay manufacturers
- PC: Provides operating platform for software, i.e., MATLAB
- Signal Conditioner: Scale the signals fed to relays and protect relays from overvoltage

8.3.1.1 Test System Overview

Figure 8.27 illustrates the system configuration including a PC used to run related software programs, a digital simulator used to generate the low-level voltage

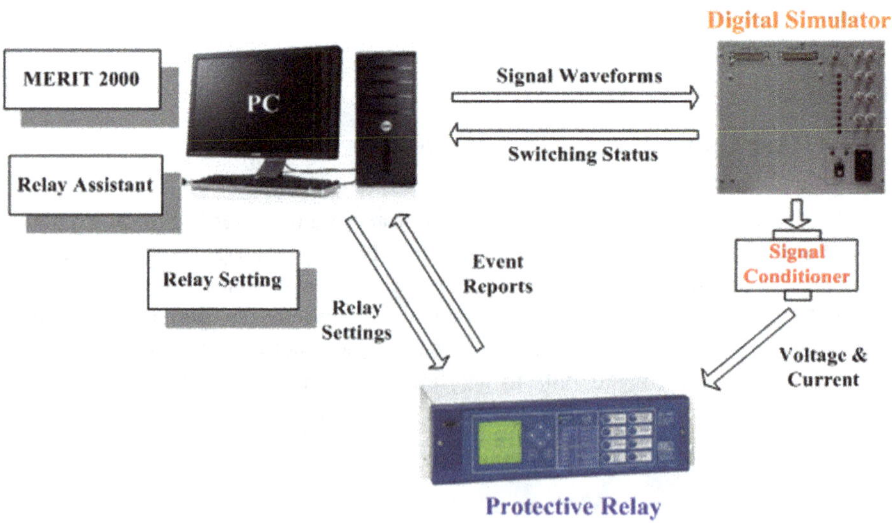

Fig. 8.27 Relay test system configuration

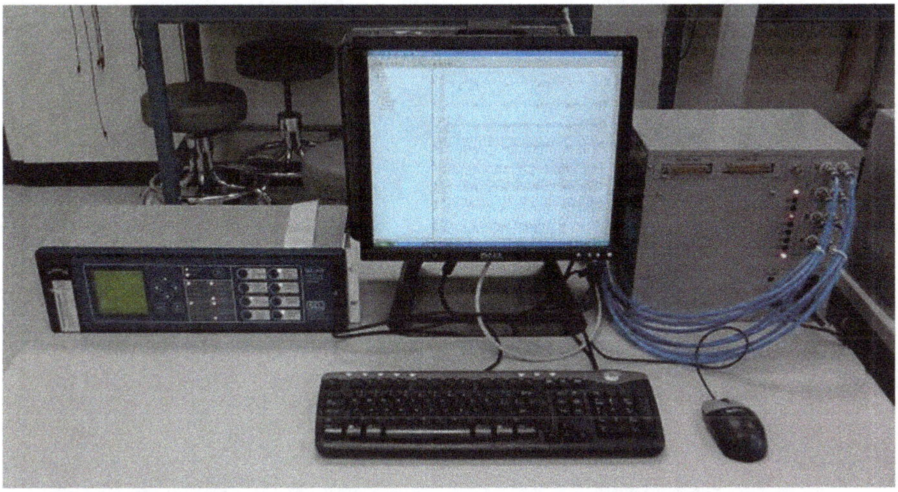

Fig. 8.28 Assembled test laboratory setup

signals, a signal conditioner used to scale the signals from the digital simulator, and the physical relay under test. A commercial software program called Relay Assistant™ residing on the PC communicates with digital simulator to send transient voltage and current data and receive contact status data. The digital simulator applies the voltage and current waveforms for the relay and records the relay trip contact status. A relay setting software program residing on the PC communicates with the relay to configure relay settings and retrieve relay event reports.

The assembled laboratory setup is shown in Fig. 8.28. The test applications are performed using a PC with commercial software Relay Assistant™ which creates test session for each test case and plays reconstructed waveforms to the relay being tested. A USB cable is used to connect the PC and the digital simulator. The simulator outputs four-channel voltage and current signals through the signal conditioning board to the relay. Coaxial cables with twin BNC connectors are used to connect the simulator and the signal conditioning board. The low-level test signals are carried by the flat ribbon cable, which comes out of the relay's front panel.

8.3.1.2 Digital Simulator

Figure 8.29 shows the front panel of digital simulator used in lab exercises. It features multi-terminal eight-channel 16-bit digital to analog converter with a high-performance waveform reconstruction system up to 40 kHz generating rate. Functional modules are described in the following:

- Two groups of coaxial twin BNC connectors are able to output four-phase voltage and current signals with magnitude up to 10 V

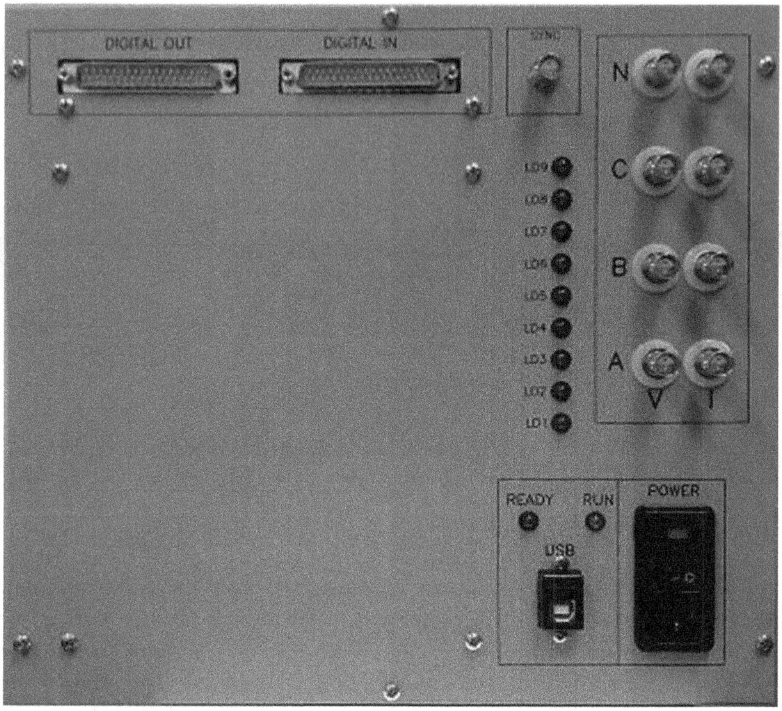

Fig. 8.29 Front panel of digital simulator

- Power switch is used to turn it on and off
- Two groups of indicators show the operating status, i.e., "Ready" and "Run", which indicate whether the data has been loaded and ready to replay
- A coaxial twin BNC connector "SYNC" is used as input signal source of synchronization (will not be used in exercises)
- Digital input connector is used to receive external control and/or state signals
- Digital output connector is used to send control and/or state signals for external devices

8.3.1.3 Protective Relays

Two types of commercial relays were used in the examples that follow: Overcurrent relay Relay-1 and impedance/distance relay Relay-2. Brief descriptions for each model are given in Table 8.1. Students can refer to the reference manuals provided by manufacturers for details.

Table 8.1 Description of relays used in lab exercises

Name	Model	Description
Overcurrent relay	Relay-1	Complete overcurrent protection. Protect lines and equipment using a sensitive and secure mix of phase, negative-sequence, and ground overcurrent elements
Distance relay	Relay-2	Protect transmission lines using combination of five zones of phase- and ground-distance and directional overcurrent elements

Table 8.2 System data of the power network

Parameter	Value
Nominal system line-to-line voltage	230 kV
Nominal relay current	5 A secondary
Nominal frequency	60 Hz
Line length	40 mile
Line impedances: Z_{1L1}, Z_{0L1}	$39\angle 84°$ primary, $124\angle 81.5°$ primary
Source Bus1 impedances: $Z_{1S1} = Z_{0S1}$	$50\angle 86°$ primary
Source Bus2 impedances: $Z_{1S2} = Z_{0S2}$	$50\angle 86°$ primary
PTR (potential transformer ratio)	230 kV:115 V = 2000
CTR (current transformer ratio)	500 A: 5 A = 100
Phase rotation	ABC

Table 8.3 Secondary impedance of the transmission line

Parameter	Value
Line impedances: Z_{1L1}, Z_{0L1}	$1.95\angle 84°$ secondary, $6.2\angle 81.5°$ secondary
Source Bus1 impedances: $Z_{1S1} = Z_{0S1}$	$2.5\angle 86°$ secondary
Source Bus2 impedances: $Z_{1S2} = Z_{0S2}$	$2.5\angle 86°$ secondary

8.3.1.4 Default System Data

The system data and secondary impedance are given in Tables 8.2 and 8.3, respectively.

8.3.1.5 Relay Assistant Software

This commercial software including following features:

- 32-Bit Windows software for automated open-loop transient testing of protective relays. It is designed to complement and improve widely accepted methods and practices for relay testing

- Load and replay transient files in a variety of file formats including DFR native format, EMTP/ATP output file format, MATLAB .mat file format, and COMTRADE format
- It can drive several different I/O hardware platforms enabling users to match simulator configuration with their testing requirement. It can also combine portable relay test sets with transient testing capabilities of modern power system simulators
- Provide user-friendly GUI through which user can create and reuse test objects of various levels of complexity to speed up the test preparation. Comprehensive set of signal processing and data handling functions is available for further increasing efficiency of relay testing

An example is given next to show the general procedures of creating test session and implementing test cases using Relay Assistant. Detailed instruction can be found in the instruction manual.

Step 1. Double click the icon on the desktop to open the main window of program as shown in Fig. 8.30.

Step 2. Click on Build—Insert Signal..., the Insert Signal window is shown in Fig. 8.31. Select option "by converting an external file" and click on Create New...; Select file type "Matlab Files (.mat)" as shown in Fig. 8.32, locate the signal file and click "Open". Nine signals with unknown information are shown on signal list in Fig. 8.33. From top to bottom, the signals are time, voltage phase A (VA), voltage phase B (VB), voltage phase C (VC), voltage phase N (VN), current phase A (IA), current phase B (IB), current

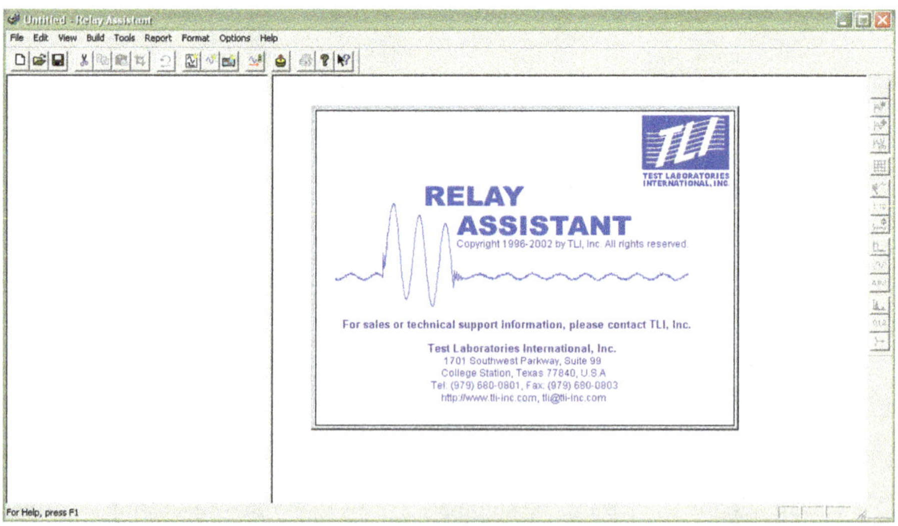

Fig. 8.30 Startup interface of the Relay Assistant software

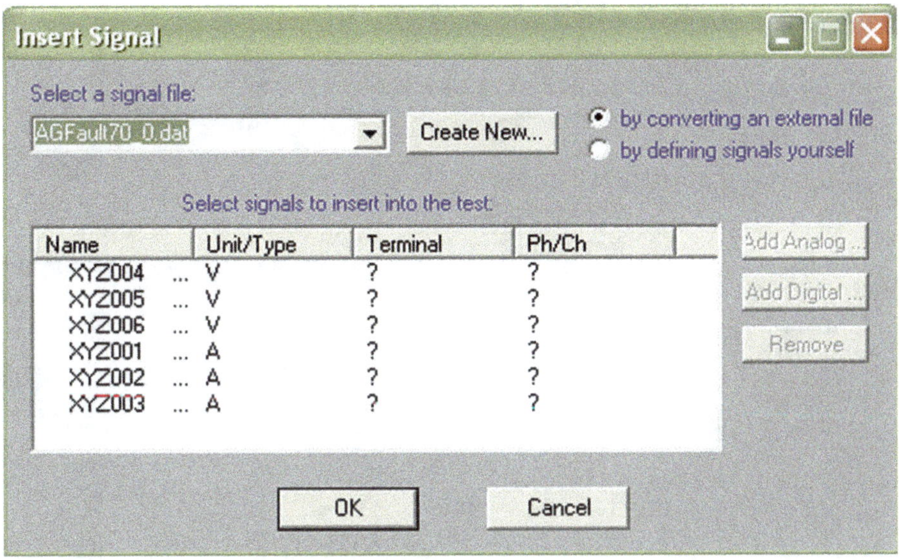

Fig. 8.31 Interface for inserting signals

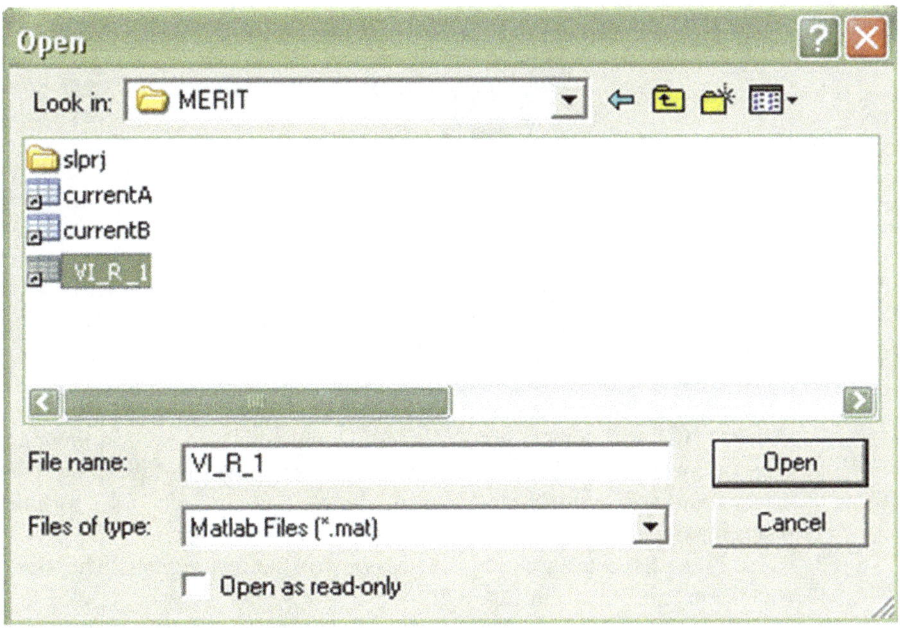

Fig. 8.32 Interface for selecting signal files

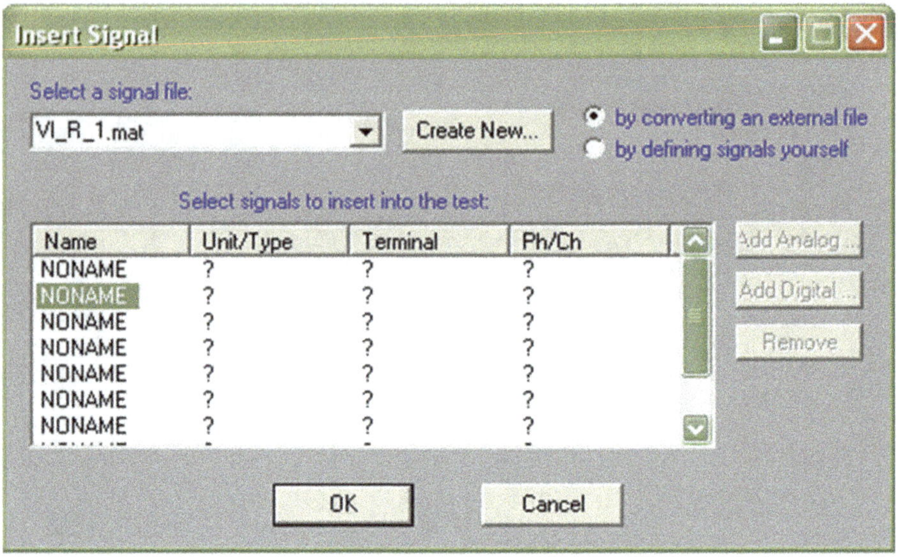

Fig. 8.33 Interface after signals are imported

Fig. 8.34 Interface for editing signal properties

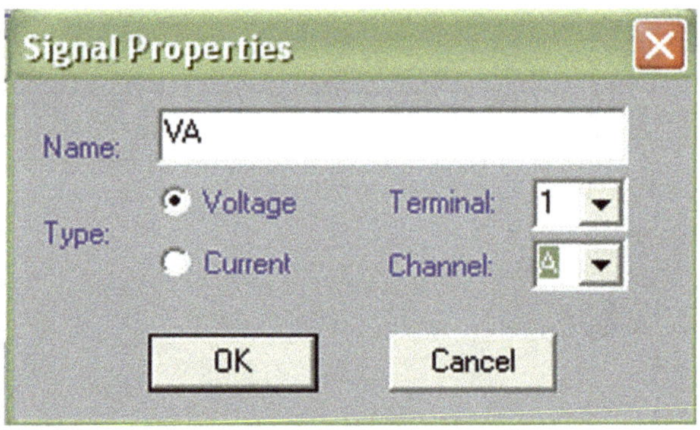

phase C (IC), and current phase N (IN). Double click on each of them and type in correct information, i.e., terminal 1 as shown in Fig. 8.34. Then select four-phase voltage and current signals and click on "Ok" to finish inserting signals from external file as shown in Fig. 8.35.

Step 3. Click File—Save to create a test session for the test signals. The signal waveforms are shown in Fig. 8.36.

Step 4. Click File—Execute... to run simulation after properly setting Device and Execution environment.

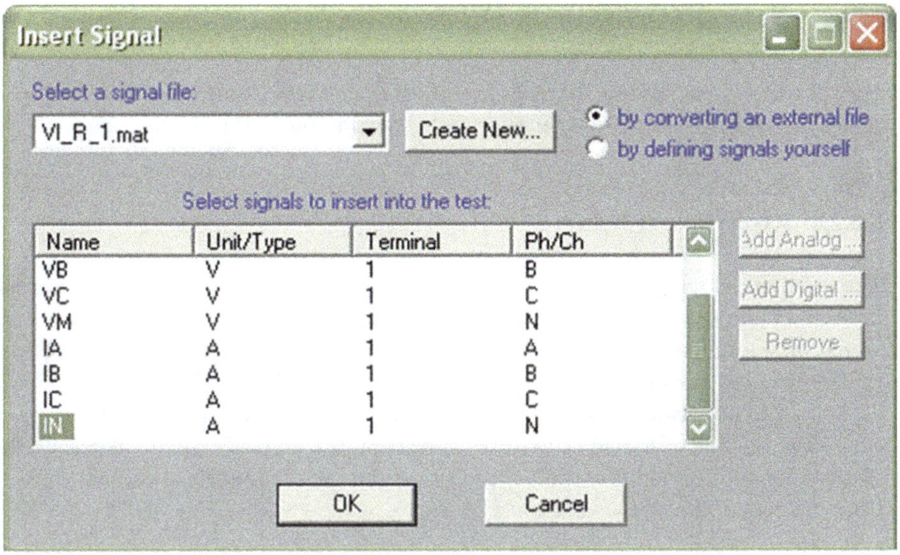

Fig. 8.35 Interface for imported signals

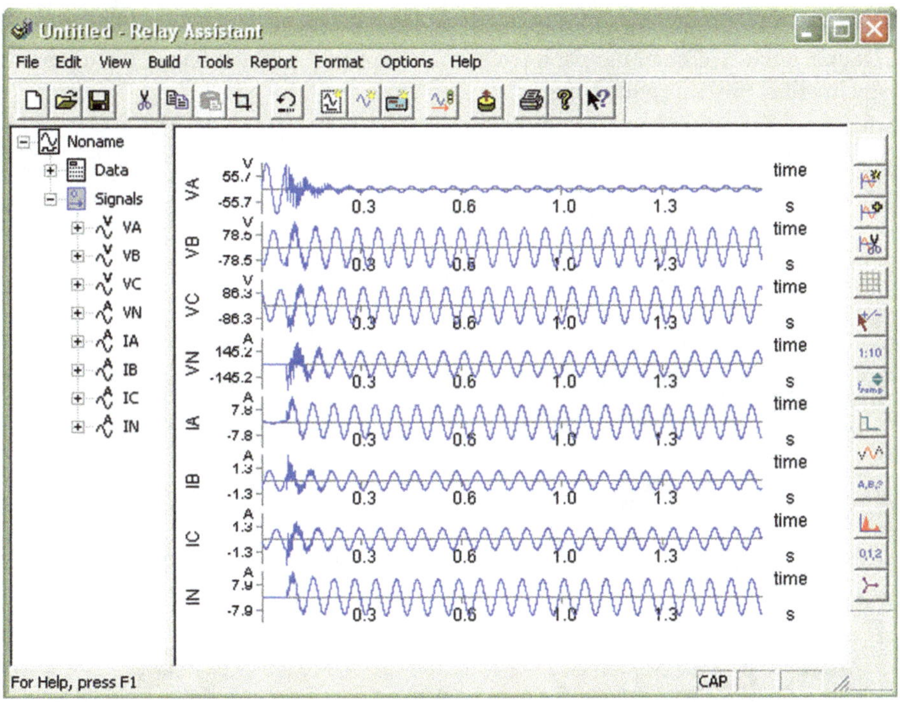

Fig. 8.36 Display of imported signal waveforms

8.3.2 System Modeling and Simulation Programs

Two power system models for overcurrent relay and impedance relay are developed in SIMULINK/MATLAB to generate test cases through simulation.

8.3.2.1 Overcurrent Relay

The off-line overcurrent relay testing program developed in SIMULINK is shown in Fig. 8.37, in which four-phase voltage and current signals at Bus P, Q, and R are stored in MAT files. These files can be used as input signals to perform relay tests on the selected commercial relay, Relay-1 through Relay Assistant software and digital simulator. Elements used in the model are described in following subsections. The models of transmission line, generator, CT and VT, and breaker, and the display for voltage and current waveforms, trip logic, and result of fault type have been described in previous sections.

8.3.2.2 To File Block

This element is used for storing output signals to MAT file as shown in Fig. 8.38. Double click will open the parameter window in Fig. 8.39 to change the file name, decimation, and sampling time.

8.3.2.3 Default System Data

Generator

- Voltage magnitude: 194,367 V (230 kV system)
- Phase angle: 0°

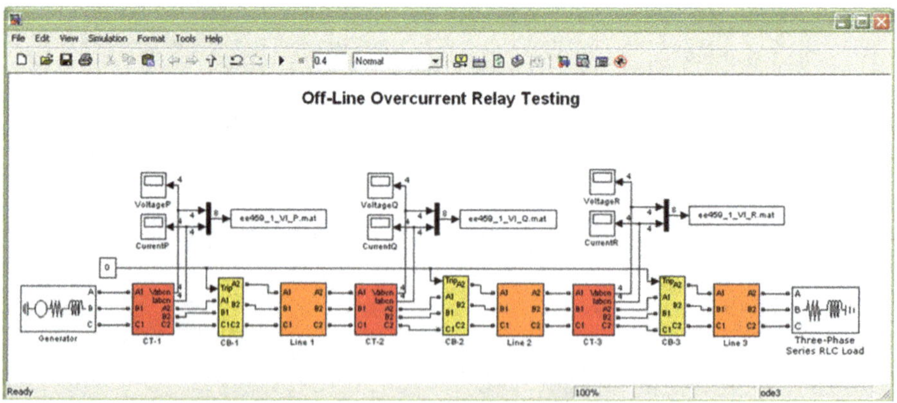

Fig. 8.37 Open-loop (off-line) overcurrent relay testing model

Fig. 8.38 Store signals to file block

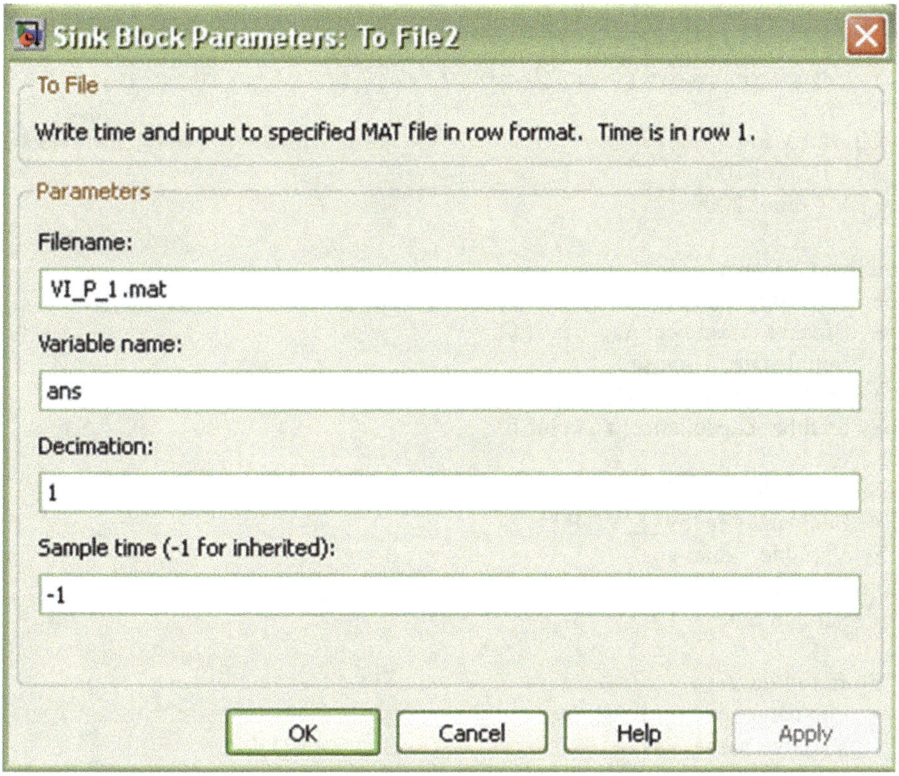

Fig. 8.39 Store signals to file block parameter box

- Frequency: 60 Hz
- Source impedance: $3.4878 + j0.1323 \times 2\pi \times 60\,\Omega$

Transmission Line 1, Line 2, and Line 3
- Resistance: $R1 = 0.0815\,\Omega$/mile; $R0 = 0.3666\,\Omega$/mile
- Inductance: $L1 = 2.1\text{e}{-3}$ H/mile; $L0 = 6.5\text{e}{-3}$ H/mile
- Capacitance: $C1 = 31.6\text{e}{-9}$ F/mile; $C0 = 12.7\text{e}{-9}$ F/mile
- Line length: 40 mile
- Frequency: 60 Hz
- Fault location: 0.9 p.u.
- Fault type: A-G
- Fault resistance: $0.01\,\Omega$
- Fault time sequence: [9 10 11 12 13] s

Load
- Configuration: Y grounded;
- Nominal phase-to-phase voltage: 230 kV
- Nominal frequency: 60 Hz
- Active power: $P = 54$ kW
- Inductive reactive power: $Ql = 26$ kvar
- Capacitive reactive power: $Q_c = 0$ var

CT and VT
- CT ratio: 100
- VT ratio: 2000

Circuit Breaker
- Operating time: 0.02 s
- Breaker resistance: $R_{on} = 0.01\ \Omega$
- Initial state: 1 (closed)
- Snubber resistance: $R_s = 1e6\ \Omega$
- Snubber capacitance: $C_s = \inf$ F

Store to File-1
- File name: ee459_1_VI_P
- Variable name: ans
- Decimation: 1
- Sample time: −1

Store to File-2
- File name: ee459_1_VI_Q
- Variable name: ans
- Decimation: 1
- Sample time: −1

Store to File-3
- File name: ee459_1_VI_R
- Variable name: ans
- Decimation: 1
- Sample time: −1

8.3.2.4 Model Activation

The model is stored as a MATLAB file named "OORT.mdl." To use the model, firstly run MATLAB (R2008b or later version). Then search the Current Directory to find where this file is stored or copy the file into MATLAB's default working directory (typically MATLAB\R2008b\work) and type "OORT.mdl." The main menu will be displayed in Fig. 8.20. Double clicking on "Off-Line Testing" icon will open the model shown in Fig. 8.26.

8.3.2.5 Impedance Relay

The off-line impedance relay testing program developed in SIMULINK is shown in Fig. 8.40, in which four-phase voltage and current signals at Bus P, Q, and R are stored in MAT files. These files can be used as input signals to perform relay tests on the selected commercial relay, Relay-2 through Relay Assistant software and digital simulator. Elements used in the model are described in following subsections. The models of transmission line, generator, CT and VT, and breaker, and the display for voltage and current waveforms, trip logic, and result of fault type have been described in previous sections.

8.3.2.6 To File Block

This element is used for storing output signals to MAT file as shown in Fig. 8.41. Double click will open the parameter window in Fig. 8.42 to change the file name, decimation, and sampling time.

8.3.2.7 Default System Data

Generator P

- Voltage magnitude: 194,367 V (230 kV system);
- Phase angle: 10°
- Frequency: 60 Hz
- Source impedance: $3.4878 + j0.1323 \times 2\pi \times 60\,\Omega$

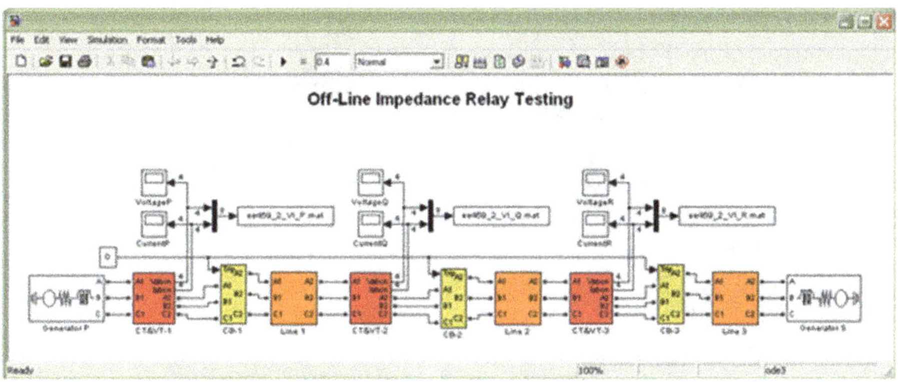

Fig. 8.40 Open-loop (off-line) impedance relay testing model

Fig. 8.41 Store signals to file block

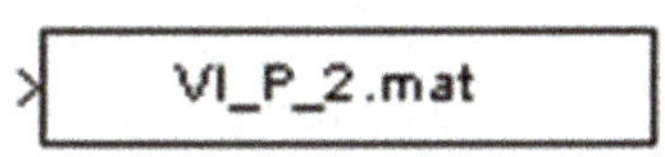

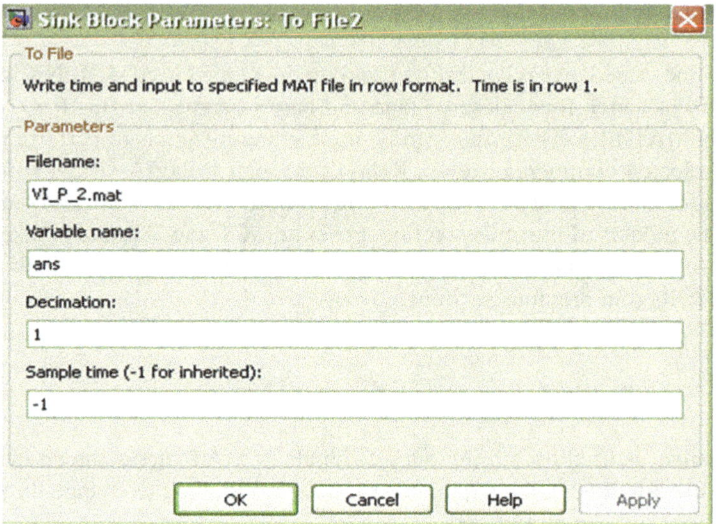

Fig. 8.42 Store signals to file block parameter box

Generator S

- Voltage magnitude: 194,367 V (230 kV system)
- Phase angle: 0°
- Frequency: 60 Hz
- Source impedance: $3.4878 + j0.1323 \times 2\pi \times 60\,\Omega$

Transmission Line 1, Line 2, and Line 3

- Resistance: $R1 = 0.0815\,\Omega$/mile; $R0 = 0.3666\,\Omega$/mile
- Inductance: $L1 = 2.1\text{e}{-3}$ H/mile; $L0 = 6.5\text{e}{-3}$ H/mile
- Capacitance: $C1 = 31.6\text{e}{-9}$ F/mile; $C0 = 12.7\text{e}{-9}$ F/mile
- Line length: 40 mile
- Frequency: 60 Hz
- Fault location: 0.9 p.u.
- Fault type: A-G
- Fault resistance: $0.01\,\Omega$
- Fault time sequence: [9 10 11 12 13] s

Load

- Configuration: Y grounded
- Nominal phase-to-phase voltage: 230 kV
- Nominal frequency: 60 Hz
- Active power: $P = 54$ kW

- Inductive reactive power: $Q_l = 26$ kvar
- Capacitive reactive power: $Q_c = 0$ var

CT and VT
- CT ratio: 100
- VT ratio: 2000

Circuit Breaker
- Operating time: 0.02 s
- Breaker resistance: $R_{on} = 0.01\ \Omega$
- Initial state: 1 (closed)
- Snubber resistance: $R_s = 1e6\ \Omega$
- Snubber capacitance: $C_s = \text{inf}$ F

Store to File-1
- File name: ee459_2_VI_P
- Variable name: ans
- Decimation: 1
- Sample time: -1

Store to File-2
- File name: ee459_2_VI_Q
- Variable name: ans
- Decimation: 1
- Sample time: -1

Store to File-3
- File name: ee459_2_VI_R
- Variable name: ans
- Decimation: 1
- Sample time: -1

8.3.2.8 Model Activation

The model is stored as a MATLAB file named "OIRT.mdl." To use the model, firstly run MATLAB (R2008b or later version). Then find the Current Directory to find where this file is stored or copy the file into MATLAB's default working directory (typically MATLAB\R2008b\work) and type "OIRT.mdl." The main menu will be displayed in Fig. 8.20. Double clicking on "Off-Line Testing" icon will open the model shown in Fig. 8.26.

8.4 Closed-Loop and Open-Loop Analysis

After completing the lab exercises, students will learn how to test overcurrent relay and impedance relay using custom digital simulator-based relay test system. The following activities are performed:

- Model various power system disturbances and create test cases for relay tests through MATLAB simulation
- Use Relay Assistant™ software to insert signals from external file, create test session, and replay test waveforms
- Set up test object relays by using controller on front panel and/or application software provided by manufacturers
- Observe relay behaviors when exposed to various disturbance and explain them
- Compare test results with closed-loop (on-line) relay test results in Sect. 8.2 and study relay operation characteristics.

8.4.1 General Procedures for Performing Tests

Step 1. Log in to computer with username and password, switch on the power of simulator and relay

Step 2. Set up calibration files for each relay being tested.

Step 3. Find the serial number (S/N) of the simulator being used printed on the left hand side of chassis. An instance given below indicates the S/N is "P806304-4".

Step 4. Copy the corresponding calibration file named "CalFile1_4_P04.cal" (the last digit of S/N on chassis is "4" as shown in Fig. 8.43) to "C:\Program Files\TLI\RA400" and rename it as "CalFile1_4.cal".

Step 5. Set up relay properly in terms of the protection scheme (refer to instruction manuals).

Step 6. Open the software "Relay Assistant" and operate tests on selected relays.

Fig. 8.43 Serial label of digital simulator

Step 7. Observe relay behaviors, retrieve event reports.

Note: Repeat Steps 2–4 when the computer is restarted. For Relay-2, signal input group IW and VY are used for the current inputs and voltage inputs.

8.5 Summary

This chapter provides description of the online and off-line testing of protective relays using digital models and low-voltage digital simulator. The software examples are available in MATLAB/SIMULINK. The protective relay models including overcurrent relay and distance relay are described. Digital simulator-based low-voltage relay test platform is described as well. Examples of laboratory exercise are provided at the end of each section.

Reference

1. Test Laboratories International, Inc. (2004) Relay assistant-digital simulator software [Online]. http://www.tli-inc.com